HANDBOOK OF RESEARCH ON FOOD PROCESSING AND PRESERVATION TECHNOLOGIES

Volume 1

Nonthermal and Innovative Food Processing Methods

Handbook of Research on Food Processing and Preservation Technologies, 5 volume set:

Volume 1: Nonthermal and Innovative Food Processing Methods

Volume 2: Nonthermal Food Preservation and Novel Processing Strategies

Volume 3: Computer-Aided Food Processing and Quality Evaluation Techniques

Volume 4: Design and Development of Specific Foods, Packaging Systems, and Food Safety

Volume 5: Emerging Techniques for Food Processing, Quality, and Safety Assurance

HANDBOOK OF RESEARCH ON FOOD PROCESSING AND PRESERVATION TECHNOLOGIES

Volume 1

Nonthermal and Innovative Food Processing Methods

Edited by

Megh R. Goyal, PhD, PE
Preeti Birwal, PhD
Monika Sharma, PhD

AAP APPLE
ACADEMIC
PRESS

First edition published 2022

Apple Academic Press Inc.
1265 Goldenrod Circle, NE,
Palm Bay, FL 32905 USA

4164 Lakeshore Road, Burlington,
ON, L7L 1A4 Canada

CRC Press
6000 Broken Sound Parkway NW,
Suite 300, Boca Raton, FL 33487-2742 USA

2 Park Square, Milton Park,
Abingdon, Oxon, OX14 4RN UK

© 2022 Apple Academic Press, Inc.

Apple Academic Press exclusively co-publishes with CRC Press, an imprint of Taylor & Francis Group, LLC

Library and Archives Canada Cataloguing in Publication

Title: Handbook of research on food processing and preservation technologies / edited by Megh R. Goyal, PhD, PE, Preeti Birwal, PhD, Monika Sharma, PhD.

Names: Goyal, Megh R., editor. | Birwal, Preeti, editor. | Sharma, Monika (Food scientist), editor.

Series: Innovations in agricultural and biological engineering.

Description: First edition. | Series statement: Innovations in agricultural and biological engineering | Includes bibliographical references and indexes. | Incomplete contents: Volume 1. Nonthermal and innovative food processing methods -- Volume 2. Nonthermal food preservation and novel processing strategies

Identifiers: Canadiana (print) 20210143002 | Canadiana (ebook) 2021014307X | ISBN 9781771889827 (v. 1 ; hardcover) | ISBN 9781774638514 (v. 1 ; softcover) | ISBN 9781774630037 (v. 2 ; hardcover) | ISBN 9781774638521 (v. 2 ; softcover) | ISBN 9781774630334 (v. 3 ; hardcover) | ISBN 9781774638538 (v. 3 ; softcover) | ISBN 9781774630341 (v. 4 ; hardcover) | ISBN 9781774638545 (v. 4 ; softcover) | ISBN 9781774630358 (v. 5 ; hardcover) | ISBN 9781774638552 (v. 5 ; softcover) | ISBN 9781003153221 (v. 1 ; ebook) | ISBN 9781003161295 (v. 2 ; ebook) | ISBN 9781003184591 (v. 3 ; ebook) | ISBN 9781003184645 (v. 4 ; ebook) | ISBN 9781003184720 (v. 5 ; ebook)

Subjects: LCSH: Food industry and trade. | LCSH: Food—Preservation.

Classification: LCC TP371.2 .H36 2021 | DDC 664/.028—dc23

Library of Congress Cataloging-in-Publication Data

Names: Goyal, Megh R., editor. | Birwal, Preeti, editor. | Sharma, Monika (Food scientist), editor.

Title: Handbook of research on food processing and preservation technologies. Volume 1, Nonthermal and innovative food processing methods / edited by Megh R. Goyal, Preeti Birwal, Monika Sharma.

Other titles: Innovations in agricultural and biological engineering.

Description: 1st edition. | Palm Bay, FL, USA : Apple Academic Press, 2021. | Series: Innovations in agricultural and biological engineering | Includes bibliographical references and index. | Contents: High-Pressure Processing: Potential Applications for Foods / K. Gomathy, R. Pandiselvam, Anjineyulu Kothakota, and S. V. Ramesh -- Ultraviolet Light Technology: Applications for Fresh Produce / Gozde Oguz Korkut and Gurbuz Gunes -- Microwave-Assisted Extraction (MAE) Technology: Potential for Extraction of Food Components / Faizan Ahmad, Sadaf Zaidi, and Z. R. A. A. Azad -- High-Pressure Assisted Freezing of Foods / Pavan Manjunath Gundu, Preeti Birwal, Chaitradeepa G. Mestri, and Gunjal Mahendra Vishram -- Microencapsulation Technology: Potential in Formulations of Probiotic Foods / Meenatai G. Kamble, Ajay Chinchkar, and Anurag Singh -- Dense Phase Carbon Dioxide (DPCD)-Aided Preservation of Foods / Lavanya Devraj, Mohan G. Naik, Nikitha Modupalli, Suka Thangaraju, and Venkatachalapathy Natarajan -- Ionic Liquids: Applications in Food Science and Food Processing / Carolina Elisa Demaman Oro, Victor De Aguiar Pedott, Marcelo Luis Mignoni, Rogério Marcos Dallago, Marcus Vinicius Tres, and Giovani Leone Zabot -- Application of Ozone in the Food Industry: Recent Advances and Prospects / Ravi Pandiselvam, Nukasani Sagarika, Yarrakula Srinivas, Anjineyulu Kothakota, Kamalakkannan Gomathy, and Kaliramesh Siliveru -- Advances in Osmotic Dehydration / Barinderjeet Singh Toor, Harinderjeet Kaur Bhullar, and Amarjeet Kaur. | Summary: "Handbook of Research on Food Processing and Preservation Technologies will be a 5-volume collection that attempts to illustrate various design, development, and applications of novel and innovative strategies for food processing and preservation. The role and applications of minimal food processing techniques (such as ozone treatment, vacuum drying, osmotic dehydration, dense phase carbon dioxide treatment, pulsed electric field, and high-pressure assisted freezing) are also discussed, along with a wide range of applications. The handbook also explores some exciting computer-aided techniques emerging in the food processing sector, such as robotics, radio frequency identification (RFID), three-dimensional food printing, artificial intelligence, etc. Some emphasis has also been given on nondestructive quality evaluation techniques (such as image processing, terahertz spectroscopy imaging technique, near infrared, Fourier transform infrared spectroscopy technique, etc.) for food quality and safety evaluation. The significant roles of food properties in the design of specific foods and edible films have been elucidated as well. The first volume in this set, Nonthermal and Innovative Food Processing Methods, provides a detailed discussion of many nonthermal food process techniques. These include high-pressure processing, ultraviolet light technology, microwave-assisted extraction, high pressure assisted freezing, microencapsulation, dense phase carbon dioxide aided preservation, to name a few. The volume is a treasure house of valuable information and will be an excellent reference for researchers, scientists, students, growers, traders, processors, industries, and others"-- Provided by publisher.

Identifiers: Canadiana (print) 20210143002 | Canadiana (ebook) 2021014307X | ISBN 9781771889827 (v. 1 ; hardcover) | ISBN 9781774638514 (v. 1 ; softcover) | ISBN 9781774630037 (v. 2 ; hardcover) | ISBN 9781774638521 (v. 2 ; softcover) | ISBN 9781774630334 (v. 3 ; hardcover) | ISBN 9781774638538 (v. 3 ; softcover) | ISBN 9781774630341 (v. 4 ; hardcover) | ISBN 9781774638545 (v. 4 ; softcover) | ISBN 9781774630358 (v. 5 ; hardcover) | ISBN 9781774638552 (v. 5 ; softcover) | ISBN 9781003153221 (v. 1 ; ebook) | ISBN 9781003161295 (v. 2 ; ebook) | ISBN 9781003184591 (v. 3 ; ebook) | ISBN 9781003184645 (v. 4 ; ebook) | ISBN 9781003184720 (v. 5 ; ebook)

Subjects: LCSH: Food industry and trade--Technological innovations. | Food--Preservation.

Classification: LCC TP370.5 .H363 2021 (print) | LCC TP370.5 (ebook) | DDC 664/.028--dc23

LC record available at https://lccn.loc.gov/2021007371

LC ebook record available at https://lccn.loc.gov/2021007372

ISBN: 978-1-77463-036-5 (5-volume set)
ISBN: 978-1-77188-982-7 (hbk)
ISBN: 978-1-77463-851-4 (pbk)
ISBN: 978-1-00315-322-1 (ebk)

ABOUT THE BOOK SERIES: INNOVATIONS IN AGRICULTURAL AND BIOLOGICAL ENGINEERING

Under this book series, Apple Academic Press Inc. is publishing book volumes over a span of 8–10 years in the specialty areas defined by the American Society of Agricultural and Biological Engineers (www.asabe. org). Apple Academic Press Inc. aims to be a principal source of books in agricultural and biological engineering. We welcome book proposals from readers in areas of their expertise.

The mission of this series is to provide knowledge and techniques for agricultural and biological engineers (ABEs). The book series offers high-quality reference and academic content on agricultural and biological engineering (ABE) that is accessible to academicians, researchers, scientists, university faculty and university-level students, and professionals around the world.

Agricultural and biological engineers ensure that the world has the necessities of life, including safe and plentiful food, clean air and water, renewable fuel and energy, safe working conditions, and a healthy environment by employing knowledge and expertise of the sciences, both pure and applied, and engineering principles. Biological engineering applies engineering practices to problems and opportunities presented by living things and the natural environment in agriculture.

ABE embraces a variety of the following specialty areas (www.asabe.org): aquaculture engineering, biological engineering, energy, farm machinery and power engineering, food, and process engineering, forest engineering, information, and electrical technologies, soil, and water conservation engineering, natural resources engineering, nursery, and greenhouse engineering, safety, and health, and structures and environment.

For this book series, we welcome chapters on the following specialty areas (but not limited to):

1. Academia to industry to end-user loop in agricultural engineering.
2. Agricultural mechanization.
3. Aquaculture engineering.
4. Biological engineering in agriculture.
5. Biotechnology applications in agricultural engineering.
6. Energy source engineering.

7.   Farm to fork technologies in agriculture.
8.   Food and bioprocess engineering.
9.   Forest engineering.
10. GPS and remote sensing potential in agricultural engineering.
11. Hill land agriculture.
12. Human factors in engineering.
13. Impact of global warming and climatic change on agriculture economy.
14. Information and electrical technologies.
15. Irrigation and drainage engineering.
16.   Micro-irrigation engineering.
17. Milk Engineering.
18. Nanotechnology applications in agricultural engineering.
19. Natural resources engineering.
20. Nursery and greenhouse engineering.
21. Potential of phytochemicals from agricultural and wild plants for human health.
22. Power systems and machinery design.
23. Robot engineering and drones in agriculture.
24. Rural electrification.
25. Sanitary engineering.
26. Simulation and computer modeling.
27. Smart engineering applications in agriculture.
28. Soil and water engineering.
29. Structures and environment engineering.
30. Waste management and recycling.
31. Any other focus areas.

Books published in the Innovations in Agricultural & Biological Engineering Series

- Biological and Chemical Hazards in Food and Food Products: Prevention, Practices, and Management
- Bioremediation and Phytoremediation Technologies in Sustainable Soil Management
  - o   Volume 1: Fundamental Aspects and Contaminated Sites
  - o   Volume 2: Microbial Approaches and Recent Trends
  - o   Volume 3: Inventive Techniques, Research Methods and Case Studies
  - o   Volume 4: Degradation of Pesticides and Polychlorinated Biphenyls

- Dairy Engineering: Advanced Technologies and Their Applications
- Developing Technologies in Food Science: Status, Applications, and Challenges
- Emerging Technologies in Agricultural Engineering
- Engineering Interventions in Agricultural Processing
- Engineering Interventions in Foods and Plants
- Engineering Practices for Agricultural Production and Water Conservation: An Interdisciplinary Approach
- Engineering Practices for Management of Soil Salinity: Agricultural, Physiological, and Adaptive Approaches
- Engineering Practices for Milk Products: Dairyceuticals, Novel Technologies, and Quality
- Field Practices for Wastewater Use in Agriculture: Future Trends and Use of Biological Systems
- Flood Assessment: Modeling and Parameterization
- Food Engineering: Emerging Issues, Modeling, and Applications
- Food Process Engineering: Emerging Trends in Research and Their Applications
- Food Processing and Preservation Technology: Advances, Methods, and Applications
- Food Technology: Applied Research and Production Techniques
- Handbook of Research on Food Processing and Preservation Technologies:
  o Volume 1: Nonthermal and Innovative Food Processing Methods
  o Volume 2: Nonthermal Food Preservation and Novel Processing Strategies
  o Volume 3: Computer-Aided Food Processing and Quality Evaluation Techniques
  o Volume 4: Design and Development of Specific Foods, Packaging Systems, and Food Safety
  o Volume 5: Emerging Techniques for Food Processing, Quality, and Safety Assurance
- Modeling Methods and Practices in Soil and Water Engineering
- Nanotechnology and Nanomaterial Applications in Food, Health, and Biomedical Sciences
- Nanotechnology Applications in Agricultural and Bioprocess Engineering: Farm to Table
- Nanotechnology Applications in Dairy Science: Packaging, Processing, and Preservation

- Novel Dairy Processing Technologies: Techniques, Management, and Energy Conservation
- Novel Strategies to Improve Shelf-Life and Quality of Foods: Quality, Safety, and Health Aspects
- Processing of Fruits and Vegetables: From Farm to Fork
- Processing Technologies for Milk and Milk Products: Methods, Applications, and Energy Usage
- Scientific and Technical Terms in Bioengineering and Biological Engineering
- Soil and Water Engineering: Principles and Applications of Modeling
- Soil Salinity Management in Agriculture: Technological Advances and Applications
- State-of-the-Art Technologies in Food Science: Human Health, Emerging Issues and Specialty Topics
- Sustainable Biological Systems for Agriculture: Emerging Issues in Nanotechnology, Biofertilizers, Wastewater, and Farm Machines
- Technological Interventions in Dairy Science: Innovative Approaches in Processing, Preservation, and Analysis of Milk Products
- Technological Interventions in Management of Irrigated Agriculture
- Technological Interventions in the Processing of Fruits and Vegetables
- Technological Processes for Marine Foods, from Water to Fork: Bioactive Compounds, Industrial Applications, and Genomics

OTHER BOOKS ON AGRICULTURAL AND BIOLOGICAL ENGINEERING FROM APPLE ACADEMIC PRESS, INC.

Management of Drip/Trickle or Micro Irrigation
Megh R. Goyal, PhD, PE, Senior Editor-in-Chief

Evapotranspiration: Principles and Applications for Water Management
Megh R. Goyal, PhD, PE, and Eric W. Harmsen, Editors

Book Series: Research Advances in Sustainable Micro Irrigation
Senior Editor-in-Chief: Megh R. Goyal, PhD, PE

  Volume 1: Sustainable Micro Irrigation: Principles and Practices
  Volume 2: Sustainable Practices in Surface and Subsurface Micro
        Irrigation
  Volume 3: Sustainable Micro Irrigation Management for Trees and Vines
  Volume 4: Management, Performance, and Applications of Micro
        Irrigation Systems
  Volume 5: Applications of Furrow and Micro Irrigation in Arid and
        Semi-Arid Regions
  Volume 6: Best Management Practices for Drip Irrigated Crops
  Volume 7: Closed Circuit Micro Irrigation Design: Theory and
        Applications
  Volume 8: Wastewater Management for Irrigation: Principles and Practices
  Volume 9: Water and Fertigation Management in Micro Irrigation
Volume 10: Innovation in Micro Irrigation Technology

Book Series: Innovations and Challenges in Micro Irrigation
Senior Editor-in-Chief: Megh R. Goyal, PhD, PE

  Volume 1: Management of Drip/Trickle or Micro Irrigation
  Volume 2: Sustainable Micro Irrigation Design Systems for Agricultural
        Crops
  Volume 3: Principles and Management of Clogging in Micro Irrigation
  Volume 4: Performance Evaluation of Micro Irrigation Management

ABOUT THE EDITORS

Megh R. Goyal, PhD, PE
Retired Professor in Agricultural and Biomedical Engineering, University of Puerto Rico, Mayaguez Campus; Senior Technical Editor-in-Chief, Biomedical Engineering and Agricultural Science, Apple Academic Press, Inc.

Megh R. Goyal, PhD, PE, is a Retired Professor in Agricultural and Biomedical Engineering from the General Engineering Department in the College of Engineering at the University of Puerto Rico–Mayaguez Campus; and Senior Acquisitions Editor and Senior Technical Editor-in-Chief in Agriculture and Biomedical Engineering for Apple Academic Press, Inc. He has worked as a Soil Conservation Inspector and as a Research Assistant at Haryana Agricultural University and Ohio State University.

During his professional career of 52 years, Dr. Goyal has received many prestigious awards and honors. He was the first agricultural engineer to receive the professional license in Agricultural Engineering in 1986 from the College of Engineers and Surveyors of Puerto Rico. In 2005, he was proclaimed as "Father of Irrigation Engineering in Puerto Rico for the Twentieth Century" by the American Society of Agricultural and Biological Engineers (ASABE), Puerto Rico Section, for his pioneering work on micro irrigation, evapotranspiration, agroclimatology, and soil and water engineering.

The Water Technology Centre of Tamil Nadu Agricultural University in Coimbatore, India, recognized Dr. Goyal as one of the experts "who rendered meritorious service for the development of micro irrigation sector in India" by bestowing the Award of Outstanding Contribution in Micro Irrigation. This award was presented to Dr. Goyal during the inaugural session of the National Congress on "New Challenges and Advances in Sustainable Micro Irrigation" held at Tamil Nadu Agricultural University. Dr. Goyal received the Netafim Award for Advancements in Microirrigation: 2018 from the American Society of Agricultural Engineers at the ASABE

International Meeting in August 2018. VDGOOD Professional Association of India awarded Lifetime Achievement Award at 12th Annual Meeting on Engineering, Science and Medicine that was held on 20–21 of November of 2020 in Visakhapatnam, India.

A prolific author and editor, he has written more than 200 journal articles and textbooks and has edited over 85 books. He is the editor of three book series published by Apple Academic Press: Innovations in Agricultural & Biological Engineering, Innovations and Challenges in Micro Irrigation, and Research Advances in Sustainable Micro Irrigation. He is also instrumental in the development of the new book series Innovations in Plant Science for Better Health: From Soil to Fork.

Dr. Goyal received his BSc degree in engineering from Punjab Agricultural University, Ludhiana, India; his MSc and PhD degrees from Ohio State University, Columbus; and his Master of Divinity degree from Puerto Rico Evangelical Seminary, Hato Rey, Puerto Rico, USA.

Preeti Birwal, PhD
Scientist (Processing and Food Engineering),
Department of Processing and Food Engineering,
College of Agricultural Engineering and
Technology, Punjab Agricultural University,
Ludhiana, Punjab, India

Preeti Birwal, PhD, is working as a Scientist (Processing and Food Engineering) in the Department of Processing and Food Engineering, College of Agricultural Engineering and Technology, Punjab Agricultural University, Ludhiana, Punjab, India.

She holds a BSc (2012) in Dairy Technology from ICAR-National Dairy Research Institute (NDRI), Karnal; MSc (2014) in Food Process Engineering and Management from NIFTEM, Haryana; and PhD (Dairy Engineering) from ICAR-NDRI, Bangalore. She is a recipient of MHRD (2008), Nestle India (2009), GATE (2012–2014), and UGC-RGN fellowships (2014–2018).

She is currently working in the area of nonthermal food preservation, fermented beverages, food packaging, and technology of millet-based beer. She has served at Jain Deemed to be University, Bangalore as a member of the board of examiners and placements. She is advising several MTech scholars in food technology. She has participated in several national and international conferences and seminars. She has delivered lectures as a resource person on doubling farmers' income through dairy technology in training sponsored by the Directorate of Extension, Ministry of Agriculture and Farmers Welfare, Government of India. She has named outstanding reviewer of the month by the *Current Research in Nutrition and Food Science Journal*. She has successfully completed AUTOCAD 2D and 3D certification.

She has 18 research papers, one edited book, five book chapters, about 28 popular articles, five conference papers, 56 abstracts, and two editorial opinions to her credit. She has successfully guided five postgraduate students for their dissertation work. She is serving as the external examiner for various Indian state agricultural universities. She is also serving as editor and reviewer of several journals. She is a life member of IDEA. Readers may contact her at: preetibirwal@gmail.com.

Monika Sharma, PhD
Scientist, Dairy Technology Division, Southern Regional Station, ICAR-National Dairy Research Institute, Bengaluru, India

Monika Sharma, PhD, is working as a Scientist, Dairy Technology Division, Southern Regional Station, ICAR-National Dairy Research Institute, Bangalore, India.

She received a BSc (2006) in Food Science and Technology from Delhi University, New Delhi; MSc (2008) in Food Technology from Govind Ballabh Pant University of Agriculture and Technology, Pantnagar; and PhD (2015) in Dairy Technology from ICAR-National Dairy Research Institute (NDRI), Karnal, Haryana, India. She started her career as a Scientist in the Indian Council of Agricultural Research in 2010; as a Scientist in ICAR-Central Institute of Postharvest Engineering and Technology, Ludhiana, Punjab, for more than five years. She is now working as a Scientist at Southern Regional Station, ICAR-National Dairy Research Institute, Bangalore, and is actively involved in teaching and research activities. She has more than ten years of research experience. She has worked in the area of convenience and ready-to-eat foods, functional foods, quality evaluation, composite dairy foods, starch modification and its application in dairy food products, etc. Presently, she is working in the area of functional and indigenous dairy foods.

She has published 25 research papers in peer-reviewed journals, three edited books, four technical bulletins, two technology inventory books, six book chapters, more than 25 popular articles, and more than 20 conference papers. She has successfully guided 6 postgraduate students for their dissertation work. She has worked as Principal Investigator in several research projects and has developed various technologies. She has also conducted entrepreneurship development programs for some of the developed technologies. She has earned several awards, such as an ICAR-JRF award and fellowship (2006–2008), first rank in all India level Agricultural Research Services (2008) examination in the discipline of Food Science and Technology, ICAR-NET (2009), conference awards, institute awards, etc. She is a life member of the Indian Science Congress and the Association of Food Scientists and Technologists (India). Readers may contact her at: sharma.monikaft@gmail.com.

CONTENTS

CONTRIBUTORS

Faizan Ahmad
Assistant Professor, Department of Postharvest Engineering and Technology,
Faculty of Agricultural Sciences, A.M.U., Aligarh – 202002, Uttar Pradesh, India,
Mobile: +91-9359230912, E-mail: f4faizahmad1989@gmail.com

Z. R. A. A. Azad
Associate Professor, Department of Postharvest Engineering and Technology,
Faculty of Agricultural Sciences, A.M.U., Aligarh – 202002, Uttar Pradesh, India,
Mobile: +91-9810810816, E-mail: zrazad@gmail.com

Harinderjeet Kaur Bhullar
PhD Research Scholar, Department of Food Science and Technology,
Punjab Agricultural University (PAU), Ludhiana – 141004, Punjab, India,
Mobile: 91-8059860824, E-mail: harinderjeet-fst@pau.edu

Preeti Birwal
Scientist, Department of Processing and Food Engineering, Punjab Agricultural University (PAU),
Ludhiana – 141004, Punjab, India, Mobile: +91-9896649633, E-mail: preetibirwal@gmail.com

Ajay Chinchkar
PhD Research Scholar, National Institute of Food Technology Entrepreneurship and Management
(NIFTEM) (Deemed to be University, Under MoFPI, Government of India), Plot No. 97,
Sector 56, HSIIDC Industrial Estate, Kundli, Sonipat – 131028, Haryana, India,
Mobile: +91-9049648454, E-mail: ajaychinchkar100@gmail.com

Rogério Marcos Dallago
Professor, Department of Food Engineering, URI Erechim, 1621 Sete de Setembro Av., Fátima,
Erechim-RS, 99709-910, Brazil, Mobile: +55-54999440174, E-mail: dallago@uricer.edu.br

Lavanya Devraj
PhD Research Scholar, Department of Food Engineering, Indian Institute of Food Processing
Technology, Pudukkottai Road, Thanjavur – 613005, Tamil Nadu, India,
Mobile: +91-9686771314, E-mail: lavanya.devraj91@gmail.com

Kamalakkannan Gomathy
Assistant Professor, VIT School of Agricultural Innovations and Advanced Learning,
Vellore Institute of Technology, Vellore – 632014, Tamil Nadu, India, Mobile: +91-9710272017,
E-mail: gomsfoodlife@gmail.com

Megh R. Goyal
Retired Faculty in Agricultural and Biomedical Engineering from College of Engineering at University
of Puerto Rico-Mayaguez Campus; and Senior Technical Editor-in-Chief in Agricultural and
Biomedical Engineering for Apple Academic Press Inc.; PO Box 86, Rincon-PR – 006770086, USA,
E-mail: goyalmegh@gmail.com

Pavan Manjunath Gundu
MTech Research Scholar, Department of Food Technology, School of Engineering and Technology,
Jain (Deemed-to-be University), Bangalore – 562112, Karnataka, India, Mobile: +91-9591007045,
E-mail: pavanmg777@gmail.com

Gurbuz Gunes
Professor, Istanbul Technical University, Faculty of Chemical and Metallurgical Engineering,
Food Engineering Department, Ayazaga Campus, Istanbul – 34469, Turkey, Mobile: +90-5365403628,
E-mail: gunesg@itu.edu.tr

Meenatai G. Kamble
PhD Research Scholar, National Institute of Food Technology Entrepreneurship and Management
(NIFTEM) (Deemed to be University, Under MoFPI, Government of India), Plot No. 97, Sector 56,
HSIIDC Industrial Estate, Kundli, Sonipat, Haryana – 131028, India, Mobile: +91-7404032936,
E-mail: meenatai1193@gmail.com

Amarjeet Kaur
Senior Milling Technologist, Department of Food Science and Technology,
Punjab Agricultural University (PAU), Ludhiana – 141004, Punjab, India,
Mobile: 91-9888466677, E-mail: foodtechak@gmail.com

Gozde Oguz Korkut
PhD Research Scholar, Istanbul Technical University, Faculty of Chemical and Metallurgical
Engineering, Food Engineering Department, Ayazaga Campus, Istanbul – 34469, Turkey,
Mobile: +90-5362409090, E-mail: korkut16@itu.edu.tr

Anjineyulu Kothakota
Scientist, Agro-Processing and Technology Division, CSIR-National Institute for Interdisciplinary
Science and Technology (NIIST), Trivandrum – 695019, Kerala, India,
Mobile: +91-9719427663, E-mail: anjineyuluk@niist.res.in

Chaitradeepa G. Mestri
Assistant Professor, College of Horticultural Engineering and Food Technology, Devihosur,
University of Horticultural Sciences, Bagalkot – 587104, Karnataka, India,
Mobile: +91-8296563842, E-mail: chaitradeepa869@gmail.com

Marcelo Luis Mignoni
Professor, Department of Food Engineering, URI Erechim, 1621 Sete de Setembro Av., Fátima,
Erechim-RS, 99709-910, Brazil, Mobile: +55-54991434000, E-mail: mignoni@uricer.edu.br

Nikitha Modupalli
PhD Research Scholar, Department of Food Engineering, Indian Institute of Food Processing
Technology, Pudukkottai Road, Thanjavur – 613005, Tamil Nadu, India,
Mobile: +91-8531081060, E-mail: nikitha.modupalli93@gmail.com

Mohan G. Naik
PhD Research Scholar, Department of Food Engineering, Indian Institute of Food Processing
Technology, Pudukkottai Road, Thanjavur – 613005, Tamil Nadu, India,
Mobile: +91-7353615709, E-mail: mohannaik023@gmail.com

Venkatachalapathy Natarajan
Professor and Head, Department of Food Engineering, Indian Institute of Food Processing Technology,
Pudukkottai Road, Thanjavur – 613005, Tamil Nadu, India, Mobile: +91-9750968403,
E-mail: venkat@iifpt.edu.in

Carolina Elisa Demaman Oro
PhD Research Scholar, Department of Food Engineering, URI Erechim, 1621 Sete de Setembro Av.,
Fátima, Erechim-RS, 99709-910, Brazil, Mobile: +55-54991106727,
E-mail: carolinae.oro@hotmail.com

Ravi Pandiselvam
Scientist, Physiology, Biochemistry, and Post-Harvest Technology Division,
ICAR-Central Plantation Crops Research Institute (CPCRI), Kasaragod – 671124, Kerala, India,
Mobile: +91-8606915321, E-mail: anbupandi1989@yahoo.co.in

Victor De Aguiar Pedott
PhD Research Scholar, Department of Food Engineering, URI Erechim,
1621 Sete de Setembro Av., Fátima, Erechim-RS, 99709-910, Brazil,
Mobile: +55-54996557286, E-mail: v.a.pedott@gmail.com

S. V. Ramesh
Scientist (Physiology), Biochemistry, and Postharvest Technology Division,
ICAR-Central Plantation Crops Research Institute, Kasaragod – 671124, India,
Mobile: +91-9993892302, E-mail: rameshsvbio@gmail.com

Nukasani Sagarika
PhD Research Scholar, Department of Food Process Engineering, College of Food Processing
Technology and Bio-Energy, Anand Agricultural University, Anand – 388110, Gujarat, India,
Mobile-+91-9265299100, E-mail:sagarikanukasani@gmail.com

Monika Sharma
Scientist (Food Science and Technology), Department of Dairy Technology, Southern Regional Station,
ICAR-National Dairy Research Institute, Bengaluru – 560030, Karnataka, India,
Mobile: +91-9915018948, E-mail: sharma.monikaft@gmail.com

Kaliramesh Siliveru
Assistant Professor, Department of Grain Science and Industry, Kansas State University, Waters Hall,
Manhattan, KS 66506, USA, Mobile: +1-6302102462, E-mail: kaliramesh@ksu.edu

Anurag Singh
Assistant Professor, National Institute of Food Technology Entrepreneurship and Management
(NIFTEM) (Deemed to be University, Under MoFPI, Government of India), Plot No. 97, Sector 56,
HSIIDC Industrial Estate, Kundli, Sonipat – 131028, Haryana, India, Mobile: +91-8199990641,
E-mail: anurag.niftem@gmail.com

Yarrakula Srinivas
PhD Research Scholar, Department of Food Process Engineering, Indian Institute of Food Processing
Technology, Thanjavur – 613005, Tamil Nadu, India, Mobile: +91-9849309525,
E-mail: srinivasyarrakula007@gmail.com

Suka Thangaraju
PhD Research Scholar, Department of Food Engineering, Indian Institute of Food Processing
Technology, Pudukkottai Road, Thanjavur – 613005, Tamil Nadu, India, Mobile: +91-9488111443,
E-mail: sukathangaraj@gmail.com

Barinderjeet Singh Toor
PhD Research Scholar, Department of Food Science and Technology,
Punjab Agricultural University (PAU), Ludhiana – 141004, Punjab, India,
Mobile: +91-8054829495, E-mail: barinderjeet-fst@pau.edu

Marcus Vinícius Tres
Professor, Laboratory of Agroindustrial Processes Engineering (LAPE),
Federal University of Santa Maria, 1040 Sete de Setembro St., Center DC, Cachoeira do Sul-RS,
96508-010, Brazil, Mobile: +55-5137248417, E-mail: marcus.tres@ufsm.br

Gunjal Mahendra Vishram
MTech Research Scholar, Department of Food Technology, School of Engineering and Technology,
Jain (Deemed-to-be) University, Bangalore – 562112, Karnataka, India, Mobile: +91-9657259325,
E-mail: mahendragunjal74@gmail.com

Giovani Leone Zabot
Professor, Laboratory of Agroindustrial Processes Engineering (LAPE), Federal University of Santa
Maria, 1040 Sete de Setembro St., Center DC, Cachoeira do Sul-RS, 96508-010, Brazil,
Mobile: +55-5137248419, E-mail: giovani.zabot@ufsm.br.

Sadaf Zaidi
Professor, Department of Postharvest Engineering and Technology,
Faculty of Agricultural Sciences, A.M.U., Aligarh – 202002, Uttar Pradesh, India,
Mobile: +91-9411212432, E-mail: sadaf63in@yahoo.com

ABBREVIATIONS

[AMPyrr][TFSI]	1-allyl-1-methylpyrrolidinium bis(trifluoromethane-sulfonyl)imide
[bmin][BF$_4$]	1-butyl-3-methylimidazolium-tetrafluoroborate
[bmin][TfO]	1-butyl-3-methylimidazolium trifluoromethanesulfonate
[bmpyrr][NTf$_2$]	1-butyl-1-methylpyrrolidinium-bis(trifluoromethylsulfonyl)imide
[C6MIM][NTf$_2$]	1-hexyl-3-methylimidazolium bis(trifluoromethylsulfonyl)-imide
[C$_8$MIM][PF$_6$]	1-octyl-3-methylimidazoliumhexafluorophosphate
[em$_2$im][NTf$_2$]	1-ethyl-2,3-dimethylimidazolium-bis(trifluoromethylsulfonyl)imide
[emin][NTf$_2$]	1-ethyl-3-methylimidazolium-bis(trifluoromethylsulfonyl)imide
[hmin][Cl]	1-hexyl-3-methylimidazolium chloride
[PMPip][TFSI]	1-methyl-1-propylpiperidinium bis(trifluoromethane-sulfonyl)imide
[PMPyrr][TFSI]	1-methyl-1-propylpyrrolidinium bis(trifluoromethanesulfonyl)imide
[pyrr][C$_7$CO$_2$]	pyrrolidinium-octanoate
ΔH	latent heat
ΔV	volume change
AA	acetic acid
AA	ascorbic acid
ABA	abscisic acid
ABS	aqueous biphasic systems
ACC	1-aminocyclopropane-1-carboxylic acid
AFILMC	ionic liquid micelle collapse capillary electrophoresis
ANS	anthocyanidin synthase
Bas	biogenic amines
BD	balanced diet
BF$_4$	tetrafluoroborate
BI	browning index

BMIM	Br1-butyl-3-methylimidazolium bromide
BMIMBr	1-butyl-3-methylimidazolium bromide
BPA	bisphenol A
C	cytosine
CA	citric acid
$CaCl_2$	calcium chloride
CCO	cytochrome C oxidase
CD	conventional hot air drying
CFU	colony forming units
CFU/ml	colony forming unit per milliliter
Chl	chlorophyll
Chl-POX	chlorophyll-degrading peroxidase
CHS	chalcone synthase
CI	chilling injury
cm	centimeter
CMs	conventional methods
CO_2	carbon dioxide
d	day
DE	dextrose equivalent
DFR	dihydroflavonol 4-reductase
DLLME	dispersive liquid-liquid microextraction
DNA	deoxyribonucleic acid
DPCD	dense phase carbon dioxide
DPPH	1,1-diphenyl-2-picrylhydrazyl
DSC	differential scanning calorimetry
DUV-LEDs	deep-ultraviolet light-emitting diodes
DWB	delignified wheat bran
E. coli	Escherichia coli
EAE	enzyme assisted extraction
EC	European Commission
EDCs	endocrine disrupting compounds
EFSA	European Food Safety Authority
EL	excimer lamps
EVA	ethylene-vinyl acetate
EVOH	ethylene-vinyl alcohol
F3H	flavanone 3-hydroxylase
FAO	Food and Agriculture Organization
FD	freeze-drying
FDA	Food and Drug Administration

FDC	fluidized bed coating
Fe_3O_4NPs	1-butyl-3-methyl imidazolium tetrafluoroborate ionic liquid, boron nitride, and magnetite nanoparticles
FOS	fructooligosaccharides
FRAP	ferric reducing/antioxidant power
g	gram
G6PDH	glucose-6-phosphate dehydrogenase
GA	gum Arabic
GCE	glassy carbon electrode
GE	gelatin
GHO	germicidal high output
GI	gastrointestinal tract
GLDH	L-galactose-1,4-lactone dehydrogenase
GO	gaseous ozone
GPa	gigapascal
GRAS	generally recognized as safe
HA	hot air
HACCP	hazard analysis and critical control point
HAV	hepatitis A virus
Hg	mercury
HHP	high hydrostatic pressure
HPAF	high-pressure assisted freezing
HPIF	high-pressure induced freezing
HPP	high-pressure processing
HPSF	high-pressure shift freezing
HS-GC	headspace gas chromatography
IB	internal browning
IL	ionic liquids
INPI-BR	Brazilian National Institute of Industrial Property
K	Kelvin
K_3PO_4	potassium phosphate
kHz	kilohertz
kV	kilovolt
L	lightness
LAB	lactic acid bacteria
LEDs	light-emitting diodes
LGaLDH	L-galactose dehydrogenase
LLE	liquid-liquid extraction

LPA	low-pressure amalgam
LPHO	low-pressure high output
LPM	low-pressure mercury
MAE	microwave-assisted extraction
MASE	microwave-assisted solvent extraction
MIL	magnetic ionic liquid
MNV	murine norovirus
MO	microorganisms
MPa	megapascal
MPM	medium pressure mercury
MSE	microwave solvent extraction
MSFE	microwave solvent-free extraction
$N(CN)_2$	dicyanamide
NCMs	nonconventional methods
nm	nanometer
NO	nitric oxide
NTf_2	bis(trifluoromethylsulfonyl)imide
NUV-LEDs	near-ultraviolet light-emitting diodes
O/W	oil in water
O_2	oxygen
O_3	ozone
OH	osmotic dehydration
P	pressure
PAL	phenylalanine ammonia-lyase
PCBs	polychlorinated biphenyls
PEF	pulsed electric field
PF_6	hexafluorophosphate
PG	polygalacturonase
pH	power of hydrogen
PL	pulsed light
PLE	pressurized liquid extraction methods
PME	pectin methylesterase
POD	peroxidase
ppb	parts per billion
ppm	parts-per-million
PPO	polyphenol oxidase
PUL	pullulan
RH	relative humidity
RI	ripening index

RNA	ribonucleic acid
ROS	reactive oxygen species
SD	spray drying
SD	surface discharge
SDH	succinate dehydrogenase
SDME	single drop microextraction
SDS	sodium dodecyl sulfate
SE	soxhlet extraction
SFE	supercritical fluid extraction
SGJ	simulated gastric juice
SIJ	simulated intestinal juice
SLE	solid-liquid extraction
SPME	solid-phase microextraction
T	thymine
TA	titratable acidity
TBA	thiobarbituric acid
TfO	trifluoromethylsulfate
Tg	glass transition temperature
TGA	thermo-gravimetric analysis
TSS	total soluble solids
TUVP	TiO_2-UV photocatalysis
UAE	ultrasound-assisted extraction
UFGT	UDP-glucose flavonoid 3-O-glucosyltransferase
USA-IL-DLLME	ultrasound-assisted ionic liquid dispersive liquid-liquid microextraction
USD	The United States Dollar
USDA	The United States Department of Agriculture
USFDA	The United States Food and Drug Administration
USPTO	The United States Patent and Trademark Office
UV	ultraviolet
UV-LEDs	UV light-emitting diodes
UV-visible	ultraviolet-visible
v	volume
VOJ	valencia orange juice
W/O	water in oil
WHC	water-holding capacity
WHO	World Health Organization
WIPO	World Intellectual Property Organization
WPC	whey protein concentrate
XeBr	xenon bromine

PREFACE

Food preservation has existed since the early times. Food is not only important for relief from hunger, but food with nutrition is also the demand of today's millennium. Food with convenience, good sensorial attributes, longer shelf life is today's challenge. The preservation in the current scenario is done by different methods like heating, cooling, drying, concentration, freezing, and cooling. All these techniques have their own pros and cons. Therefore, now a challenge to save the food from deteriorating its quality in terms of loss of nutrients, texture, sensorial characteristics with longer shelf life has been taken by the food scientists and industrialist. Several new food processing, preservation, and quality assessment technologies have been investigated by researchers to preserve the quality and design of specific nutrient-rich food.

Handbook of Research on Food Processing and Preservation Technologies is a bouquet of various emerging techniques in the food processing sector and quite relevant for the food industry and academic community. Nonthermal techniques (such as high-pressure processing [HPP], pulsed electric field [PEF], pulsed light [PL], ultraviolet [UV], microwave, ohmic heating, electrospinning, nano, and microencapsulation) are a few novel processing techniques that are being investigated thoroughly. The role and application of minimal processing techniques (such as ozone treatment, vacuum drying, osmotic dehydration, dense phase carbon dioxide [DPCD] treatment, and high pressure-assisted freezing) have also been included with a wide range of applications. Literature has reported successful applications on juices, meat, fish, fruits and vegetable slices, food surfaces, purees, milk, and milk products, extraction, drying enhancement, and encapsulation of micro-macro nutrients.

Furthermore, this handbook also covers some computer-aided techniques emerging in the food processing sector, such as robotics, radio frequency identification (RFID), three-dimensional food printing, artificial intelligence, etc. Enough emphasis has also been given on nondestructive quality evaluation techniques (such as image processing, terahertz spectroscopy imaging technique, near-infrared [NIR], Fourier transform infrared [FTIR] spectroscopy technique, etc.), for food quality and safety evaluation. A significant role of food properties in the design of specific food and edible films have been elucidated. Thus, this handbook would have significant scope in food processing, preservation, safety, and quality evaluation. The handbook is organized under five volumes.

Volume 1: Nonthermal and Innovative Food Processing Methods attempts to illustrate various applications of novel nonthermal food processing techniques. This book will serve many professionals working in the area of food science technology and engineering around the world. The book will also act as a reference book for researchers, students, scholars, industries, universities, and research centers. Each chapter covers major aspects pertaining to principles, design, and applications of various novel and advanced food processing techniques. This book is divided into two parts. Part I: Emerging Nonthermal Approaches of Food Science and Engineering discusses various techniques viz. HPP, a UV light for fresh produce, high-pressure assisted freezing (HPAF) of foods, novel microwave application for extraction of food components, and application of microencapsulation for the development of probiotic foods. Part II: Prospects of Innovative Food Preservation Strategies emphasizes DPCD-aided preservation of foods, applications of ionic liquid in food science and food processing, application of ozone in the food industry, and recent advances in osmotic dehydration.

This book volume will serve as a valuable resource of information and excellent reference for researchers, scientists, students, growers, traders, processors, industries, and others for food preservation, processing, and quality aspects of food products.

This book has taken its present shape due to the excellent contribution by the contributing authors, who have been the soul of this compendium. We have mentioned their names in each chapter and also in the list of contributors. We are indeed indebted to them for their knowledge, dedication, and enthusiasm. We expect this book to prove a helpful resource for all the food processing and engineering academicians, food processors, and students.

We also extend our sincere thanks to Apple Academic Press, Inc., for their immense facilitation and suggestions right through this assignment.

We take this opportunity to thank: (1) our families for their motivation, moral support, and blessings in counteracting every obstacle coming in the way; (2) our spouses for their understanding, patience, and encouragement throughout this project; (3) our institutes: PAU, Ludhiana, Punjab;

ICAR-NDRI, Karnal; Southern Regional Station, ICAR-NDRI, Bangalore; and UPRM, Mayaguez, for their support during the compilation of this publication.

— Editors

PART I

Emerging Nonthermal Approaches in Food Science and Engineering

CHAPTER 1

HIGH-PRESSURE PROCESSING: POTENTIAL APPLICATIONS FOR FOODS

K. GOMATHY, R. PANDISELVAM, ANJINEYULU KOTHAKOTA, and S. V. RAMESH

ABSTRACT

High-pressure processing (HPP) is a non-thermal technology that involves the application of 100–600 MPa to food products. The food is sealed hermetically inside a flexible packaging material, and the high-pressure is transmitted to the food through a pressure transfer medium. The pasteurization of food products by uniform application of pressure is based upon Le Chatelier's principle, the principle of microscopic ordering, and the isostatic rule. With modern technical advancements, HPP has gained wide applications in the food industry, including the processing of fruits and vegetables, ready-to-eat foods, meat-based products (such as sausages, frankfurters, etc.). The application of high-pressure targets the cell membranes, ribosomes and denatures the protein of the microbial cell leading to its inactivation and lysis. HPP guarantees food safety and quality without altering much of its nutritional, sensorial, and functional properties.

1.1 INTRODUCTION

The never-ending consumer demand for safe, convenient, and quality food products has urged the food industries to hunt for novel food processing technologies. The consumer's preference for minimally processed foods is witnessing exponential growth. This demand has diverted the research and developments in public-funded research institutes as well as in industries towards the development of novel non-thermal technologies, which tend to maintain the fresh-like properties of the food products along with its extended shelf-life. Some of the non-thermal technologies are high-pressure processing (HPP), PEF, irradiation, ozone, cold plasma, and ultrasound [8, 16, 37].

The efficient and economical production of quality food products are now becoming the top priority of food industries. At the same time, the industries have started adopting innovative techniques, which are eco-friendly and utilize minimal energy and water. One such emerging non-thermal food processing technique is high-pressure processing (HPP) [6, 27, 29, 46, 64, 71, 74, 102]. Though the technology was introduced a century back, yet the commercial and consumer acceptance has happened only during recent years. The technology has been widely adopted by the food industries in United States, South Korea, Japan, and Spain, accounting for sales of about US $10 billion [47].

HPP technology allows the extension of the shelf life of the product without altering much of its physical, biochemical, and nutritional characteristics. Since the food quality is directly related with the biochemical, nutritional, and microbial attributes of the foods, HPP guarantees the production of high-quality food products [82]. The consumer's perception of high-quality foods depends on the fresh-like properties of the food even after processing. HPP is often termed as cold pasteurization technology, wherein the pre-packed food material either in the solid or liquid state is exposed to 300–600 MPa for a shorter holding time [35, 83]. However, the process does not entail any heating source, yet the increase in pressure results in heat generation in a minimal amount that does not affect the food quality. This attribute has made the HPP processed products comparable to the unprocessed products [51, 96].

High-pressure food processing was initially experimented in the University of West Virginia, USA [44] for extending the shelf life of milk for a period of 4 days. The high-pressure of about 600 MPa was applied to milk for one hour at room temperature. During the 1980s, the research conducted at the University of Delaware (USA) showed the reduction of microbial and enzymatic activities of high-pressure processed foods without much alteration of basic natural compositions. Unlike the thermal treatments, HPP deactivates microorganisms of the food and has the ability to produce fresh-like products with better functional properties [2]. This unique property of HPP led to the widespread development of this technology by food industries across many countries. Thereon from the 1990s, the technology started gaining momentum, and many Japanese programs on HPP created awareness among the research institutes, corporate of different countries.

During 1988–1993, a research consortium was formed by the Japanese food industry for the commercialization of HPP. It was followed by the US Army initiative to conduct research on HPP during 1993–1995 by signing a joint contract with Professor Hoover of the University of Delaware and Professor Farkas of Oregon State University. HPP studies were widened for the processing of various value-added fruits and vegetable products (like juices, salad dressings, sauces, jams, and jellies) [43, 70]. In addition, innovations and interventions were made in HPP towards raw ingredients aiming at advanced and fresh-like sensorial and functional properties [12].

The high-pressure processing technology entails sealing the sample within the packaging material and placing it within a high-pressure chamber. The packaging material must withstand the sample being exposed to higher pressures of about 300–600 MPa. The adhesion between the multi-layered pouches needs to be highly stable to contain the sample during processing. Even hairline leakages in the package could cause a threat to the microbial contamination of the product [50].

Packaging materials selected for HPP should be highly impermeable to water vapor, oxygen, and moisture and are able to retain the color, fresh flavor, and aroma of the contained sample. The heavy capital cost of the pressure vessel and the allied components of the HPP set up had resulted in the limited commercial success of this technology [10]. However, with the advancement of science and technology, the equipment cost has been reduced, and commercialization has started up in different countries. More than 230 HPP machines have been installed globally for producing HPP food products during 2012. The first commercially processed HPP food was released by Japan in 1990 by Meidiya [97]. Currently, commercial HPP food products, namely ready-to-eat meats and meat products, fruit juices, and jams are available in U.S.A., France, Italy, Portugal, Spain, and Poland.

The US Department of Agriculture (USDA) and US Food and Drug Administration (FDA) has recognized HPP as a safe post-packaging pasteurization technology for shelf-stable high acid foods provided the processing is achieved targeting the *E. coli* O157:H7 as an indicator strain. FDA in 2009 had approved HPP for the manufacturing of low acid foods and formulated suitable regulations. Since 2004, the Canadian markets have witnessed the HPP processed foods. Since then, Health Canada assessed various HPP processed foods, such as applesauce, fruit juices, ready-to-eat meat products, etc. Initially, the European Commission (EC) had considered the available legislation on novel foods for HPP food products, but later after repeated assessments, EC declared that HPP is not to be considered as a novel food and has done away with the need for pre-notification (Division 28 Part B of FDA).

The manufacturer is solely responsible for ensuring that the packaging material containing the HPP product do not cause any health hazard (Division 23 Part B of FDA). In addition, the EC has formulated a research project on HPP to unlock the practical feasibility of HPP for commercialization. The third-party purchasing the HPP processed product as their raw material is to be informed prior to consider the HPP treatment given to the material in order to follow their HACCP (hazard analysis and critical control point) assessments accordingly. In India, consumer awareness on HPP foods is very minimal. The first HPP system in India has been established in the Defense Food Research Laboratory, Mysore-Karnataka, with a maximum operating pressure of 900 MPa and with a capacity of 2 L [34].

This chapter focuses on the construction of batch and continuous type HPP equipment, principles of microbial destruction through HPP, and various applications in the field of food processing.

1.2   EQUIPMENTS FOR HIGH-PRESSURE PROCESSING (HPP)

The HPP system consists of the high-pressure vessel, enclosure valves, plugs, and wedges for closing the vessel, intensifier pumps, data monitoring and acquisition system, thermocouples, and temperature logger, product handling, and conveying system [84, 105]. Among all the components, the pressure vessel is the most critical element as the design of the pressure vessel should withstand high-pressure in a safe manner. The commercial production capacity of about 100 L, pilot-scale HPP vessels of 10–25 L, and lab-scale pressure vessels of 0.1 to 2 L capacity are commercially available in the market.

1.2.1   PRESSURE VESSEL

Hite [44] used a thick-walled built-in high-pressure forged monolithic cylindrical vessel. These high-pressure vessels can be mounted either in horizontal, vertical, or even in tilted mode. The thickness of the pressure vessel depends upon the maximum operating pressure required for the process, the capacity of vessel, method of loading and unloading, temperature rise, diameter of the vessel, and frequency of pressurizing cycles. Decreasing of working pressure prolongs the durability of the vessel. High tensile steel alloy monoblocs are used upon applications of pressures from 400–600 MPa. Modern processing uses HPP vessels using concentric cylinders of very high mechanical strength. The inner cylinder contains the food product and is often chosen to be food grade and non-corrosive stainless steel. The outer cylinder compressing the inner cylinder is usually wire-wound. Wounding high tensile tensed wires around the thin-walled vessel enhances the strength of the cylinder against the developing internal pressure. This wounding puts the outer cylinder under constant compression, thereby reducing the hoop and longitudinal stresses acting inside the cylinder. This type of pre-stressing improves the number of loading cycles of about 100,000 at a pressure of 680 MPa and more. For higher-pressure applications, the HPP vessel is fabricated using the pre-stressed vessel techniques. This confirms the safety and reliability of the equipment upon operation. The common techniques for the fabrication of pre-stressed cylindrical vessels include autofrettage, heat-shrink, and wire-wounded techniques [54, 67].

1.2.1.1 AUTOFRETTAGE TECHNIQUE

It is similar to the process of pre-stressing concrete for building bridges. The technique strains the material artificially, creating a stress higher than the elastic limit dragging the inner cylinder into a plastic effect. It is a technique used for improving the fatigue and durability of the cylindrical pressure vessel by means of building a locked in residual compressive stress at the bore. Autofrettage creates a high-pressure, which deforms the central bore plastically without cracking it down. This process creates a very little stress on the outside cylinder causing it to expand elastically. When the pressure is removed, the outer wall tends for the elastic recovery, but the plastic deformation of the inner cylinder resists this recovery. The elastic stress on the outer cylinder tries to overcome the inner cylindrical bore, and it instills a hoop stress of compressive nature. This ultimately causes the continuous residual compressive stress conditions for the outer cylinder.

Autofrettage minimizes the adverse effects of pressure fluctuations on the inner cylinder of the pressure vessel. Autofrettage can increase the pressure bearing capacity of the cylindrical pressure vessel by 30% to 200%, depending on the intensity of the initial stress provider. This implies that increasing the pressure carrying capacity of the vessel enhances the magnitude of operating cycles.

The various methods to incorporate autofrettage are peenage, swage autofrettage, and hydraulic autofrettage. Peenage involves bombarding or hammering the surface of the material with metallic spheres called shots by the cold working process. This results in surface yielding, but resistance is offered by the bore, thereby inducing residual compressive stress.

Swage autofrettage technology was introduced by Davidson [7]. The inner cylinder is plasticized and deformed by introducing the interference of an oversized mandrel into the cylinder forcefully. The mandrel usually consists of two interconnected conical sections. Three types of swaging used for creating deformation within the inner cylinder include mechanical push, mechanical pull, and hydraulic push swaging. A ram is forced against the mandrel in mechanical push swaging. In the case of mechanical pull swaging, an overhead crane is used for pulling the mandrel vertically. In hydraulic push swaging, hydraulic fluids (such as mixture of water and glycerin) are used to create the pressure to force against the mandrel. During the swaging processes, high amounts of sliding frictional forces are developed between the mandrel and the cylinder. Hence lubrication using a mixture of molybdenum disulfide oil and copper plating are carried to prevent any frictional losses. Figure 1.1 represents the pre-stressing of pressure vessels using the swaging method.

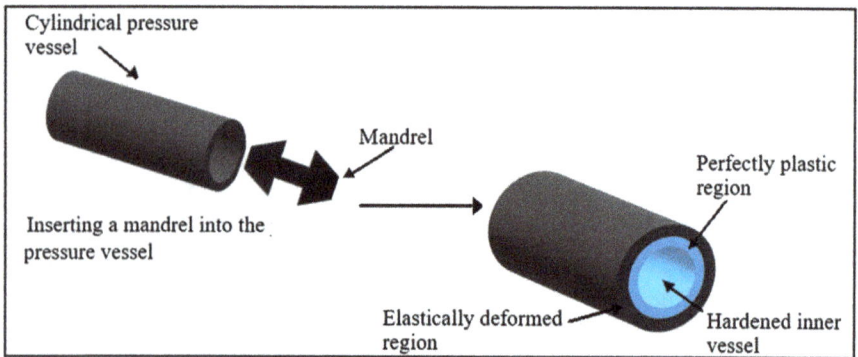

FIGURE 1.1 Designing a pre-stressed pressure vessel using autofrettage technique: swage method.

The advantages of using the autofrettage technique include increased fatigue life, reduction in cylindrical vessel wall thickness without compromising on the working pressure. When the working pressure is not altered, the cost of equipment could be reduced. The finishing for boring and surfaces could be eliminated, and the service life of the vessel can be improved for the given working pressure.

In hydraulic autofrettage, a cylindrical vessel is subjected to a hydrostatic pressure by means of hydraulic oil pumped through the pump. This process results in plastic deformation of the inner cylinder as the yield strength of the material becomes comparatively lower than the equivalent stress of the inner cylinder. During this process, a solid spacer is inserted into the inner cylinder to reduce the volume of hydraulic oil pumped into the cylinder. Upon the release of the pressurized oil, the outer cylinder tries to return to its original form, but the plastically deformed inner cylinder prevents this from occurring as it had already undergone deformation. A cylinder hydraulic autofrettage is represented in Figure 1.2.

The hydraulic autofrettage is a time-consuming process, and care must be taken to prevent the spillage of the highly pressurized oil. The technique is expensive because of the costly hydraulic power pack for pressurization.

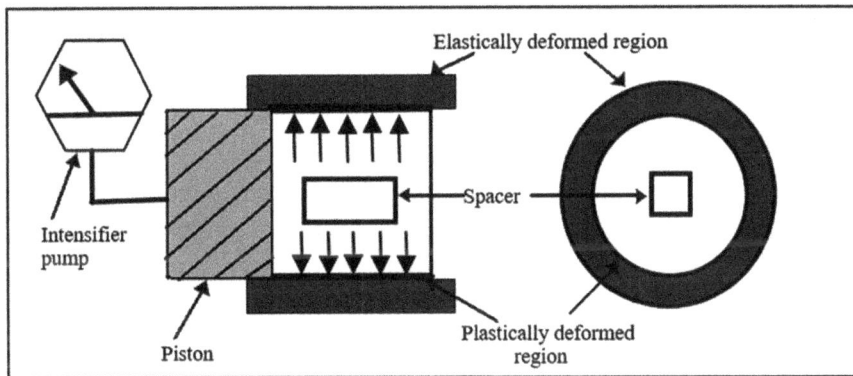

FIGURE 1.2 Designing of a pre-stressed pressure vessel using autofrettage technique: hydraulic method.

1.2.1.2 *HEAT SHRINK TECHNIQUE*

This technique is applied to both concentric outer and inner cylinders. This technique involves the application of heat to the external cylinder until it expands and alternate cooling of the inner cylinder until it shrinks. The outer layer is then arranged on the inner cylinder, and the entire assembly is cooled to room temperature, which creates hoop stress to develop and maintain the inner cylinder under compression. Figure 1.3 represents the pre-stressed pressure vessel using the heat shrink technique.

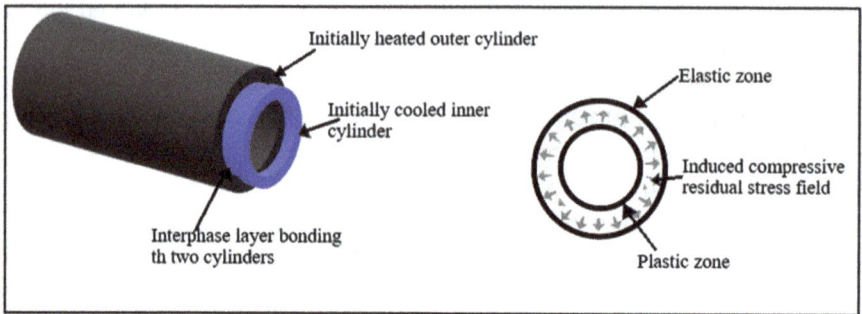

FIGURE 1.3   Designing of a pre-stressed pressure vessel using autofrettage technique: heat shrink technique.

1.2.1.3   WIRE WINDING

The wire winding technique involves winding of wire in a helical manner in pretension around the outer wall of the inner cylinder (Figure 1.4). With the increase in the number of windings around the cylinder, compressive stresses pile up in the cylinder. The internal wire windings experience lesser stress due to the radial compressive stresses created by the last wound layer. The pressure vessel with the inner cylinder now experiences compressive stress, whereas the wire-wound block undergoes tensile stress. During operation, compressive stress on the inner cylinder decreases, whereas the tensile strength of the outer wire wounding increases.

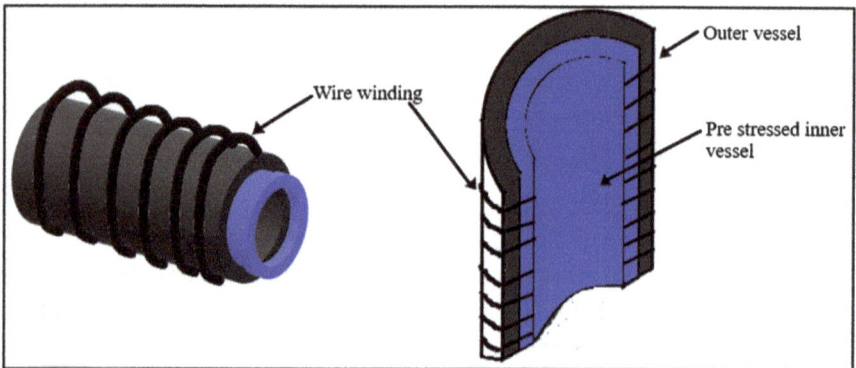

FIGURE 1.4   Designing of a pre-stressed pressure vessel using the wire-winding technique.

The wire-wound pressure vessels are safe compared to other techniques as it has a leak system prior to the breaking set up. The wire-wound pressure vessels are lighter in weight, durable, and could withstand millions of cycles of working pressure. During operation, the inner diameter of the cylinder will undergo shrinkage as the wire bonding induces high-pressure leading to the residual compressive stresses within the vessel wall. Hence, even under high internal pressure, the tensile stresses would not increase. The wire-wound technique increases the strength to weight ratio and fatigue life of the pressure vessel [54, 91, 108].

1.2.2 PRESSURIZING PUMP

There are two important means by which pressure is generated within the high-pressure processing system. They are direct and indirect compression. Direct compression is accomplished using a low-pressure pump. The closure of the vessel is connected to a piston filled with fluid. The principle of increasing the pressure is achieved by reducing the specific volume within the inner cylinder of the pressure vessel. Figure 1.5 depicts the direct and indirect compression system.

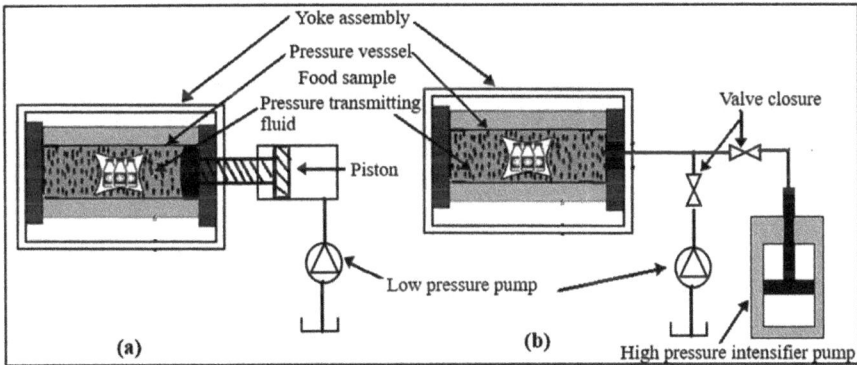

FIGURE 1.5   (a) Direct and (b) indirect compression systems.

The compression is obtained at a faster rate within the pressure vessel, but the utilization of the dynamic sealing principle between the mobile pressure vessel and the stationary piston makes it unsuitable for pilot-scale applications [7]. In the indirect compression method, a high-pressure intensifier pump is employed. The pressurizing fluid is stored in the reservoir tank from where it is pumped through the high-pressure tubes into the pressure vessel. The required holding time can be set up using this intensifier pump. The compression time within the pressure vessel is a function of the horsepower of the pump. In general, a 100 hp pump can produce an operating pressure of about 680 MPa in a 50 L capacity vessel in 3–4 minutes.

1.2.3   PRESSURE TRANSMITTING FLUID

Samples, which are loaded within the packaging materials, are pressurized using the pressuring fluids, whereas the liquid food samples are pumped directly into the pressure vessels, and they themselves act as the pressurizing medium. The pressure transmitting fluid transmits pressure uniformly throughout the entire product instantaneously. The factors involved in the selection of the pressure transmitting fluid are type of HPP equipment, range of process temperature, viscosity, and corrosion-resistant ability of the fluid [45]. Water is the most commonly used pressure transmitting fluid in the food industry. Other food grade fluids include ethanol, castor oil, glycol, sodium benzoate, and silicone oil.

1.2.4   BATCH AND CONTINUOUS TYPE HPP SYSTEMS

The food product required to be HPP processed is packaged into a flexible container so as to withstand the pressure of processing. The general processing of the HPP unit is represented in Figure 1.6. The packaging pouches may be multi-layered, polymeric bottles, or even tray-lid packages. The important factors are good transparency and visual integrity, gas, and water vapor barrier properties, moisture barrier property, heat withstanding ability, headspace gas mixture, printability, flexibility, high seal strength of the laminated adhesive [28]. Polymer-based materials allow excellent visual transparency, while aluminum and metals lack this property. The adhesive used between the layers of the pouches is altered during the HPP treatment. If the integrity of the adhesive layer is lost, then it can pose a threat to the microbial safety of the HPP processed product. The polymers of the packaging material should tolerate the volume changes of the product resulting from the pressurized compressions. Due to this salient feature materials, such as cardboard, rigid packaging materials (such as glasses and metals) are not preferred for HPP treatment [17, 69].

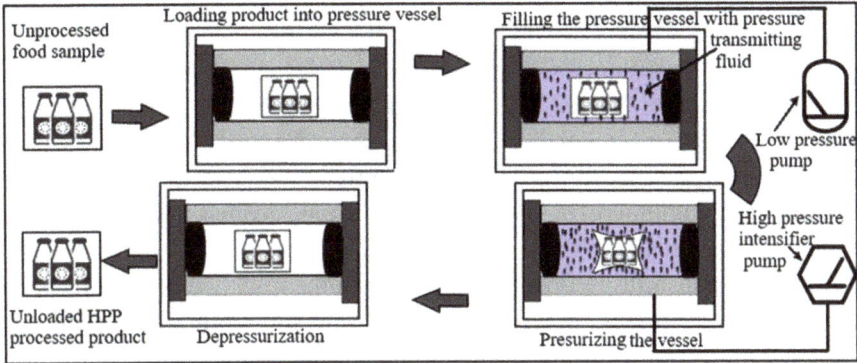

FIGURE 1.6 General processing and workflow in a HPP unit.

The integrity of the packaging material should prevent delamination, holes, and other such defects. Gas and water vapor barrier properties vary with the nature of the food sample being processed. Industrially 12% relaxation from the standard oxygen and water barrier levels are permitted. In addition, the changes in tensile strength, elongation, elasticity modulus, as well as seal strength should be within 25% [57].

The packaging material should have good thermal conductivity to enhance fast heating as well as cooling. Foil laminated materials tend to have a better heating rate compared to polymeric materials owing to the aluminum layers within the pouch [55]. The copolymer of ethylene and vinyl alcohol, namely EVOH can maintain the physical barrier properties as well as in maintaining the visual integrity of the packaging film under high-pressure applications. In addition, intelligent, active, and convenient packaging parameters could also be incorporated into the packaging films.

The packaged food within any of the selected packaging material is then loaded into the pressure vessel, and the required pressure is applied. According to the Universal Gas Equation (PV = MRT), the process of pressurization results in a decrease in volume. However, during decompression, expansion occurs; hence the packaging material should withstand a 10–20% decrease in its volume decrease during pressurization and should be able to regain its original volume upon the release of the applied pressure. There are two methods of applying HPP technology for food systems, namely, batch and continuous processing.

1.2.4.1 BATCH TYPE SYSTEM

Batch type HPP systems are utilized for processing solid-state and liquid state food products in any kind of packaging materials, namely, pouches, cups, or even bags. In batch type HPP system (Figure 1.7), the air within the pressure vessel is displaced before the treatment process to reduce the compression pumping cost. The food pouches are loaded into the pressure vessel and then sealed tightly by means of closures and yoke threads. Then the pressure transmitting fluid is pumped from the reservoir tank and passed into the pressure vessel. Widely the pressure transmitting fluid being selected for many industrial applications is water only. The vessel is then pressurized by the intensifier pump either under direct compression or indirect compression mode.

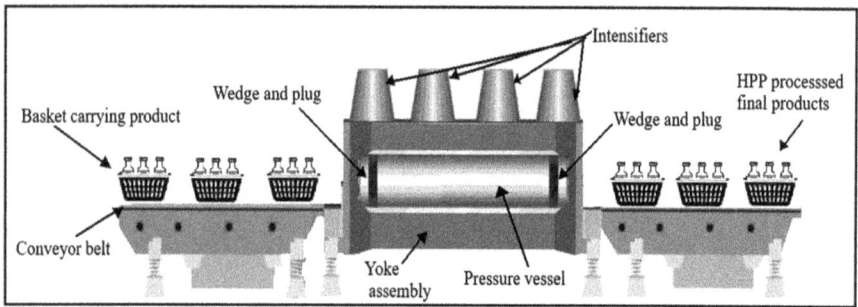

FIGURE 1.7   Batch type HPP processing unit.

The rate of pressurization is proportional to the horsepower of the pump selected. Pressurization of 100 MPa causes a 4% reduction in volume, whereas a 15% reduction in volume is obtained at a pressure of 680 MPa [103]. Once the desired pressure is achieved for a given time (holding time), the depressurization of the vessel is performed by removal of the closures, and the pressure transmitting fluid is pumped back into the reservoir tank [33]. The HPP processed product is then unloaded from the vessel and then goes for tertiary packing and ready for sales and distribution.

1.2.4.2 CONTINUOUS HPP SYSTEM

The continuous HPP system is applicable only for liquid foods [9]. A valve is opened for the inlet of the food inside the inner cylinder of the pressure vessel. A floating cylinder divides the liquid food sample from the pressure transmitting fluid. The high-pressure intensifiers pressurize the transmitting fluid leading to the compression of the liquid food sample. The product is held in the desired pressure level for a predetermined period of time, and then the pressurizing fluid is depressurized and released from the bottom valve of the pressure vessel. The pressurized product is passed through a pressure release component, wherein it is subjected to shear, cavitational, and other frictional changes. This shearing action leads to a rise in temperature within the product. Though the increase in temperature contributes to microbial destruction the textural, yet functional properties of the food product may get altered. This depressurizing reduces the excessive shearing and increase in temperature of the product. The liquid food sample is decanted through a separate valve located at the top of the vessel. A balance tank assembled along with the system constantly maintains the flow level of the product into the vessel. Thus, there is a continuous outflow of HPP processed liquid food sample.

1.2.4.3   SEMI-CONTINUOUS HPP SYSTEM

The high shearing and heating effect of continuous systems can be overcome by means of the semi-continuous process system, which consists of series of pressure vessels, low-pressure pumps, free piston, high pressurizing intensifiers, holding, surge tanks, and control valves [21]. The food sample is pumped into the pressure vessel by means of the low-pressure pumps. A free piston floating within the pressure vessel gets relocated as the vessel is filled with the sample. The inlet valve is then sealed, and next, the pressure transmitting fluid is pumped into the vessel, which compresses the free piston against the food sample. After achieving the necessary holding time, the vessel is depressurized, and the piston is again relocated to its initial place. The HPP food product is then transferred into the surge tank through the steel pipes. The processed product is then aseptically packed and distributed.

1.3   PRINCIPLE OF OPERATION OF HPP

According to the ideal gas law, the increase in pressure is accompanied by a decrease in volume and an increase in temperature and free energy called Gibbs energy. The preservation of food products through HPP [107] is explained based on the three important principles, namely Le Chatelier's principle, the principle of microscopic ordering, and isostatic principle:

1. **Le Chatelier's Principle:** Here, any chemical system in an equilibrium condition undergoing a reaction change is complemented with a reduction in volume when enhanced by pressure [49, 60].
2. **Isostatic Principle:** It states that the pressure applied is transmitted instantaneously and uniformly throughout the entire sample, and there exists no pressure gradient. Since the pressure transmitted is not influenced by the attributes of the food product (such as size, shape, and geometry, etc.), the uniform transmission of pressure occurs [107]. Once the pressure is released, the food returns back to its original shape.
3. **Microscopic Ordering Principle:** At a given temperature, a corresponding increase in the degree of ordering of molecules is observed with the increase in pressure. Both pressure and temperature exert an incompatible reaction over the molecular structure and chemical reactions [14].

1.4 PRINCIPLE OF MICROBIAL DESTRUCTION

The effectiveness of HPP on the population of microorganisms depends upon the chemistry of the food product and the type of microorganism [3]. Understanding the basic mechanism of cell damage and repair mechanism is helpful in achieving a greater preservation effect of HPP [26, 40]. The food sample is thoroughly analyzed prior to the HPP treatment. Yeast and molds are highly sensitive to HPP, and most of the vegetative species are inactivated at a pressure range of 300–400 MPa at room temperature. Research studies have proven that the inactivation of yeast inoculated within orange juice is achieved at a high-pressure of 200–300 MPa [56]. However, the efficacy of HPP in the inactivation of microorganisms under some conducive conditions has been proven the potential of the microbes to recover to its original condition [50].

The pressure of about 300 MPa may damage the intracellular and cell envelope of *L. lactis* subsp. *cremoris*. Microbial cells treated at lower pressure of 100 MPa exhibited an increase in reducing sugars, which clearly indicated that lower pressures could potentially activate the cell wall hydrolases. At pressures >400 MPa, the cell wall hydrolase activity was reduced [99].

High-pressure applications of above 300 MPa generally cause a significant reduction of vegetative microorganisms [108].

The vegetative microorganisms are inactivated in HPP treatment due to the structural changes in the cell membrane and interfering with the enzymatic metabolism of the microorganism [53]. High-pressure treatments may denature the enzymes required for carrying certain metabolic functions (Figure 1.8).

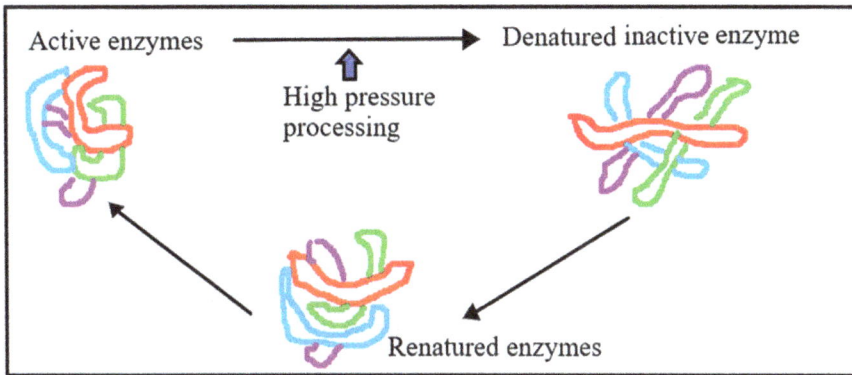

FIGURE 1.8   Effect of HPP on the microbial enzymes.

The lipids and proteinaceous membranes are the primary sources of the target in vegetative microbes. A reversible phase change and macromolecular changes in the protein units are documented when high-pressure of 100–200 MPa are applied to the food product. When the pressure exceeds 220 MPa, interface separation, and protein unfolding occur, causing irreversible destruction to the microbial membranes [81, 104]. Bacterial spore formers are highly resistant to higher pressures. The DNA binding proteins, which are acid-soluble spore proteins contributing to about 10–20% of the total spore proteins; protect the bacterial spore against the physical, chemical, or even pressure changes [92]. Spore inactivation could be achieved by combining other techniques along with HPP.

Ritz et al. [86] studied the proteins of *S. typhimurium* by electrophoretic profiles and found that pressure treatment destroyed major and minor proteins in the bacteria. From initial presence of three major protein only two were observed after treatment where as minor 12 proteins were completely disappeared. In addition, conformational changes of the proteins are observed under stronger pressure resistance. In the case of *L. mesenteroides*, HPP treatment of 345 MPa for 5 min at 25°C inhibited the cells from synthesizing

ATP, resulting in degradation of enzymes in the cell walls. This was due to the formation of pores in the cell walls and cell membranes, making it permeable and thereby reducing the potential gradients across the membrane [52]. Figure 1.9 represents the pore formation in the cell wall of microbes with HPP. Temperature (80–90°C) in combination with high pressures above 1000 MPa has been found to inactivate the bacterial spores successfully [85].

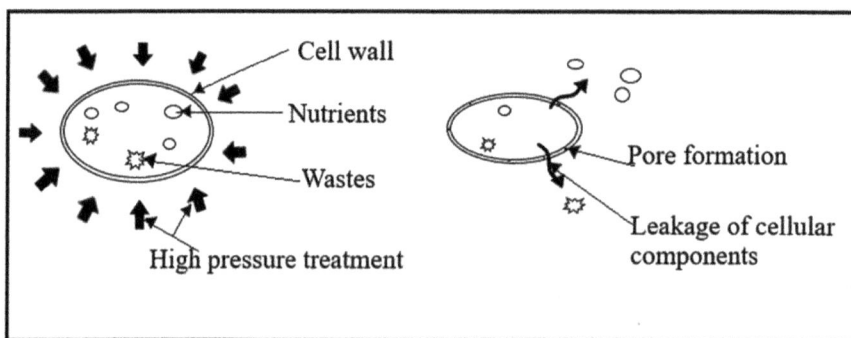

FIGURE 1.9   Pore formations on the microbial cell by HPP.

The clostridium spores were found to be significantly reduced at a pressure-temperature combination of 690 MPa, 80–90°C, at a 20 min holding time period [30]. Even at a moderate temperature of 70°C along with pressure, *Bacillus* strains were deactivated [41]. When bacterial cells experience high pressures, the membranes are damaged, and the mechanism of transport phenomena involved in nutrient intake is affected. Analyzing the cell suspending fluids obtained after application of pressure revealed much of the intracellular fluid components suggesting the cell wall leakages upon high-pressure applications [94]. Though the bacterial spore destruction has been successful following the combination of heat treatment and high pressure, yet the efficacy also depends upon the adiabatic heating process, pH of the sample, and its initial temperature [19, 70].

1.5   APPLICATIONS OF HPP FOR DIFFERENT FOOD PRODUCTS

High-pressure processing has overcome conventional thermal processing by means of high nutrition retention and guaranteed microbial safety. HPP was applied to the strawberry puree, and it was found to enhance the nutritional and sensorial qualities of the puree in comparison with the thermal pasteurization technique [66]. Many studies have been conducted to examine the changes in the physical, biochemical, and sensorial properties of various fruits and vegetable-based products with the application of HPP.

Table 1.1 lists some of the research findings on HPP processed fruits and vegetable products. Most of the physicochemical changes happen during the storage period due to partial enzyme inactivations and microbial growth. HPP treated pomegranate juice at a pressure of 350–550 MPa for a period of 30–50 s showed significant variations from the initial pH, Brix, and acidity value after being stored in refrigerated conditions for a period of 15 days [100]. The effect of HPP on the color retention of products depends not only on the innate pigments but also on the reaction of the browning pigments under different pressure treatments. HPP at room temperature retains carotenoids, whereas anthocyanin remains unstable [36].

TABLE 1.1   Application of HPP on Fruits and Vegetable Products

Fruit and Vegetable Products	Processing Conditions	Outcome	References
Apple Juice	500 MPa, 1 min, 25°C	A synergistic effect of TiO_2-UV photocatalysis (TUVP), high-pressure, and temperature in microbial inactivation of commercially available apple juice was observed.	[93]
Apple Puree	400–600 MPa, 15 min, 20°C	High-pressure of 600 MPa dint affect the nutritional and instrumental quality of the apple puree compared to the pressure of 400 MPa.	[58]
Apple, Orange Juices	200–600 MPa, 10 min, 20–60°C	High-pressure combined with heat treatment resulted in the inactivation of acidoterrestris successfully in the commercial samples of apple and orange juices.	[39]
Banana Smoothie	550 MPa, 2–10 min, 20°C	HPP in combination with N_2 infusion into banana smoothies resulted in no significant changes in the microbiological and physicochemical qualities, even after a storage period of 15 days at 4°C	[62]
Cantaloupe Puree	300–500 MPa, 5 min, 8–15°C	HPP resulted in guaranteed microbial destruction of microorganisms in Cantaloupe Puree	[72]

TABLE 1.1   *(Continued)*

Fruit and Vegetable Products	Processing Conditions	Outcome	References
Cherimoya Pulp	600 MPa, 8 min, 20°C	The HPP in combination with the effect of enterocin AS was studied on the cherimoya pulp. The combined effect was effective in preserving the microbial quality of the pulp compared with the HPP treatment alone.	[79]
Guava puree	600 MPa, 3 min	No significant change in the color value of the high-pressure processed guava puree when it was stored at 4°C for a period of 45 days	[48]
Keiskei Juice	550 MPa, 1.5 min	HPP inhibited the activity of yeast, mold, coliforms, as well as *Pseudomonas,* but the *Bacillus cereus* in the Keiskei Juice could not be controlled. *Bacillus* exhibited its growth within 8 days of refrigerated storage at 4°C	[22]
Mandarin Juice	150–450 MPa, 0–1 min, 15–45°C	HPP was found to be the best alternative preservation techniques for the preservation of mandarin juice as it inhibited the microbe *Lactobacillus plantarum* in the juice in comparison with thermal and high-pressure homogenization techniques	[20]
Mango Nectar	600 MPa, 1 min, 20°C	HPP did not create any change in viscosity of the mango nectar immediately after treatment but exhibited a gradual decrease in viscosity during storage	[63]
Orange Juice	400 MPa, 1 min	The color, Brix value, pH of the orange juice remained unaltered after high-pressure treatment. A 35.1% decrease in ascorbic acid (AA) concentration and 34.7% decrease in pectin methylesterase activity of the juice were observed	[106]

TABLE 1.1 *(Continued)*

Fruit and Vegetable Products	Processing Conditions	Outcome	References
Orange juice mixed with milk	100–400 MPa, 20–42°C, 2–9 min	High-pressure dint has a significant effect on the pH and Brix of the juice, but the turbidity of the juice was reduced as the pressure increased beyond 200 MPa	[11]
Pear Nectar	103–241 MPa, 2 s–15 min, 25°C	The microbes inoculated in pear nectar, namely *E. coli*, *S. cerevisiae*, and *L. innocua* were successfully inactivated by HPP without the aid of thermal treatments	[38]
Pomegranate juice	350–550 MPa, 30–50 s	Pressure-treated pomegranate juice exhibited changes in pH, Brix, and acidity after being stored in refrigerated conditions for a period of 15 days	[11]
Red grape, white, grape juices	150–250 MPa, 5–15 min, 20–40°C	HPP inhibited the formation of 5-hydroxymethylfurfural under all pressures in both red and white grape juices	[68]
Sauerkraut fermented cabbage	300 MPa, 10 min, 40°C	HPP was found to be an alternative technique in enhancing the microbial safety and shelf life of sauerkraut even after storing up to a period of 3 months under refrigerated conditions	[78]
Sour Chinese cabbage	400–600 MPa, 10–30 min	HPP at 600 MPa was found to be a highly acceptable method for preserving Sour Chinese Cabbage as it inactivated the total aerobic bacteria, lactic acid bacteria (LAB) much within the standard permitted levels	[61]
Soy smoothie	550–650 MPa, 3 min, 20°C	HPP of 450 MPa resulted in the smoothie with very high acceptability in organoleptic evaluation similar to that of fresh like smoothies	[4]
Strawberry Puree	300–500 MPa, 1–15 min, 50°C	HPP enhanced the nutritional and sensorial qualities of the strawberry puree in comparison with the thermal pasteurization technique	[66]
Watermelon juice	300–900 MPa, 60°C, 5–60 min	As the pressure increased beyond 600 MPa, the browning in watermelon juice was prevented	[7]

When strawberry was treated with high-pressure (200–700 MPa, 80–130°C), the anthocyanin was degraded at a faster rate. The degradation followed the first-order kinetics [101]. An insignificant effect of HPP on pigments at ambient temperature has been reported in the literature [1, 74]. There was no significant change in color index at high-pressure (600 MPa, 3 min) processing of guava puree, when it was stored at 4°C for a period of 45 days [48].

However, HPP applied at high temperatures may result in loss of color. HPP at 400 MPa for 1 min resulted in 35.1% decrease in ascorbic acid (AA) concentration and 34.7% decrease in pectin methylesterase (PME) activity of orange juice [106]. HPP at 100–400 MPa along with of 30–60°C was able to retain 91% ascorbic acid in the orange juice-milk beverage [11]. With the increase in pressure levels from 400 to 600 MPa and the subsequent maintenance at room temperature also preserved the ascorbic acid successfully [15, 98]. However, when at temperatures above 65°C, oxidation started to occur. Hence removing oxygen from packages prior to processing can prevent the degradation of ascorbic acid to a greater extent [75]. In addition, HPP treated orange juices retained vitamin C contents [80].

HPP caused some changes in the biopolymers of the plant-based foods, which could be due to either enzymatic or non-enzymatic reactions. The applied pressure causes changes in the polysaccharides of the cells, which indirectly affects the integrity and texture [18]. The pressure treatment disrupts the cell membrane and causes a loss in turgor pressure of the plant-based foods, such as fruits, and vegetables even at 100 MPa. This results in a loss of firmness of many fruits and vegetables. It was also found that the Apple was very sensitive at a higher pressure of about 200 MPa compared to the pear [13].

Furthermore, different ions, enzymes, and substrates could be released from different compartments of the cell leading to different reactions after or during the pressure treatments. Generally, pressure treatments do not target and affect the non-covalent bonds and hence are regarded to maintain the flavor of most of the food products. However, due to some enzymatic chemical reactions, HPP could indirectly aid in changing the intensity of some flavor compounds, specifically during storage.

Rancid flavor development in tomato related products at high pressures could catalyze the oxidation of fatty acids [87]. The HPP and 68°C were performed at a pressure of 650 MPa for 2 min in orange juice, and it was concluded that 12 different aroma compounds were lost significantly for every 20°C rises in temperature [89]. Since the appearance of the first fruit juice product in Japan market, the growth of such fruit-based products has grown all over the world in recent years.

HPP has been used for different meat products and has been investigated in detail for texture, color, flavor, nutrients retention, and microbial safety. Some of these products include cod muscles, deboned turkey meat, cured ham, sausages, pastrami, cooked ham, minced meat, Strasburg beef, etc. Since meat products are a high source of microorganisms, the processing of these products involves stringent techniques, which could inactivate the microbes after processing as well as storage. It was found that in certain meat products at a high-pressure of 600 MPa, 20°C for 180 s, several species of microorganisms were observed at mild levels during a storage period of 98 days at 4°C [42].

Studies on the effect of HPP on ready to eat salads are very limited. High-pressure of 500–800 MPa was sufficient for the inactivation of *Lactobacillus fructivorans* and *Zygosaccharomyces bailli*, which were inoculated in different salad dressings [73]. In minimally processed products, HPP is considered as an efficient technique in inactivating salmonella and *Listeria monocytogenes* [65].

HPP causes partial denaturation of myoglobin of meats, which is an important protein responsible for the color of the meat [32]. The other possible reason for discoloration of meat includes heme displacement and oxidation of myoglobin at pressures above 400 MPa. It is concluded that though high-pressure processing is acceptable with cured meat yet the same is not acceptable in terms of undesirable color changes in fresh meat [25]. HPP causes changes in the enzymes depending upon the processing pressure, temperature, pH, ion concentration, and availability of substrate. The changes induced on the enzymes indirectly reflect on the texture of the final product. During HPP, the pH of the product decreases due to alteration in the acidic and basic ions [95].

HPP interacts at the molecular level causing denaturation of tissue proteins, aggregation, gelation, and release of proteases, thereby causing the conditioning of muscles [76, 77]. The denaturation of myosin within the meat due to the pressure leads to the formation of gels. The texture of both heat-treated and pressure-treated meats are different, as the pressure (200–800 MPa, 20 min) yields hydrogen-bonded structures that are heat-labile; whereas the heat treatment (40–80°C, 20 min) yields structures that are stabilized by hydrophobic interactions [5].

Application of HPP may catalyze lipid oxidation in different meat products. The denatured proteins also contribute towards the lipid oxidation.

The initiation of oxidation takes place during pressurization, wherein non-heme iron compounds are released, and membranes are damaged. The critical pressure condition inducing lipid oxidation in pork was within the range of 300–400 MPa [23, 24]. With respect to the effect of HPP on meat aroma, only limited studies are available. The volatile compounds released from the packages of raw chicken and beef meat that were stored for 14 days under 400–600 MPa were analyzed for their aroma profile and were subjected to sensorial analysis by a trained panel. Mild typical meat odors evolved from HPP processed and off-flavor evolved from the unprocessed meat [90]. The consumer acceptance of processed meat products depends upon the color, odor, aroma, taste, degree of lipid and protein oxidation [88].

1.6 SUMMARY

Consumer's demand for fresh and safe food is a never-ending task of the food industry. HPP inhibits microbial activity even at room temperature without creating any detrimental changes in the nutritional, color, taste, and flavor of the food product. Heat sensitive products are highly suitable for being processed and preserved for a longer time by HPP. The development of HPP would benefit both the consumers as well the environment as it uses a non-thermal eco-friendly technology. This technique can also be combined with other existing technologies to overcome the hurdles faced during processing and extension of shelf life. The magnitude of HPP on the consumer's health and safety are to be established to ensure its full-scale processing in the food industry. The history of HPP dates back to several centuries through it garnered its attention only during the 20th century. The commercialization of HPP products in the world market has been taken up from then on.

The high cost involved in manufacturing the high-pressure withstand pressure vessels makes the technology very expensive. The initial high cost of investment, lower workload, and production lead to an increased production cost. This cost concern and lack of regulatory approvals have held back many emerging industries from adopting this technology.

It is forecasted that technological developments and commercialization of the eco-friendly HPP technology would benefit both the consumers as well as the environment. With the increase in awareness and improved efficiency, and reduced investment cost of HPP technology, many entrepreneurs would enter into this field and popularize the utility of this technology.

KEYWORDS

- batch HPP system
- continuous HPP system
- food quality
- food safety
- high-pressure processing (HPP)
- pasteurization

REFERENCES

1. Ahmed, J., Ramaswamy, H. S., & Hiremath, N., (2005). The effect of high-pressure treatment on rheological characteristics and color of mango pulp. *International Journal of Food Science and Technology, 40*(8), 885–895.
2. Akhmazillah, M. F. N., Farid, M. M., & Silva, F. V. M., (2013). High-pressure processing (HPP) of honey for the improvement of nutritional value. *Innovative Food Science and Emerging Technologies, 20,* 59–63.
3. Alpas, H., Kalchayanand, N., Bozoglu, F., Sikes, A., Dunne, C. P., & Ray, B., (1999). Variation in resistance to hydrostatic pressure among strains of food-borne pathogens. *Journals American Society for Microbiology, 65*(9), 4248–4251.
4. Andrés, V., Villanueva, M. J., & Tenorio, M. D., (2016). Influence of high-pressure processing on microbial shelf life, sensory profile, soluble sugars, organic acids, and mineral content of milk and soy-smoothies. *LWT-Food Science and Technology, 65,* 98–105.
5. Angsupanich, K., Edde, M., & Ledward, D. A., (1999). Effects of high-pressure on the myofibrillar proteins of cod and turkey muscle. *Journal of Agricultural and Food Chemistry, 47*(1), 92–99.
6. Aubourg, S. P., Rodríguez, A., Sierra, Y. S., Tabilo, G. T., & Perez, M., (2013). Sensory and physical changes in chilled farmed Coho salmon (*Oncorhynchus kisutch*): Effect of previous optimized hydrostatic high-pressure conditions. *Food and Bioprocess Technology, 6,* 1539–1549.
7. Aurelio, L. M., Barry, G. S., Gustavo, V. B. C., & Enrique, P., (2007). High-pressure treatment in food preservation. In: Rahman, M. S., (ed.), *Handbook of Food Preservation* (2nd edn., pp. 815–853). Boca Raton, Florida - USA: CRC Press.
8. Awad, T. S., Moharram, H. A., Shaltout, O. E., Asker, D., & Youssef, M. M., (2012). Applications of ultrasound in analysis, processing, and quality control of food: A review. *Food Research International, 48,* 410–427.
9. Balasubramaniam, V. M., Martínez, S. I., & Gupta, R., (2015). Principles and application of high-pressure technologies in the food industry. *Annual Review of Food Science and Technology, 6,* 435–462.

10. Balci, A. T., & Wilbey, R. A., (1999). High-pressure processing of milk: The first 100 years in the development of new technology. *International Journal of Dairy Technology, 52,* 149–155.
11. Barba, F. J., Cortes, C., Esteve, M. J., & Frigola, A., (2011). Study of antioxidant capacity and quality parameters in an orange juice milk beverage after high-pressure processing treatment. *Food and Bioprocess Technology, 5,* 2222–2232.
12. Bárcenas, M. E., Altamirano-Fortoul, R., & Rosell, C. M., (2010). Effect of high-pressure processing on the wheat dough and bread characteristics. *LWT-Food Science and Technology, 43*(1), 12–19.
13. Basak, S., & Ramaswamy, H. S., (1998). Effect of high-pressure processing on the texture of selected fruits and vegetables. *Journal of Texture Studies, 29,* 587–601.
14. Benet, G. U., (2005). *High-Pressure Low-Temperature Processing of Foods: Impact of Metastable Phases on Process and Quality Parameters* (p. 218). PhD Thesis; Berlin University of Technology, Berlin.
15. Bull, M. K., Zerdin, K., Howe, E., & Goicoechea, D., (2004). The effect of high-pressure processing on the microbial, physical, and chemical properties of Valencia and navel orange juice. *Innovative Food Science and Emerging Technologies, 5*(2), 135–149.
16. Caminiti, I. M., Noci, F., Munoz, A., & Whyte, P., (2011). Impact of selected combinations of non-thermal processing technologies on the quality of an apple and cranberry juice blend. *Food Chemistry, 124,* 1387–1392.
17. Caner, C., Hernandez, R. J., & Harte, B. R., (2004). High-pressure processing effects on the mechanical, barrier and mass transfer properties of food packaging flexible structures: A critical review. *Packaging Technology Science, 17*(1), 23–29.
18. Cano, M. P., & De Ancos, B., (2005). Advances in use of high-pressure to processing and preservation of plant foods. In: Barbosa-Canovas, G. V., Tapia, M. S., & Cano, M. P., (eds.), *Novel Food Processing Technologies* (pp. 283–309). Boca Raton, FL-USA: CRC Press.
19. Capellas, M., Mor, M., Gervilla, R., Yuste, J., & Guamis, B., (2000). Effect of high-pressure combined with mild heat or nisin on inoculated bacteria and mesophiles of goat's milk fresh cheese. *Food Microbiology, 17,* 633–641.
20. Carreño, J. M., Gurrea, M. C., Sampedro, F., & Carbonell, J. V., (2001). Effect of high hydrostatic pressure and high-pressure homogenization on *Lactobacillus plantarum* inactivation kinetics and quality parameters of mandarin juice. *European Food Research and Technology, 232,* 265–274.
21. Cavender, G. A., (2011). *Continuous High-Pressure Processing of Liquid Foods: An Analysis of Physical, Structural and Microbial Effects* (p. 229). PhD Thesis; University of Georgia.
22. Chai, C., Lee, J., Lee, Y., Na, S., & Park, J., (2014). A combination of TiO_2-UV photocatalysis and high hydrostatic pressure to inactivate *Bacillus cereus* in freshly squeezed *Angelica keiskei* juice. *LWT-Food Science and Technology, 55,* 104–109.
23. Cheah, P. B., & Ledward, D. A., (1997). Catalytic mechanism of lipid oxidation following high-pressure treatment in pork fat and meat. *Journal of Food Science, 62,* 1135–1139.
24. Cheah, P. B., & Ledward, D. A., (1996). High-pressure effects on lipid oxidation in minced pork. *Meat Science, 43,* 123–134.
25. Cheftel, J. C., & Culioli, J., (1997). Effect of high-pressure on meat: A review. *Meat Science, 46,* 211–234.

26. Chilton, P., Isaacs, N. S., Manas, P., & Mackey, B. M., (2001). Biosynthetic requirements for the repair of membrane damage in pressure-treated *Escherichia coli*. *International Journal of Food Microbiology, 71*, 101–104.

27. Chung-Yi, W., Hsiao-Wen, H., Chiao-Ping, H., & Binghuei, B. Y., (2015). Recent advances in food processing using high hydrostatic pressure technology. *Critical Reviews in Food, Science and Nutrition, 56*(4), 527–540.

28. Juliano, P., Richter, T., & Buckow, R., (2015). Food packaging for high-pressure processing. In: Cirillo, G., Gianfranco, S. U., & Iemma, F., (eds.), *Functional Polymers in Food Science* (pp. 253–280). Scrivener Publishing LLC.

29. Clouston, J. G., & Wills, P. A., (1969). Initiation of germination and inactivation of *Bacillus pumillus* spores by hydrostatic pressure. *Journal of Bacteriology, 97*, 684–690.

30. Crawford, Y. J., Murano, E. A., Olson, D. G., & Shenoy, K., (1996). Use of high hydrostatic pressure and irradiation to eliminate clostridium sporogenes spores in chicken breast. *Journal of Food Protection, 59*, 711–715.

31. Davidson, T. E., Barton, C. S., Reiner, A. N., & Kendall, D. P., (1962). New approach to the autofrettage of high strength cylinders. *Experimental Mechanics, 2*, 33−40.

32. Defaye, A., Ledward, D. A., MacDougall, D. A., & Tester, R. F., (1995). Renaturation of metmyoglobin subjected to high isostatic pressure. *Food Chemistry, 52*, 19–22.

33. Farkas, D. F., & Hoover, D. G., (2000). High-Pressure processing. *Journal of Food Science, 65*, 47–64.

34. Ferstl, C., (2013). *High-Pressure Processing Insights on Technology and Regulatory Requirements* (Vol. 10, pp. 67–80). Food for Thought Topical Insights from the National Food Lab White Paper Series.

35. Flick, G. J., (2003). *High Hydrostatic Pressure Processing has Potential: Commercial Fish and Shellfish Technology (CFAST)* (Vol. 6, No. 1, pp. 88, 89). Originally Appearing in Global Aquaculture Advocate, Middletown, Ohio; Avure Technologies, Inc.

36. Garcia-Palazon, A., Suthanthangjai, W., Kajda, P., & Zabetakis, I., (2004). The effects of high hydrostatic pressure on β-glucosidase, peroxidase and polyphenol oxidase in red raspberry (*Rubus idaeus*) and strawberry (*Fragaria* x *ananassa*). *Food Chemistry, 88*(1), 7–10.

37. Geveke, D. J., & Torres, D., (2012). Pasteurization of grapefruit juice using a centrifugal ultraviolet light irradiator. *Journal of Food Engineering, 111*, 241–246.

38. Guerrero-Beltrán, J., Barbosa-Cánovas, G. V., & Welti-Chanes, J., (2011). High hydrostatic pressure effect on *Saccharomyces cerevisiae*, *Escherichia coli*, and *Listeria innocua* in pear nectar. *Journal of Food Quality, 34*, 371–378.

39. Hartyáni, P., Dalmadi, I., & Knorr, D., (2013). Electronic nose investigation of *Alicyclobacillus acidoterrestris* inoculated apple and orange juice treated by high hydrostatic pressure. *Food Control, 32*, 262–269.

40. Hauben, K. J. A., Wuytack, E. Y., Soontjens, C. C. F., & Michiels, C. W., (1996). High-pressure transient sensitization of *Escherichia coli* to lysozyme and nisin by disruption of outer-membrane permeability. *Journal of Food Protection, 59*, 350–355.

41. Hayakawa, I., Kanno, T., Yoshiyaina, K., & Fujio, Y., (1994). Oscillatory compared with continuous high-pressure sterilization on *Bacillus stearothermophilus* spores. *Journal of Food Science, 59*, 164–167.

42. Hayman, M. M., Baxter, I., O'Riordan, P. J., & Stewart, C. M., (2004). Effects of high-pressure processing on the safety, quality, and shelf life of ready-to-eat meats. *Journal of Food Protection, 67*, 1709–1718.

43. Hite, B. H., Giddings, N. J., & Weakly, C. E., (1914). The Effects of pressure on certain microorganisms encountered in the preservation of fruits and vegetables. *Virginia Agricultural Experiment Station Bulletin, 146*, 1–67.
44. Hite, B. H., (1899). The effect of pressure in the preservation of milk. *Virginia Agricultural Experiment Station Bulletin, 58*, 15–35.
45. Hogan, E., Kelly, A., & Sun, D., (2005). High-pressure processing of foods: An overview. *Food Science and Technology, 1*, 1–27.
46. Hoover, D. G., Metrick, C., Papineau, A. M., Farkas, D. F., & Knorr, D., (1989). Biological effects of high hydrostatic pressure on food microorganisms. *Food Technology, 43*(3), 99–107.
47. Hsiao-Wen, H., Sz-Jie, W., Jen-Kai, L., Yuan-Tay, S., & Chung-Yi, W., (2017). Current status and future trends of high-pressure processing in food industry. *Food Control, 72*, 1–8.
48. Jacobo-Velazquez, D. A., & Hernandez-Brenes, C., (2010). Biochemical changes during the storage of high hydrostatic pressure processed avocado paste. *Journal of Food Science, 75*(6), 264–270.
49. Jaeger, H., Reineke, K., Schoessler, K., & Knorr, D., (2012). Effects of emerging processing technologies on food material properties. In: Bhandari, B., & Roos, Y. H., (eds.), *Food Materials Science and Engineering* (pp. 222–262). New York, USA: Wiley-Blackwell.
50. Jofré, A., Aymerich, T., Bover-Cid, S., & Garriga, M., (2010). Inactivation and recovery of *Listeria monocytogenes, Salmonella enterica,* and *Staphylococcus aureus* after high hydrostatic pressure treatments up to 900 MPa. *International Microbiology, 13*, 105–112.
51. Jolie, R. P., Christiaens, S., De Roeck, A., & Fraeye, I., (2012). Pectin conversions under high pressure: Implications for the structure related quality characteristics of plant-based foods. *Trends in Food Science and Technology, 24*, 103–118.
52. Kalchayanand, N., Frethem, C., Dunne, P., Sikes, A., & Ray, B., (2002). Hydrostatic pressure and bacteriocin-triggered cell wall lysis of *Leuconostoc mesenteroides*. *Innovative Food Science and Emerging Technologies, 3*, 33–40.
53. Knorr, D., & Heinz, V., (2001). Development of non-thermal methods for microbial control. In: Block, S. S., (ed.), *Disinfection, Sterilization, and Preservation* (5th edn., pp. 853–877). Philadelphia-USA: Lippincott Williams & Wilkins, A Wolters Kluwer Company.
54. Koutchma, T., (2014). *Adapting High Hydrostatic Pressure (HPP) for Food Processing Operations* (pp. 11, 12). San Diego, California-USA: Academic Press.
55. Koutchma, T., Song, Y., Setikaite, I., & Juliano, P., (2010). Packaging evaluation for high-pressure, high-temperature sterilization of shelf-stable foods. *Journal of Food Process Engineering, 33*(6), 1097–114.
56. Kuldiloke, J., & Eshtiaghi, M. N., (2008). Application of non-thermal processing for preservation of orange juice. *KMITL Science and Technology Journal, 8*(2), 64–74.
57. Lambert, Y., Demazeau, G., & Largeteau, A., (2000). Packaging for high-pressure treatments in the food industry. *Packaging Technology Science, 13*(2), 63–71.
58. Landl, A., Abadias, M., Sárraga, C., Viñas, I., & Picouet, P. A., (2010). Effect of high-pressure processing on the quality of acidified granny smith apple purée product. *Innovative Food Science and Emerging Technologies, 11*, 557–564.

59. Lange, J., & Wyser, Y., (2003). Recent Innovations in barrier technologies for plastic packaging: A review. *Packaging Technology Science, 16*(4), 149–158.
60. Le Chatelier, H. L., (1884). A general statement of the laws of chemical equilibrium. *Comptes Rendus, 99*, 786–789.
61. Li, L., Feng, L., Yi, J. Y., Hua, C., Chen, F., Liao, X., Wang, Z., & Hu, X., (2010). High hydrostatic pressure inactivation of total aerobic bacteria, lactic acid bacteria, yeasts in sour Chinese cabbage. *International Journal of Food Microbiology, 142*, 180–184.
62. Li, R., Wang, Y., Wang, S., & Liao, X. A., (2015). Comparative study of changes in microbiological quality and physicochemical properties of N_2-infused and N_2-degassed banana smoothies after high-pressure processing. *Food Bioprocess Technology, 8*, 333–342.
63. Liu, F., Wang, Y., Bi, X., Guo, X., Fu, S., & Liao, X., (2013). Comparison of microbial inactivation and rheological characteristics of mango pulp after high hydrostatic pressure treatment and high temperature short time treatment. *Food and Bioprocess Technology, 6*, 2675–2684.
64. Ma, H., & Ledward, D. A., (2013). High-pressure processing of fresh meat: Is it worth it? *Meat Science, 95*(4), 897–903.
65. Marco, C., (2010). High-pressure processing of meat, meat products, and seafood. *Food Engineering Reviews, 2*, 256–273.
66. Marszałek, K., Mitek, M., & Skapska, S., (2015). The effect of thermal pasteurization and high-pressure processing at cold and mild temperatures on the chemical composition, microbial and enzyme activity in strawberry purée. *Innovative Food Science and Emerging Technologies, 27*, 48–56.
67. Mathur, A., (2009). Pressure vessels and heat exchangers. In: Grote, K. H., & Antonsson, E. K., (eds.), *Springer Handbook of Mechanical Engineering* (pp. 947–966). Berlin Heidelberg-Germany: Springer.
68. Mert, M., Buzrul, S., & Alpas, H., (2013). Effects of high hydrostatic pressure on microflora and some quality attributes of grape juice. *High-Pressure Research, 33*, 55–63.
69. Mertens, B., (1993). Packaging aspects of high-pressure food processing technology. *Packaging Technology Science, 6*, 31–36.
70. Meyer, R. S., Cooper, K. L., Knorr, D., & Lelieveld, H. L. M., (2000). High-pressure sterilization of foods. *Food Technology, 54*(11), 67–72.
71. Mozhaev, V., Heremans, K., Frank, J., Masson, P., & Balny, C., (1996). High-pressure effects on protein structure and function. *Proteins, Structure, Function and Genetics, 24*, 81–91.
72. Mukhopadhyay, S., Sokorai, K., Ukukub, D., Fan, X., & Junej, V., (2017). Effect of high hydrostatic pressure processing on the background microbial loads and quality of cantaloupe puree. *Food Research International, 91*, 55–62.
73. Neinaber, U., Arora, A., & Shellhammer, T. H., (2001). *High-Pressure Processed Salad Dressings: Microbiological and Rheological Aspects.* Oral presentation No. 28-4, Annual meeting of the Institute of Food Technologists, New Orleans, LA.
74. Oey, I., Lille, M., Van, L. A., & Hendrickx, M., (2008). Effect of high-pressure processing on color, texture, and flavor of fruit-and vegetable-based food products: A review. *Trends in Food Science and Technology, 19*(6), 320–328.

75. Oey, I., Van, D. P. I., Loey, A. V., & Hendrickx, M., (2008). Does high-pressure processing influence nutritional aspects of plant-based food systems. *Trends in Food Science Technology, 19*, 300–308.

76. Ohmori, T., Shigehisa, T., Taji, S., & Hayashi, R., (1991). Effect of high-pressure on the protease activity in meat. *Agricultural and Biological Chemistry, 55*(2), 357–361.

77. Orlien, V., Hansen, E., & Skibsted, L. H., (2000). Lipid oxidation in high-pressure processed chicken breast muscle during chill storage: Critical working pressure in relation to oxidation mechanism. *European Food Research Technology, 211*, 99–104.

78. Peñas, E., Frias, J., Gomez, R., & Vidal-Valverde, C., (2010). High hydrostatic pressure can improve the microbial quality of sauerkraut during storage. *Food Control, 21*, 524–528.

79. Pérez, P. R., Toledo, J., Grande, M. J., Gálvez, A., & Lucas, R., (2015). Analysis of the effect of high hydrostatic pressure treatment and enterocin AS-48 addition on the bacterial communities of cherimoya pulp. *International Journal of Food Microbiology, 196*, 62–69.

80. Plaza, L., Sanchez-Moreno, C., & Elez-Martinez, P., (2006). Effect of refrigerated storage on vitamin c and antioxidant activity of orange juice processed by high-pressure or pulsed electric fields with regard to low pasteurization. *European Food Research and Technology, 223*(4), 487–493.

81. Powalska, E., Janosch, S., & Kinne-Saffran, E., (2007). Fluorescence spectroscopic studies of pressure effects on Na^+, K^+- ATPase reconstituted into phospholipid bilayers and model raft mixtures. *Biochemistry, 46*(6), 1672–1683.

82. Queiroz, C., Moreira, C. F. F., & Lavinas, F. C., (2010). Effect of high hydrostatic pressure on phenolic compounds, ascorbic acid, and antioxidant activity in cashew apple juice. *High-Pressure Research, 30*, 507–513.

83. Ramaswamy, H. S., Zaman, S. U., & Smith, J. P., (2008). High-pressure destruction kinetics of *E. coli* (O157: H7) and *Listeria monocytogenes* (Scott A) in a fish slurry. *Journal of Food Engineering, 87*(1), 99–106.

84. Rastogi, N. K., Raghavarao, K. S. M. S., & Balasubramaiam, V. M., (2007). Opportunities and challenges in high-pressure processing of foods. *Critical Reviews in Food Science and Nutrition, 47*(1), 69–112.

85. Reineke, K., (2012). *Mechanisms of Bacillus Spore Germination and Inactivation During High-Pressure Processing* (p. 180). PhD Thesis; Technischen University, Berlin.

86. Ritz, M., Freulet, M., Orange, N., & Federighi, M., (2000). Effects of high hydrostatic pressure on membrane proteins of *Salmonella typhimurium*. *International Journal of Food Microbiology, 55*, 115–119.

87. Rodrigo, D., Cortes, C., Clynen, E., & Schoofs, L., (2006). Thermal and high-pressure stability of purified polygalacturonase and pectin methylesterase from four different tomato-processing varieties. *Food Research International, 39*(4), 440–448.

88. Rubio, B., Martinez, B., Garcia-Cachan, M. D., Rovira, J., & Jaime, I., (2007). Effect of high-pressure preservation on the quality of dry-cured beef "Cecina de Leon". *Innovative Food Science and Emerging Technologies, 8*(1), 102–110.

89. Sampedro, F., Geveke, D. J., Fan, X., & Zhang, H. Q., (2009). Effect of PEF, HHP and thermal treatment on PME inactivation and volatile compounds concentration of an orange juice milk-based beverage. *Innovative Food Science and Emerging Technologies, 10*, 463–469.

90. Schindler, S., Krings, U., Berger, R. G., & Orlien, V., (2010). Aroma development in high-pressure-treated beef and chicken meat compared to raw and heat-treated. *Meat Science, 86*(2), 317–23.
91. Sedighi, M., & Jabbari, A. A., (2013). New analytical approach for wire-wound frames used to carry the loads of pressure vessel closures. *Journal of Pressure Vessel Technology, 135*(6), Article ID: 061206-1-7.
92. Setlow, B., Setlow, C. A., & Setlow, P., (1997). Killing bacterial spores by organic hydroperoxides. *Journal of Industrial Microbiology, 18*, 384–388.
93. Shahbaz, H. M., Yoo, S., Seo, B., Ghafoor, K., Un Kim, Y., Lee, D. U., & Park, J., (2016). Combination of TiO$_2$-UV photocatalysis and high hydrostatic pressure to inactivate bacterial pathogens and yeast in commercial apple juice. *Food Bioprocess Technology, 9*, 182–190.
94. Shimada, S., Andou, M., Naito, N., Yamada, N., Osumi, M., & Hayashi, R., (1993). Effects of hydrostatic pressure on the ultrastructure and leakage of internal substances in the yeast *Saccharomyces cerevisiae*. *Applied Microbiology and Biotechnology, 40*, 123–131.
95. Stippl, V. M., Delgado, A., & Becker, T. M., (2005). Ionization equilibria at high pressure. *European Food Research and Technology, 221*, 151–156.
96. Tao, Y., Sun, D., Górecki, A., & Błaszczak, W., (2012). Effects of high hydrostatic pressure processing on the physicochemical and sensorial properties of a red wine. *Innovative Food Science and Emerging Technologies, 16*, 409–416.
97. Thakur, B. R., & Nelson, P. E., (1998). High-pressure processing and preservation of foods. *Food Review International, 14*, 427–447.
98. Torres, B., Tiwari, B. K., Patras, A., Cullen, P. J., Brunton, N., & O'Donnell, C. P., (2011). Stability of anthocyanins and ascorbic acid of high-pressure processed blood orange juice during Storage. *Innovative Food Science and Emerging Technologies, 12*(2), 93–97.
99. Torres, J. A., & Velazquez, G., (2005). Commercial opportunities and research challenges in the high-pressure processing of foods. *Journal of Food Engineering, 67*, 95–112.
100. Varela-Santos, E., & Ochoa, E. A., (2012). Effect of high hydrostatic pressure processing on physicochemical properties, bioactive compounds, and shelf life of pomegranate juice. *Innovative Food Science and Emerging Technologies, 13*, 13–22.
101. Verbeyst, L., Oey, I., Van, D. P. I., Hendrickx, M., & Van, L. A., (2010). Kinetic study on the thermal and pressure degradation of anthocyanins in strawberries. *Food Chemistry, 123*(2), 269–274.
102. Viljanen, K., Lille, M., Heinio, R., & Buchert, J., (2011). Effect of high-pressure processing on volatile composition and odor of cherry tomato puree. *Food Chemistry, 129*, 1759–1765.
103. Wael, M., Elamin, Johari, B., Endan, Yus, A. Y., & Rosnah, S., (2015). High-pressure processing technology and equipment evolution: A review. *Journal of Engineering Science and Technology Review, 8*(5), 75–83.
104. Winter, R., & Jeworrek, C., (2009). Effect of pressure on membranes. *Soft Matter, 5*, 3157–3173.
105. Yaldagard, M., Mortazavi, S. A., & Tabatabaie, F., (2008). The principles of ultra-high-pressure technology and its application in food processing/preservation: A review of microbiological and quality aspects. *African Journal of Biotechnology, 7*(16), 2739–2767.

106. Yoo, S., Ghafoor, K., Un Kim, J., Kim, S., Jung, B., Lee, D. U., & Park, J., (2015). Inactivation of *Escherichia coli* O157:H7 on orange fruit surfaces and in juice using photocatalysis and high hydrostatic pressure. *Journal of Food Protection, 78*, 1098–1105.
107. Yordanov, D. G., & Angelova, G. V., (2010). High-pressure processing for preserving foods. *Biotechnology and Biotechnological Equipment, 24*(3), 1940–1945.
108. Zhang, H. Q., Barbosa-Cánovas, G. V., & Balasubramaniam, V. B., (2011). *Non-thermal Processing Technologies for Food* (p. 345). New Delhi, India: John Wiley & Sons.

CHAPTER 2

ULTRAVIOLET LIGHT TECHNOLOGY: APPLICATIONS FOR FRESH PRODUCE

GOZDE OGUZ KORKUT and GURBUZ GUNES

ABSTRACT

Ultraviolet (UV) light is a non-thermal process with a high potential for fresh produce to inactivate microorganisms and to maintain food quality. The effectiveness of the UV treatment depends on UV dose (J/m^2), UV intensity (W/m^2), exposure time (s), surface properties, initial microorganism load, and microorganisms. In addition to microbial inactivation, investigation of UV application on the attributes of fresh products and quality changes throughout the storage life of fresh products is also considered for the optimization of UV treatment.

2.1 INTRODUCTION

World Health Organization (WHO), Food and Agriculture Organization (FAO), US Department of Agriculture (USDA), and European Food Safety Authority (EFSA) suggest increasing fruit and vegetable consumption as it reduces the possibility of cancer and cardiovascular disease, and protects health [6, 76]. Processes applied to food products are carried out to provide high quality and safety of foods. Thermal technologies and chlorine-based chemical disinfectants are the most commonly used methods to achieve these goals, but these treatments may have adverse effects on food quality, consumer health, and the environment. The development of new technologies to minimize microbial risk while maintaining the freshness of fruits and vegetables is crucial to meet the increasing demand. Therefore, many non-thermal novel technologies, including ultraviolet (UV) light, are being investigated.

Recently, comprehensive research on the use of UV light has shown that this technology holds significant promise as an alternative to conventional methods (CMs) to extend the storage life of fresh produce. UV light has been found to be safe for the inactivation of microorganisms on food surfaces and has been approved by FDA [28]. There has been an increased interest in this technology due to the advantages of UV light application, such as being easy to apply, eco-friendly, low-cost of equipments, a legal and safe method, a process that does not leave chemical residues and does not produce by-products that can change the sensory properties of the final product.

This chapter presents (1) a summary of the basic principles of UV technology and includes the applications of UV in fresh and fresh-cut fruit and vegetable products;(2) Factors related with the product and the process affecting UV treatment are emphasized; (3) microbial inactivation mechanisms are explained; and (4) the effects of UV light on the fresh produce quality and composition are discussed; (5) The challenges and future trends for the development of UV technology are also discussed.

2.2 BASIC PRINCIPLES OF UV-LIGHT TECHNOLOGY

Although UV light is a non-ionizing radiation, yet the term "irradiation" is also used for UV treatment [30, 61]. The UV light is located between the X-rays and visible light in the electromagnetic spectrum and has a wavelength ranging from 100 to 400 nm [37]. UV light is grouped as vacuum UV, UV-C, UV-B, and UV-A with wavelength ranges of 100–200 nm, 200–280 nm, 280–315 nm, and 315–400 nm, respectively.

Sun is the natural source of UV light. While UV-A and some UV-B reach us from the sun, UV-C is totally absorbed by the ozone layer [96]. UV-A can damage human skin and is responsible for tanning, UV-B cause's skin burns that can lead to skin cancer. UV-C is a germicidal agent that inactivates bacteria, mold, yeast, viruses, algae, and protozoa [50]. UV-A is not used much in food applications as it is less absorbed by living cells and affects the microbial cells less than UV-C. UV-A applications in the food industry are often used for quality control, especially for the determination of aflatoxins that occur during storage [14]. UV-B is usually used in the field or garden without harvesting fruits and vegetables. In the food industry, most of the research on preserving fresh produce involves UV-C applications.

For the interaction between light and matter, photons must be absorbed by atoms or molecules. Atoms and many ions consist of electrons rotating in a certain orbit around a nucleus consisting of protons and neutrons. The electron closest to the nucleus has the lowest energy level, while the outer electron has the highest energy level. The emission of photons is related to the transition of electrons from the high-energy state (E_2) to the low energy state (E_1) [49]. Based on Plank's Law, each photon contains certain energy described by Eqn. (1).

$$E = E_2 - E_1 = \frac{hc}{\lambda} \qquad (1)$$

where; E is photon energy (kJ/Einstein) or (eV); h is the Plank coefficient (6.23×10^{-34} Js); c is light velocity ($2,998 \times 10^8$ m/s); and λ is radiation wavelength (m) [49].

The emission of light from UV lamps takes place through gas discharge. A mixture of excited atoms, non-excited atoms, cations, and free electrons, which occurs when a sufficient amount of high voltage flows through a certain volume of gas, is referred to as gas discharge [30]. UV light diffused by gas discharge interacts with materials through absorption, reflection, refraction, and scattering [49]. It is the absorbed light that drives microbial inactivation or chemical interactions on the materials. Scattered, reflected, or refracted light will be still available to interact with the material until being absorbed. The effects of UV light on any matter depends on dose (also known as fluency), which is the amount of energy sheded on a unit surface area (J/m²). The dose of UV light is the product of UV intensity (W/m²) and exposure time (s), where the former is the amount of light power per unit area and is dependent on the power of the light source used.

2.3   SOURCES OF UV LIGHT

Choosing the right UV source is important for the effectiveness of UV applications. There is no single effective source for all photochemical reactions, thus a suitable source must be selected to perform the desired photochemical change for each application. There are many UV sources, such as mercury lamps operating at low and medium pressures (LPM, MPM), excimer lamps (EL), pulsed light lamps (PL), and light-emitting diodes (LEDs). However, LPM and MPM lamps are generally used in UV applications because of their low cost, quality, and high performance [49]. This section summarizes the most important UV sources.

2.3.1   MERCURY LAMPS

Mercury lamps contain a small amount of elemental mercury (Hg) and an inert gas. They are called low-pressure and medium-pressure lamps according to the vapor pressure of the mercury during operation [51]. Low or medium mercury lamps produce UV light continuously [45].

LPM (low-pressure mercury) lamps are cylindrical and predominantly release monochromatic light at a wavelength of 253.7 nm, which is very close to the most germicidal wavelength of 260 nm [50]. They operate at temperatures from 30 to 50°C and total gas pressure between 10^2 and 10^3 Pa [49]. The LPM lamps are produced with output powers in the range of 4 to 120 W and they have a lifetime of about 12000 h. [45, 50, 51, 66]. The UV-C efficiency of LPM lamps is 30–40% [84].

Medium pressure mercury (MPM) lamps are polychromatic lamps producing light in the wavelength band of 250–600 nm. They operate at higher temperatures (more than 400°C) and higher mercury vapor pressures (50 to 300 kPa) than LPM lamps. Compared to LPM lamps, the UV intensity is higher, but the germicidal power is less (15–20%). The lifetime of MPM is 5000 h [45, 49, 51]. The UV-C efficiency of MPM lamps is 5–15% [84].

2.3.2   EXCIMER LAMPS (EL)

Excimer lamps (EL) are mercury-free UV technologies. The word "Excimer" is an abbreviation for Excited Dimer, an excited Xe_2 molecule [84]. These exciters may be a rare gas and a halogen. Radiation is produced by the dissociation of excited molecules [50]. Depending on the gas mixture, they emit UV light at wavelengths ranging from 120 to 380 nm [49]. The source of the lamp is quasimonochromatic. UV yield ranges from 10 to 40%. The lifetime of EL is 1000 h. It can work at low surface temperatures compared to other UV sources [49–51]. The UV-C efficiency of Excimer XeBr (xenon bromine) lamps is 8% [84].

2.3.3   PULSED UV LAMPS

One of the mercury-free UV lamp technologies is pulsed lamps (PL). Pulsed lamps have a wavelength in the range of 170–1000 nm [51]. Pulsed lamps include xenon, argon, krypton or other inert gases or mixtures thereof [15, 45]. A pulse of light emission is generated by releasing energy from a high-speed switch [51]. The flash lamp and surface discharge (SD) lamp are two different lamp types that produce pulsed UV light. The pulses are generated in the flash lamp by a rare gas between the two electrodes in the envelope [45, 83], whereas in the SD lamp, a plasma discharge is produced on the surface of the dielectric substrate [15].

The UV efficiency and intensity of the SD lamps are much higher than conventional mercury lamps and flash lamps [45]. The UV-C efficiency and intensity of flash lamps are respectively 9% and 600 W/cm^2, while the UV-C efficiency and intensity of SD lamps are respectively 17% and 30,000 W/cm^2. The dose for virus inactivation of SD lamps is 22 mJ/cm^2 [49]. Recently, lot of research has been done with this innovative-pulsed lamp technology [16, 26, 36, 39, 57, 62, 88, 97].

2.3.4   LOW-PRESSURE AMALGAM (LPA) LAMPS

The low-pressure amalgam lamps (LPA) are often named as low-pressure high output (LPHO) LAMPS (LPHOs) or antiseptic high output lamps (GHO-lamps). Amalgam lamps feature Long Life Technology. It has a life-time of over 16,000 h and shows a constant UV output of over 90% during this time. The UV-C efficiency of LPA lamps is 35% [84]. LPA lamps have a UV spectrum in the range of 185–254 nm [49].

2.3.5   MICROWAVE UV LAMPS

An innovative UV-lamp technology-enabled to eliminate the use of electrodes in typical mercury lamps with microwave power. The microwave UV lamps operate at temperatures from 30 to 60°C like LPM lamps but have a quicker warm-up response. The lifetime of this lamp is about three times higher than electrode-powered UV lamps [49].

2.3.6   UV LIGHT-EMITTING DIODES (UV-LEDS)

UV light-emitting diodes (UV-LEDs) lamps produce UV light in the wave-length band of 200–400 nm. They are generally grouped as near-ultraviolet light-emitting diodes (NUV-LEDs with 300–400 nm emission wavelength), and deep-ultraviolet LEDs (DUV-LEDs with 200–300 nm emission wave-length). UV LED lamps are mercury-free. Compared to other UV lamps, the energy efficiency is higher, the lifetime is longer, and the UV intensity is more constant. The control of temperature and heat is easier [90]. Because of these advantages, it is one of the recently investigated topics [35, 46, 48, 91, 107].

2.4 FACTORS AFFECTING THE UV TREATMENT

The effectiveness of UV treatment depends on a lot of factors including UV light source and UV dose, UV treatment chamber design, microorganism type, and load, physical, and optical properties of surface to be applied, composition of food, and UV light transmission.

2.4.1 PRODUCT PARAMETERS

Foods contain many nutrients such as water, vitamins, minerals, carbo-hydrates, proteins, and lipids that may be sensitive to UV light [50]. The effectiveness of UV light varies based on the chemical, physical, and optical characteristics of the product. For instance, UV light activity varies in various product matrices, such as air, water, liquid, or solid foods [51]. UV light penetrates into solids at lower rates than liquids. In the case of liquid materials, UV light penetration varies depending on whether they are transparent, opaque, or translucent [50, 61].

Surface properties of the sample (such as the color of the surface, being dirty and rough) change the effectiveness of UV light. Roughness and pollution create a shadow effect, preventing direct access to UV light, thereby reducing the effectiveness of the process. Depending on the color of the surface on which UV is applied, the amount of light absorbed is affected. For example, the absorption of light is high in dark foods and thus the energy required is reduced [65]. The composition of the food is also important for UV light to penetrate into the food. For example, pH, water activity, °Brix, acidity or basicity of products, dissolved solids; opacity, absorptivity, transmittance, viscosity, and turbidity are food parameters that can alter UV activity [13, 50]. Dissolved solids, suspended particles, organic solvents, macromolecules (such as proteins and fat globules) can limit the penetration of the light and thus the effectiveness of light [50, 65].

In terms of their sensitivity to UV light, Gram-positive bacteria, and spore-forming microorganisms are more resistant than Gram-negatives and non-spore formers. Yeasts and molds are generally more durable to UV than bacteria due to the presence of less pyrimidine bases in their structures and morphological and physiological differences of the cell membranes. The most resistant microorganisms are viruses [8]. Therefore, the required UV dose for each microorganism is different [61].

2.4.2 PROCESS PARAMETERS

UV light interacts with substances through absorption, reflection, refraction, and scattering and these properties affect the penetration of UV light [51, 65]. UV dose (Eqn. (2)) describes the amount of energy sheded on a unit area, and is the most important factor in the effectiveness of the UV treatment [8, 14].

$$UV\ Dose\left(\frac{J}{m^2}\right) = UV\ Intensity\left(\frac{W}{m^2}\right) x\ Exposure\ Time(s) \qquad (2)$$

UV intensity is expressed as irradiance and the unit is W/m². The UV intensity can be measured using a radiometer and indicates the amount of UV light power applied to a unit area [51]. UV intensity or irradiance can be changed by using lamps of different power or by varying the distance between the lamp and the product. As the distance from the lamp increases or as the power of the lamp decreases, UV intensity decreases. However, the difference in UV intensity due to lamp power may not be significant at short distances. Increasing the lamp power by 2-fold does not mean that the intensity would increase by 2-fold, it depends on the distance between the product and the lamp.

Because UV light has no or very limited penetration to the materials, therefore, effective and homogenous exposure of the whole surface of the materials to the light during the treatment is very important. This is only being possible by proper design of the treatment chamber to achieve a uniform treatment of the material. Therefore, the treatment chamber design, number, and power of UV source used, the distance from the sample and the exposure time are important factors for determining the homogeneity of the effectiveness of the UV application [51, 65]. The design of the treatment chamber can also affect the UV intensity. For instance, inclusion of reflectors in the interior of the UV treatment chamber would result in higher intensity than the chamber without reflectors.

2.5   MECHANISM OF MICROBIAL INACTIVATION

UV light must be in the appropriate wavelength range to inactivate micro-organisms. The most effective UV light for inactivating microorganisms is UV-C followed by UV-B light and UV-A, which has almost no effects [45]. The highest germicidal effect is observed in the wavelength range of 250 to 270 nm [33]. Majority of the commercial UV lamps emit UV rays at a wavelength of 254 nm.

UV-C affects specific target molecules, and the applied UV dose causes inactivation by directly altering microbial deoxyribonucleic acid (DNA) or ribonucleic acid (RNA) through dimer formation [7]. The mechanism of microbial inactivation is due to the formation of pyrimidine dimers (T-T, T-C) by causing cross-linking of the nucleotide bases of the adjacent pyrimidine (Thymine (T) and cytosine (C)) on DNA or RNA and the fact that these dimers disrupt the genetic configuration. As a result, DNA transcription and translation are inhibited, replication stops, cell functions are compromised, repair, and reproduction is prevented, and finally cell death occurs [8, 45, 49, 80, 86].

The UV resistance is determined by the ability of microorganisms to repair UV damage (photoreactivation). Photoreactivation is a natural phenomenon that is usually effective in bacteria and spores, repairing DNA damage by allowing pyrimidine dimers to return to free pyrimidine [52]. When the microorganisms are exposed to wavelengths greater than 330 nm, light-repair enzymes (photolyases) are activated, and they repair damaged cells by allowing the dimer to split back into free forms, which is called monomerization [8, 52, 61].

Once UV is applied, a first-order reaction kinetics model (Eqn. (3)) is often used to identify microbial inactivation.

$$\frac{-dN}{dt} = kN^n \tag{3}$$

where; N is the number of microorganisms surviving after time (or dose) dependent UV-C treatment; k is the inactivation rate constant; and n is the degree of inactivation.

Many mathematical models (such as: Gompertz, Weibull, Fermi or Baranyi) have been proposed to describe the microbial inactivation curves. Among these models, the Weibull distribution (Eqn. (4)) is the most appropriate and resilient model to describe nonlinear inactivation kinetics of microorganisms [13, 71].

$$N = N_0 \exp\left[-\left(\frac{t}{\alpha}\right)^{\gamma}\right] \qquad (4)$$

where; N is the count of microorganisms surviving after dose-dependent UV-C treatment for duration of t; N_0 is the initial microorganism load; α is the scale factor; and γ is the shape parameter that determines the shape of the curve.

The value $\gamma > 1$ demonstrates the survival curve with a convex shape, while, $\gamma < 1$ yields a concave shape with a higher microbial resistance. For $\gamma = 1$, the Weibull model is equivalent to the first order. In Eqn. (4), α and γ values are determined by nonlinear optimization [93].

2.6   EFFECTS OF UV LIGHT ON FOOD QUALITY

UV light should ensure adequate microbial inactivation in foods, but the applied doses should not cause undesirable quality changes. While the majority of UV studies on fresh fruits and vegetables consist of short-wavelength UV-C light, studies investigating the effects of medium-wavelength UV-B and long-wavelength UV-A lights have also been reported in the literature. Effects of UV light on ripening and senescence, storage disorder, texture, color, bioactive compounds, and microorganisms have been investigated in many studies that are discussed in this section.

2.6.1   EFFECT ON RIPENING AND SENESCENCE OF FRESH PRODUCE

Fresh and fresh-cut fruit and vegetable products are metabolically active, and biochemical reactions associated with ripening and senescence take place during storage. Senescence is generally associated with the oxidation of proteins, lipids, and nucleic acids and causes irreversible disorders in plant cells [102]. The energy required for these metabolic reactions is supplied by respiration. It is important to slow down these metabolic reactions and thus the respiration rate to increase the storage life of fresh produce product. CCO and succinic dehydrogenase (SDH) are two prominent enzymes that play a role in respiratory activity. UV treatment decreases respiration rate by reducing the activity of SDH and CCO, and thus delays ripening and senescence [104].

In general, UV-C light inhibits senescence at low doses, while UV-B light enhances lipid oxidation that causes senescence [86]. The plant hormone, ethylene, regulates many changes, such as, growth, development, ripening, and softening [1, 11, 78]. Ethylene is known to be produced in response to biotic and abiotic stress factors, such as injury, ozone, UV light, cooling or freezing [99]. However, UV application does not always increase ethylene production. Because in some cases, UV light causes irreversible membrane degradation and results in low ethylene production by affecting the activity of 1-Aminocyclopropane-1-carboxylic acid (ACC) oxidase [63]. Polyamines can suppress ethylene production by reducing the synthesis of ACC synthase that causes ethylene formation and thereby delay ripening and senescence [63]. Some studies to determine the effectiveness of UV application on senescence and ripening are summarized in Table 2.1.

Peaches were treated with UV-C at 3.0 kJ/m^2 and stored for 8 days at 20°C [102]. It was found that UV-C application inhibited the respiration rate by reducing succinic dehydrogenase (SDH) and cytochrome C oxidase (CCO) activity after 5 days, and there was a decrease in the development of senescence. The UV-C treatment also blocked the accumulation of reactive oxygen species (ROS) after 5 days and delayed senescence by reducing oxidative damage.

In another study, UV-C treatments at 1.5, 2.5, 4.9, 7.4, 9.8 kJ/m^2 were applied to peaches before storing them at 5°C for 14 or 21 days, and rates of respiration and ethylene production at 20°C for up to 7 days were evaluated [31]. The respiration rate of fruits after 14 days of cold storage was stimulated at all doses immediately after transferring to 20°C, but except for 1.5 kJ/m^2 UV-C, all the treated fruits had a lower respiration rate than the control beyond 10 h of the transfer. On the other hand, the respiration rate of fruits after 21 days of cold storage showed different trends. The fruits treated with 1.5 and 4.9 kJ/m^2 had higher respiration while those treated with 2.5 and 7.4 kJ/m^2 had lower respiration rate than the control beyond 10 h of the transfer to 20°C. Among all samples, the UV dose of 7.4 kJ/m^2 resulted in the lowest respiration rate. All the UV-C treatments resulted in a significant increase in ethylene production measured after transferring the fruit to 20°C for up to 7 days.

Stevens et al. [92] also treated peaches with 7.5 kJ/m^2 UV-C and reported that the treatment inhibited ripening, decreased ethylene production, and stimulated phenylalanine ammonia-lyase (PAL) enzyme activity. The UV-C treatment caused up to a 25% reduction of ethylene production during the first 24 h, but increased the ethylene production after 48 h of the treatment.

Strawberries were subjected to 3 different UV-C conditions (single-step, two-step, multi-step) and then were stored for 13 days at 0°C [9]. In the single-step, researchers used 4 kJ/m^2 UV-C light before storage. Two-step UV-C application was carried out as two 2 kJ/m^2 treatments after 0 and 4 day-storage, and the multi-step UV-C application was carried out as five 0.8 kJ/m^2 treatments after 0, 2, 4, 6 and 8 days of storage. All the treatments had no effect on respiration rate during storage up to 10 days, but the treated fruits had lower respiration rate after 13 days of storage. There was no difference among the different UV-C treatment patterns on respiration rate. Firstly, strawberries were applied with concentrations of 0.01, 0.1 and 1.0 mM abscisic acid (ABA) and UV-C radiation doses of 2.0, 4.1 and 8.0 kJ/m^2 [55].

TABLE 2.1 Studies on the Effects of UV Applications on Ripening and Senescence

Fresh Produce	Treatment Conditions	Results	References
Blueberry (*Vaccinium spp.* Berkeley)	15 W lamp power 4.0 kJ/m^2 UV-C Stored for 8 days at 4°C.	• A significant reduction in respiratory rate.	[101]
Fresh-cut watermelon (*C. lanatus* cv 'Fashion')	36 W lamp power 1.6, 2.8, 4.8 and 7.2 kJ/m^2 Stored for 11 days at 5°C. 15 cm of distance.	• Increased respiration rate. • No effect on ethylene production.	[10]
Peach (*Prunus persica* Batsch cv. Yulu)	30 W lamp power 3.0 kJ/m^2 UV-C 200 µW/cm^2 25 cm distance. Storage for 8 days at 20°C.	• Lower SDH activity on 3–6 days of storage. • Lower CCO activity on 7–8 days of storage. • Inhibited the respiration rate on 3–5 days of storage. • Inhibited ROS after 5 days of storage. • Rate of reduction of mitochondrial membrane fluidity by storage decreased. • Membrane permeability transition pore opening during storage was decreased. • Delayed senescence.	[102]

TABLE 2.1 *(Continued)*

Fresh Produce	Treatment Conditions	Results	References
Peach (*Prunus persica* cv Jefferson)	15 W lamp power 1.5, 2.5, 4.9, 7.4, 9.8 kJ/m² 3, 5, 10, 15, 20 min UV-C treatments 8220 mW/m² 15 cm distance. Storage for 14 or 21 days at 5°C + 7 days at 20°C.	• Increased respiration rate after 14 and 21 days of cold storage. • Respiration was higher with 3 min treated samples after transferring to 20°C compared to control. • Respiration was lower with 15 min treated samples after transferring to 20°C compared to control. • Ethylene production was markedly higher in all UV-treated samples after transferred to 20°C compared to control.	[31]
Peach (Elberta)	30 W lamp power 7.5 J/m² UV-C. 1.24 mW/cm² 10 cm distance	• Depression of ethylene production within the first 24 h after treatment, but stimulation of it after 48 h of the treatment. • Delayed ripening.	[92]
Strawberry (*Fragaria x ananassa cv* Camarosa)	30 W lamp power 25 cm distance; 4 kJ/m² UV-C treatment before storage. Two 2 kJ/m² consecutive UV-C treatments after on day 0 and day 4. Five 0.8 kJ/m² consecutive UV-C treatments on day 0, 2, 4, 6, and 8. Storage for 13 days at 0°C.	• Reduction of respiration rate only after 13 days of storage.	[9]
Strawberry (*Fragaria ananassa* Duch. cv. Akihime)	4.1 kJ/m² UV-C. 4.1 kJ/m² UV-C+1.0 mM ABA. Storage for 4 days at 20°C in the dark.	• UV-C application stimulated ethylene production. • Combined treatments decreased the increase of ethylene production.	[55]

TABLE 2.1 *(Continued)*

Fresh Produce	Treatment Conditions	Results	References
Sweet cherry (cv. Sweetheart)	1.2, 3.0 or 6.0 kJ/m² 20 W/m² 50 cm distance. Storage for 10 days at 0°C.	• Respiration rate decreased. • Ripening index increased. • TSS unchanged. • TA decreased.	[67]
Tomato (*S. lycopersicum* cv. Flavortop)	3.7 kJ/m² for 16 min. 60 cm distance.	• Stimulated ethylene production. • Increased polyamine content. • Delayed ripening.	[95]
Tomato (*Lycopersicon esculentum* Mill cv. Capello)	30 W lamp power 3.7 and 24.4 kJ/m² UV-C 20.3 W/m² 30 cm distance; Storage for 35 days at 16°C in the dark	• Slight initial stimulation followed by decreased respiration and ethylene production during storage. • Delayed climacteric peak and ripening. • Increased putrescine levels.	[63]

Li et al. [55] found that UV-C at 4.1 kJ/m² stimulated ethylene production during storage, but when it was applied in combination with 1.0 mM ABA then ethylene production was lower than the control during storage. UV-C application is generally known to stimulate ethylene synthesis as a stress factor. Therefore, combinations of multiple interventions may be necessary to reduce the negative effects on fruits and vegetables. The blueberries were treated with UV-C at 4.0 kJ/m² (3 min) before storing them for 8 days at 4°C [101]. The respiration rate was significantly lower in treated the fruits than the control samples.

Tiecher et al. [95] treated tomatoes with 3.7 kJ/m² UV-C light and found that the treated fruits had about two-fold higher ethylene production than the control. The UV-C treatment maintained higher levels of polyamide contents and delayed ripening of the tomatoes during storage. In another study, tomatoes were applied UV-C treatment (3.7 and 24.4 kJ/m²) and stored at 16°C for 35 days [63]. It was reported that 3.7 kJ/m² dose was more effective in delaying ripening and senescence by decreasing the respiratory rate and ethylene production on tomatoes. The UV treatment delayed the emergence of ethylene peaks by at least 9 days and the occurrence of respiratory peaks by at least 7 days compared to the control. In addition, the putrescine level of the control fruit was 39 μmole/kg, while these values were 452 μmole/kg in 3.7 kJ/m^2 and 293 μmole/kg in 24.4 kJ/m^2 at the end of the 21-day storage period. Polyamines (such as: putrescine and spermidine) are important compounds to control ripening (the higher levels, slower ripening) and also reduce the activity of enzymes causing softening [31].

Sweet cherry was applied with 1.2, 3.0 or 6.0 kJ/m^2 UV-C treatment and storage for 10 days at 0°C [67]. It was found that UV-C treatment decreased respiration rate in a dose-dependent manner. Ripening index (RI) was calculated as the ratio of total soluble solids (TSS) to titratable acidity (TA), and it was 12 for the control fruits. The RI of the samples treated with 1.2 and 3 kJ/m^2 UV-C was similar to that of the control samples, but 6 kj/m^2 of UV-C treatment increased the RI to about 15, which was associated with the reduced TA and the unchanged TSS. On the other hand, the UV treatment decreased softening of the fruits. In spite of the higher RI induced by UV treatment, the authors concluded that the UV treatment delayed ripening and senescence through reduced respiration and softening.

Watermelon cubes were treated with UV-C at doses of 1.6, 2.8, 4.8 and 7.2 kJ/m^2 and stored for 11 days at 5°C [10]. While UV-C treatments increased respiration rate up to 2-folds, and no effect on ethylene production was observed.

2.6.2   EFFECT ON STORAGE AND SHELF LIFE

Low temperature storage is a physical method used to control ripening and to prolong shelf life during post-harvest storage of fruits and vegetables. However, storage above freezing point (0°C) and below some critical temperatures between 5 and 13°C causes chilling injury (CI) in tropical and subtropical fruits [42]. Symptoms of the CI include surface and internal discoloration, pitting or submerged lesions on the peel, water soaking, failure to mature, irregular ripening, formation of off-flavor compounds and increased susceptibility to decay [42, 77].

Superficial scald is an important physiological storage disorder that causes brown or black spots on the peel of the fruit. It usually occurs in apple and pear varieties [23]. Scalding is caused by oxidative reactions. The α-farnesene is oxidized to conjugated trienes, ketones, and alcohol radicals, which ultimately cause the skin browning [81].

Internal browning (IB) is one of the important physiological disorders that occur during cold storage. IB is also called black heart or endogenous brown spot [85]. IB occurs first as the formation of water sports on the underside of the fruit. Then these spots turn black and spread over time to neighboring tissues [100]. IB cannot be visually observed from the outside as it occurs in fleshy tissues inside the fruit [29]. Another physiological disease affecting the quality of the fruits during storage is pitting due to metabolic changes and mechanical injury. Table 2.2 presents some studies to determine the effectiveness of UV application on storage disorder.

TABLE 2.2 Studies on the Effects of UV Applications on Storage Disorder

Fresh Produce	Treatment Conditions	Results	References
Banana	8 W lamp power 0.03 kJ/m² UV-C 2 W/m² 50 cm distance Stored at 8°C	• Decreased CI incidence and severity. • Reduced polyphenol oxidases (PPO) activity.	[73]
Cucumber (*Cucumis sativus* L. cv. Silor)	30 W lamp power 3, 5, 10 or 15 min UV-C 7166 mW/m² 30 cm distance.	• Decreased electrolyte leakage and CI until 10 days of storage at 5°C.	[44]
Eggplant and Cucumber	11 W lamp power 3, 9, and 15 kj/m² UV-C	• 15 kj/m² in 17% less CI on eggplants than the control. • 3 kj/m² caused slightly less CI on cucumbers.	[53]
Golden Bell sweet pepper (*Capsicum annum* L.)	30 W lamp power 2.2 kJ/m² UV-C for 30 min 4.4 kJ/m² UV-C for 60 min 6.6 kJ/m² UV-C for 90 min 70 cm distance Storage for 15 days at 4±1°C	• Alleviation of CI. • The most effective dose 6.6 kJ/m²	[74]
Mango (*Mangifera indica*, L. cv. Tainong)	1, 5, 10 kJ/m² UV-B Storage at 6°C for 10 days	• Reduced CI. • Increased endogenous nitric oxide levels. • The most effective dose 5 kJ/m²	[77]
Peach (*Prunus persica* cv Jefferson)	15 W lamp power 1.5, 2.5, 4.9, 7.4, 9.8 kJ/m² 3, 5, 10, 15, 20 min UV-C treatments; 8220 mW/m² 15 cm distance Storage for 14 or 21 days at 5°C and 7 days at 20°C	• Significantly reduced CI The 1.5 kJ/m² was the most efficient in reducing CI. • Reduced decay at doses up to 7.4 kJ/m²	[31]

TABLE 2.2   *(Continued)*

Fresh Produce	Treatment Conditions	Results	References
Peach (*Prunus persica* L. Batsch)	30 W lamp power 1.5 kJ/m² UV-C for 20 min HA; at 40 °C and 90% relative humidity for 4 h; Stored at 1°C for 35 days + 3 days at 20°C	• Suppressed CI. • Lower IB index.	[106]
Peach (*Prunus persica* L. Batsch)	30 W lamp power 1.5 kJ/m² UV-C for 20 min; Stored at 1°C for 35 days	• Delayed CI.	[105]
Pepper (*Capsicum annum* L. cv. Zafiro)	30 W lamp power 1, 3, 7, 14 kJ/m² UV-C Storage for 15 and 22 days at 0°C + 4 days at 20°C 30 cm distance	• Decreased decay and CI index. • The most effective dose 7 kJ/m²	[98]
Pineapple 'Phulae'	13.2 kJ/m² UV-C for 10 min 26.4 kJ/m² UV-C for 20 min 39.6 kJ/m² UV-C for 30 min 2.2 mW/cm 25 cm distance storage for 28 days at 10°C	• Reduced disease incidence and internal browning.	[79]
Sweet cherry (cv. Sweetheart)	1.2, 3.0 or 6.0 kJ/m² 20 W/m² 50 cm distance Storage for 10 days at 0°C	• Reduced disorder pitting. • The most effective dose 6.0 kJ/m²	[67]
Tomato (*Lycopersicon esculentum* cv. Zhenfen 202)	30 W UV-C lamp power 20 W UV-B lamp power 4 kJ/m² UV-C 20 kJ/m² UV-B; approximately 20 cm distance; Stored at 2°C for 20 days + at 20°C for 10 days	• CI index is markedly lower than control.	[59]

UV treatments reduce the symptoms of CI by stimulating the antioxidant system and triggering the defense system [104]. Peaches were applied UV-C treatments (1.5, 2.5, 4.9, 7.4, 9.8 kJ/m²) [31]. It was found that the UV-C at 1.5, 2.5, 4.9 kJ/m² doses significantly alleviated CI during storage for 21 days at 5°C followed by 7 days at 20°C. These applications markedly decreased skin browning and internal deterioration. As storage time increased, the reduction of these symptoms became more pronounced. Among all applications, 1.5, 2.5, 4.9 kJ/m^2 UV-C treatments were most efficient after 14 days at 5°C followed by 7 days at 20°C, while 1.5 and 2.5, kJ/m^2 UV-C treatments were the most effective after 21 days at 5°C plus 7 days at 20°C. In the control samples, the indication of intense CI and decay were observed. In UV-C treated fruits, browning or fungal decay were seen similar to the control only in samples treated with 9.8 kJ/m^2 UV-C. Decay symptoms were minimal in 1.5 and 7.4 kJ/m^2 UV-C treatments. UV-C treatments at 1.5, 2.5, 4.9 kJ/m^2 considerably diminished the browning index (BI) while the treatments at 7.4, 9.8 kJ/m^2 increased the browning of the skin. The 1.5 kJ/m^2 UV-C treatment provided the most effective results in all evaluations. In another study, UV-C light has been found to alleviate IB in peaches.

Hot air (HA) treatment at 40°C and 90% relative humidity (RH) for 4 h, and 1.5 kJ/m^2 UV-C treatment was applied to peaches and the peach fruits were stored at 1°C for 35 days plus 3 days at 20°C [106]. The first signs of IB appeared on day 14 for control (non-treated samples) and on day 21 for both applications (for groups treated with HA and UV-C) and BI continued to increase after this time. However, the UV-C and the HA treatments decreased IB index during the entire storage compared to the control. Besides, UV-C application was more efficient than HA to inhibit IB index at 21, 28, and 35 days. Peaches were treated with UV dose of 1.5 kJ/m^2 and HA treatment at 40°C and 90% RH for 4 h [105]. It was reported that these applications significantly delayed the emergence of CI on peaches. The first signs of CI appeared on day 28 for the control fruit; on the other hand, it was seen on the 35th day for both applications with a 1-week delay. CI index was 0.54 for control samples and 0.11 for both HA and UV-C applications after 35-day storage.

Eggplant and cucumber were treated with 3, 9 and 15 kJ/m^2 UV-C [53]. The UV treatment at 15 kj/m^2 dose resulted in 17% less CI than the control for eggplants. The UV treatment at 3 kJ/m^2 brought about a 10% lower CI than the control. Kasim and Kasim [44] applied 3, 5, 10 or 15 min UV-C light (7166 mW/m^2) to cucumber. Electrolyte leakage of UV-C treated samples was less than the control on day 10 at 5°C.

In another study, tomatoes were treated with 4 kJ/m^2 UV-C or 20 kJ/m^2 UV-B and stored for 20 days at 2°C followed by 10 days at 20°C [59]. At the end of this period, the CI index was 1.47 for UV-C applied tomatoes and 1.31 for UV-B. The CI value was 2.33 in the control samples. When both applications are compared with the control, the CI index was markedly lower.

Mango fruits were treated with 1, 5, and 10 kJ/m² UV-B and then stored for 10 days at 6°C [77]. It was observed that the CI index in the treated fruits after storage was lower than the control. Treatments at 5 and 1 kJ/m² UV-B doses decreased CI index, while 10 kJ/m² UV-B dose slightly intensified the symptoms of CI. The most effective dose was 5 kJ/m². It was emphasized that endogenous nitric oxide (NO) level that was increased by UV-B was effective in reducing the CI index.

Peppers were treated with 7 kJ/m² UV-C and stored at 0°C for up to 22 days followed by 4 days at 20°C [98]. The treatment decreased decay (reducing *Alternaria* and *Botrytis*), CI index, and severity of the symptoms. After 12 days of storage at 10°C, 40% of the control fruits were decayed, whereas the UV-C treated fruits had no rotting. Similar results were reported by Promyou and Supapvanich [74], who applied 2.2, 4.4, and 6.6 kJ/m² UV-C treatments on sweet peppers. The UV-C treatment at 6.6 kJ/m² alleviated the CI with CI index values of lower than 4.

UV-C treatment at a dose of 0.03 kJ/m² decreased CI incidence and severity in banana during storage at 8°C [73]. The UV-C treatment significantly diminished the development of peel browning symptoms during storage, and markedly reduced PPO activity up to day 14.

Exposure to UV-C light (13.2, 26.4, 39.6 kJ/m²) has been demonstrated to reduce IB and disease index (reducing mold growth) in pineapple during storage at 10°C for 28 days [79]. The IB in fruits treated with 26.4 kJ/m² and 39.6 kJ/m² UV-C was significantly lower compared to the control while the treatment at 13.2 kJ/m² had no effect compared to the control. While the BI was about 10% at high doses on the 21st day, this value was 20% in the control and low dose samples. After 28-day storage, BI was 55% in the fruits treated with high doses, but it was 90% in the fruits treated at the lower doses.

Sweet cherry treated with 6.0 kJ/m² UV-C had significantly lower incidence of surface pitting compared to control during storage at 0°C for 10 days [67].

2.6.3   EFFECT ON FOOD TEXTURE

The texture of fruits and vegetables is a critical quality feature that affects consumer acceptability of foods. During metabolic processes, such as, ripening, and senescence, the texture of the fruits and vegetables changes, firmness decreases and softening occurs. UV-C application can increase the firmness and delay the softening of fresh produce by slowing the activity of the enzymes degrading the cell wall [75]. Polygalacturonase (PG) and PME are the most important enzymes that cause softening of fruits and vegetables, as well as other enzymes, such as, β-d-galactosidase, xylanase, cellulase, and protease [18]. Some studies to determine the effectiveness of UV application on texture are summarized in Table 2.3.

TABLE 2.3   Studies on the Effects of UV Applications on Texture

Fresh Produce	Treatment Conditions	Results	References
Apple Fresh-cut 'Pink Lady'	15 W lamp power 1.2, 6, 12, 24 kJ/m^2 1, 5, 10, 20 min; 20 W/m^2 Storage for up to 14 days at 6°C	• No considerable difference in firmness.	[64]
Cherry tomato (*S. lycopersicum* L. cv zhenzhu L.)	30 W lamp power 4.2 kJ/m^2 UV-C for 8 min; 15 cm distance; Stored at 18°C for 35 days in the dark;	• Reduced loss of firmness during storage.	[18]
Cucumbers (*Cucumis sativus* L. cv. Silor)	30 W lamp power 3, 5, 10 or 15 min UV-C 7166 mW/m^2 30 cm distance Storage at 5°C or 10°C for 15 days	• Reduced firmness at 5°C. • Slightly higher firmness after 10 days at 10°C.	[44]
Fresh-cut pineapple (*Ananas comosus* L. Merr. cv. Comte de Paris)	4.5 kJ/m^2 UV-C 60 s and 90 s 20 cm distance Stored at 10°C	• Reduce loss of firmness during storage.	[69]
Peach (*Prunus persica* Batsch cv. Yulu)	30 W lamp power 3.0 kJ/m^2 UV-C 200 µW/cm^2 25 cm distance Storage for 8 days at 20°C	• Effectively inhibited loss of firmness.	[102]
Pepper (*Capsicum annum* L. cv. Zafiro)	30 W lamp power 1, 3, 7, 14 kJ/m^2 UV-C Storage for 15 and 22 days at 0°C + 4 days at 20°C 30 cm distance	• Maintained the firmness. • The lower weight loss after 12-day storage. • No difference weight loss after 18 days storage.	[98]

TABLE 2.3   *(Continued)*

Fresh Produce	Treatment Conditions	Results	References
Shiitake mushrooms (*Lentinus edodes*)	20 W lamp power 4 kJ/m² 2.82 ± 0.44 mW/cm² Storage for 15 days at 1±1°C + 3 days at 20°C	• Higher firmness.	[41]
Strawberry (*Fragaria* × *ananassa* Duch., cv. Aroma)	30 W lamp power 4.1 kJ/m² UV-C for 8 min 30 cm distance Stored at 20°C for 96 h	• Delayed fruit softening. • Diminished PG and PME activities.	[72]
Tomato (*Lycopersicon esculentum* Mill cv. Trust)	3.7 kJ/m² Stored at 16°C for 25 days in the dark	• Delayed softening. • The higher firmness. • Lower PG, PME, and cellulase activities.	[12]

Barka et al. [12] applied 3.7 kJ/m² UV-C treatments on tomatoes and the fruits were stored for 25 days at 16°C in the dark. The UV-C treatment significantly slowed the loss of firmness during storage compared to the control. This was associated with the lower activities of cell wall degrading enzymes such as PG, PME, cellulase in the treated fruits than the control. Treatment of cherry tomatoes with 4.2 kJ/m² UV-C resulted in no effects on the firmness immediately, but the treated fruits were firmer than the control after 5 days of storage at 18°C [18]. This was associated with considerably lower PG and PME activities in the treated fruits. Similarly, when 4.2 kJ/m² UV-C doses were applied to strawberry fruit, PG, and PME activities were diminished [72]. After 2–4 days storage, the UV-C treated fruits preserved their firmness better than the control.

Yang et al. [102] found that the UV-C treatment at 3.0 kJ/m² prevented loss of firmness observed in the control for peaches during storage. Jiang et al. [41] found that the UV-C treatment at 4 kJ/m² preserved the firmness of the mushrooms during storage at 1°C for 15 days, and delayed the loss of firmness after the following 3 days storage at 20°C. Fresh-cut pineapple treated with 4.5 kJ/m² UV-C had a slower reduction in firmness than the control during storage at 10°C [69].

Manzocco et al. [64] applied UV-C (1.2, 6.0, 12.0 and 24.0 kJ/m²) on fresh-cut apples and found that UV-C applications had no effect on firmness.

Kasim and Kasim [44] tested 3, 5, 10 or 15 min UV-C light at 7166 mW/m² on cucumber that was stored at 5°C or 10°C for 15 days. The UV-C application for 10 min resulted in higher firmness after 10 days of storage at 10°C. On the other hand, UV-C applied for 5 min or higher had negative effects on firmness at 5°C after 5 and 15 days of storage.

Vicente et al. [98] treated peppers with 7 kJ/m² UV-C and found that the treatment did not have an immediate effect on the firmness. However, the UV-C treated fruits maintained their firmness better, while the firmness of control fruits gradually decreased after 22-day storage at 0°C.

2.6.4 EFFECT ON FOOD COLOR

Some natural pigments (anthocyanins, carotenoids, chlorophyll, etc.), are sensitive to UV light. Studies to determine the effectiveness of UV application on color of fresh produce products are summarized in Table 2.4.

TABLE 2.4 Studies on the Effects of UV Applications on Color

Fresh Produce	Treatment Conditions	Results	References
Apple Fresh-cut (*Malus domestica* L cv. Golden delicious) Pear fresh-cut (*Pyrus communis* L. cv. Abate Fétel, Decana)	UV-A LEDs 2.43×10^{-3} W/m² 10, 20, 30, 40, 50 and 60 min	• UV-A application decreased total color change on the fruits. • The approximately 60% reduction in total color change in apples, 25% reduction in total color change in pears after 60 min exposure to UV-A.	[54]
Apple Fresh-cut 'Pink Lady'	15 W lamp power 1.2, 6, 12, 24 kJ/m² UV-C 1, 5, 10, 20 min 20 W/m² Storage for up to 14 days at 6°C	• Reduce the rate of browning reaction at doses up to 12 kJ/m². • Nevertheless, increased browning at 24.0 kJ/m².	[64]
Apple Fresh-cut 'Golden Delicious'	25 W lamp power 0.5 and 1.0 kJ/m² UV-C 15 cm distance Storage for 15 days at 4C	• No effects on L*, a*, b* values.	[32]

TABLE 2.4   *(Continued)*

Fresh Produce	Treatment Conditions	Results	References
Broccoli (*Brassica oleracea* L. Italica Group)	4.5 and 9.0 kJ/m² UV-A 4.4, 8.8, 13.1, 17.5, and 26.3 kJ/m² UV-B 15 cm distance Stored at 15°C in the dark for 6 days	• UV-A doses did not prevent discoloration during storage. • UV-B at 8.8 kJ/m² or above had higher hue angle and higher chlorophyll contents after 4 days of storage compared to control.	[4]
Carrot Fresh-cut (*Daucus carota* L.cv. Nantes)	15 W lamp power 0.78 kJ/m² UV-C	• No difference between control and UV-C treated samples.	[5]
Leek Spinach Cabbage	15 W lamp power Single UV-C: only once at the beginning; Multiple UV-C: once every day; 2.46 kJ/m² for 5 min; 20 cm distance; Stored at 4°C.	• Markedly higher Chlorophyll a contents in samples treated with multiple-UV-C.	[58]
Pepper (*Capsicum chinense*) Habanero	30 W lamp power 0, 0.5 and 1 min UV-C 11.3 W/m²	• No significant effect on the L*, a*, b*, Hue, and Chroma values.	[70]
Tahitian lime (*C. latifolia* Tan.)	19.0 kJ/m² UV-B for 20 min 15 cm distance Stored at 25°C in the dark	• Delayed loss of chlorophyll a content. • Decreased reduction in hue angle during storage. • Suppressed the increases of Chllase, Mg-dechelase, and Chll-POX enzyme activity during storage.	[43]
Watermelon fresh-cut (cv. Raspa and/or Sangría)	30 W lamp power 1.4, 4.1, 6.9, 13.7 kJ/m² UV-C 4.61 W/m² 15 cm distance Stored at 2.8°C for 7 days	• Loss of red color associated with higher L* value and lower a*, b*, hue, and chroma at 13.2 kJ/m². • No significant effect of UV-C doses up to 6.9 kJ/m².	[27]

TABLE 2.4   *(Continued)*

Fresh Produce	Treatment Conditions	Results	References
Watermelon Fresh-cut (*C. lanatus* cv 'Fashion')	36 W lamp power 1.6, 2.8, 4.8 and 7.2 kJ/m² UV-C 15 cm of distance Stored for 11 days at 5°C	• No effects on color parameters.	[10]

Leek, spinach, and cabbage samples were irradiated once every day (multiple UV-C treatment) or once at the beginning (single UV-C treatment) with a 2.46 kJ/m² UV-C and before storing them at 4°C for 5 days [52]. The multiple UV-C treatment resulted in significantly higher retention of chlorophyll-a in all samples than the control or single UV-C treatments. No difference on the chlorophyll-a between the single UV-C treatment and the control was detected.

Aiamla et al. [4] applied 4.5 and 9.0 kJ/m² UV-A and 4.4, 8.8, 13.1, 17.5, and 26.3 kJ/m² UV-B to the broccoli before storing them at 15°C for 6 days. The UV-A treatments did not affect color, but the UV-B treatments at 8.8 kJ/m² or higher markedly reduced the loss of Chl and retained color, and caused a higher hue angle after 4 days of storage.

Similar effects of UV-B on retention of chlorophyll in lime fruit was reported by Kaewsuksaeng et al. [43], who applied 19.0 kJ/m² UV-B treatments to lime and stored at 25°C in the dark. While the control fruits' hue angle values were reduced considerably and the fruits began to turn yellow, and little changes occurred in the UV-B treated fruits. The reduction of chlorophyll-a was delayed in the UV-B treated fruits, whereas the chlorophyll-a content in the control samples decreased rapidly during storage. UV-B administration decreased Chlase, Mg-dechelase, and Chlorophyll-degrading peroxidase (Chl-POX) enzyme activities during storage.

Fresh-cut watermelon were treated with different UV-C light doses (1.4, 4.1, 6.9 and 13.7 kJ/m²) and stored at 2.8°C for 7 days [27]. The red color of the fruit decreased upon treating with 13.7 kJ/m² UV-C, which was associated with higher L* and lower a*, b*, hue, and chroma values. The authors concluded that UV-C treatment up to 6.9 kJ/m² had no substantial impact on the color values of the watermelon compared to the control during storage. A similar observation was made by Artés-Hernández et al. [10], who found that UV-C treatments at 1.6, 2.8, 4.8, and 7.2 kJ/m² doses did not change the color of watermelon cubes during 11 days at 5°C.

Fresh-cut apple (Golden Delicious, Fuji, and Granny Smith) and fresh-cut pear (Abate, Decana) were treated with UV-A LED at an irradiance of 2.43×10^{-3} W/m^2 with varying exposure times from 10 to 60 min [54]. The 60 min UV-A treatment caused approximately 60% and 25% reduction in color change (%RΔE) in apple and pear samples, respectively. The researchers argued that although UV-A light application is less potent than UV-C, yet the browning can be controlled by UV-A without adversely affecting the nutritional and organoleptic quality characteristics of fresh-cut fruits. Fresh-cut apples ('Pink Lady') treated with UV-C up to 12.0 kJ/m^2 had significantly lower enzymatic browning than the untreated fruits [64]. On the other hand, the UV-C treatment at 24 kJ/m^2 caused a higher enzymatic browning compared to control after 4 days of storage at 6°C.

In contrast, Graça et al. [32] treated fresh-cut apples ('Golden Delicious) with UV-C at 0.5 and 1.0 kJ/m^2 and reported that the treatment did not affect the color compared to the control. These results illustrate that the effects of UV-C on color of fruit tissues can be different depending on the dose and cultivar of fruits, and the treatment can cause quality defects beyond some dose levels. Therefore, optimum dose level must be determined for each cultivar and each quality attributes of fruits.

UV-C treatment at 11.3 W/m^2 for 0.5 and 1 min had no significant effect on the L*, a*, b*, Hue, and Chroma values of Habanero pepper during storage at 4–5°C for 30 days [70]. Similarly, the color of fresh-cut carrots was not affected by UV-C treatments at 0.78 kJ/m^2 [5].

2.6.5   EFFECT ON BIOACTIVE COMPOUNDS IN FRESH PRODUCE

Fresh produce is an excellent source of antioxidant bioactive compounds, such as vitamins, carotenoids, and phenolic compounds. It has utility effects on human health due to its antimutagenic and anticarcinogenic properties, the ability to prevent cardiovascular, neurodegenerative, and chronic diseases [38]. Bioactive compounds are also secondary metabolites with antimicrobial activity in plant defense systems [3]. These compounds control oxidation reactions alleviating free radicals and ROS in the cells. It is known that UV application enhances the secondary metabolites biosynthesis in fresh produce and increases antioxidant activity [21]. The studies focusing on the effectiveness of UV application on bioactive compounds are summarized in Table 2.5.

TABLE 2.5 Studies on the Effects of UV Applications on Bioactive Compounds

Fresh Produce	Treatment Conditions	Results	References
'Yali' pear (*Pyrus bretschneideri Rehd.*)	5 kJ/m^2 15 min 5.55 W/m^2 20 cm distance	• Increased PAL activity after 24 h and onwards. • Higher levels of superoxide dismutase, catalase, and glutathione reductase activities.	[56]
Fresh-cut carrot (*Daucus carota* L.cv. Nantes)	15 W lamp power 0.78 kJ/m^2 UV-C Storage at 5°C for 10 days.	• Higher phenolic content during storage. • Immediate reduction in total carotenoid content but more retention of carotenoids after 7 days.	[5]
Fresh-cut honey pineapple, guava, banana "pisang mas."	2.16 J/m^2 UV-C 0, 10, 20, 30 min 10 cm distance	• Decreased ascorbic acid in pineapple and guava upon 20 and 30 min exposures. • Increased total polyphenols and flavonoids. • Higher FRAP and DPPH in banana. • No effects on FRAP and DPPH in pineapple. • A significant increase FRAP after 30 min in guava.	[7]
Fresh-cut watermelon (*C. lanatus* cv 'Fashion')	36 W lamp power 1.6, 2.8, 4.8 and 7.2 kJ/m^2 UV-C 15 cm of distance stored for 11 days at 5°C	• Lycopene content was not affected by 1.6 and 2.8 kJ/m^2, but slightly decreased by 4.8 and 7.2 kJ/m^2. • No effects on total phenolic and vitamin C content. • Increased antioxidant capacity on day 0 and day 8 by 4.8 and 7.2 kJ/m^2.	[10]
Habanero pepper (*Capsicum chinense*)	30 W lamp power 0, 0.5 and 1 min UV-C 11.3 W/m^2	• Stimulated the synthesis of bioactive comp ounds. • Enhanced antioxidant capacity.	[70]
Minimally processed rocket (*Eruca vesicaria* ssp. *sativa*)	36 W lamp power 10, 20 and 30 kJ/m^2 21 cm of distance Stored at 5°C for 12 days.	• Total phenolic content increased in the first 4 days of storage by 20 and 30 kJ/m^2. • Higher antioxidant activity at all doses was detected only at the end of 12 days. • Decreased carotenoid content at all dose only after 1 day of storage.	[34]

TABLE 2.5   *(Continued)*

Fresh Produce	Treatment Conditions	Results	References
Peach (*Prunus persica* L. Batsch)	30 W lamp power 1.5 kJ/m² UV-C during 20 min stored at 1°C for 35 days	• Enhanced accumulation of anthocyanins. • Up-regulation of gene expression for anthocyanin synthesis Decreased citric and malic acid degradation during storage.	[105]
Strawberry (*Fragaria x ananassa* Duch., cv 'Aromas')	30 W lamp power 4.35 kJ/m² UV-C 30 cm distance Stored for 6 days at 20°C	• Increased total anthocyanin, gallic acid, ρ-hydroxybenzoic acid, ρ-coumaric acid, quercetin, (+)-catechin after 2 days. • Increased PAL and ANS transcript acculumation after 6 days. • Increased LGalDH (L-galactose dehydrogenase) and GLDH (L-galactono-1,4-lactone dehydrogenase) transcript accumulation after 2 days. • L-ascorbic acid content was lower before 2 days but higher after 4 days.	[87]
Table grapes (*V. vinifera×V. labrusca* cv. Summer Black)	UV-B or UV-C 3.6 kJ/m² 5.27 W/m² Storage at 4°C for 28 days.	• Increased total phenolic, flavonoid, flavonol, and anthocyanin contents, and increased antioxidant activities by both UV-C and UV-B. • Increased expression of phenylpropanoid, flavonoid, and stilbene pathway genes by both UV-C and UV-B. • UV-C effect was more pronounced than UV-B.	[89]
Table grapes 'Emperor' (*Vitis vinifera* L.)	UV-A or UV-B with 15 W lamp power 30 min exposure followed by storage at 20°C at low light of 20 μmol/m²s photon flux between 400 and 700 nm	• No significant change in caftaric acid content and antioxidant activity. • Decreased the amount of quercetin-glycoside and quercetin glucuronide after 2 h, but no difference after 48 h compared to control.	[22]

TABLE 2.5 *(Continued)*

Fresh Produce	Treatment Conditions	Results	References
Tomato on vine (*S. lycopersicum* cv. Pera variety)	15 W lamp power 1.0, 3.0, 12.2 kJ/m² UV-C 0.282 W/m² Stored 2 days at ambient temperature	• Increased the total lycopene contents except 3.0 kJ/m². • Decreased β-carotene content. • Increased total phenolic contents and some individual phenolics by 3 and 12 kJ/m².	[17]

Pérez-Ambrocio et al. [70] applied 11.3 W/m² UV-C light (0, 0.5, 1 min) to Habanero pepper and stored in the sealed package at 5°C for 30 days. UV-C light applications showed a considerable increase in total flavonoids, total phenolic compounds compared with the control, and enhanced antioxidant activity.

Severo et al. [87] found that the total amount of anthocyanins and phenolic compounds including hydroxybenzoic acid, gallic acid, ρ-coumaric acid, (+)-catechin and quercetin were increased by UV-C application (4.35 kJ/m²) in strawberry. The contents of total phenolic and total anthocyanin of the UV-C treated fruits were always higher than the control. The ascorbic acid (AA) levels were higher in UV-C treated fruits after 4 days of storage, while it was lower before 2 days of storage compared to control. The UV-C treatment resulted in higher L-galactose dehydrogenase (LGaLDH), L-galactose-1,4-lactone dehydrogenase (GLDH) transcript accumulation after 2 days, which are involved in ascorbic acid synthesis. Moreover, phenylalanine ammonia-lyase (PAL) and anthocyanin synthase (ANS) transcript accumulation involved in anthocyanin and phenolic synthesis were higher in UV-C treated fruits after 6 days of storage compared to the control.

Alegria et al. [5] applied UV-C light at 0.78 kJ/m² to fresh-cut carrot and found a higher content of total phenolic in the UV-C-treated carrots than that in control during 10 days of storage at 5°C. Total carotenoid content was initially decreased by the UV-C treatment, but more carotenoid was retained in the treated samples after 7 days of storage.

Alothman et al. [7] applied UV-C treatment (0, 10, 20, and 30 min) to fresh-cut honey pineapple, banana, and guava. The samples received an average UV radiation dose of 2.16 J/m². The total phenolic and flavonoid contents of guava and banana were enhanced markedly by the treatments. While total phenolic contents of pineapple decreased, it's content of total flavonoid increased by the UV-C treatments. The amount of Vitamin C of all fruits reduced with UV-C exposures of 20 and 30 min. Antioxidant activities (FRAP and DPPH) of fresh-cut pineapple were not affected, while higher

FRAP and DPPH values were observed in fresh-cut banana and guava after the UV-C treatments.

Tomatoes were treated with 1.0, 3.0 and 12.2 kJ/m² UV-C light [17]. Compared to the control, UV-C treatments (1.0 and 12.2 kJ/m²) increased the total lycopene (*E+Z* isomers) contents by approximately 20% while decreased the amount of β-carotene. The effects of UV-C doses on isomers of lycopene varied. *E-lycopene* was increased only by 1 kJ/m² UV-C applications, while the *Z-lycopene* increased markedly with increasing doses of UV-C. UV-C applications of 3 and 12 kJ/m² increased total phenolic compounds.

Table grapes were treated with UV-C and UV-B light at a dose of 3.6 kJ/m² and stored for 28 days at 4°C [89]. Both UV-C and UV-B applications increased total phenolic, flavonoid, flavonol, and anthocyanin contents, and antioxidant activities during storage. The UV-C effect was found to be more pronounced than UV-B. The UV applications increased the gene expression of phenylpropanoid, flavonoid, and stilbene pathways.

In another study, red table grapes were treated with UV-A and UV-B light for 30 min followed by storage at 20°C at low light intensity (20 μmol/m²s photon flux between 400 and 700 nm) for 48 h [22]. The UV treatments did not affect the antioxidant capacity and caftaric acid content. Both UV-A and UV-B treatments caused a reduction in the amount of quercetin-3-O-glycoside and quercetin-3-O-glucuronide after 2 h compared to the control, where UV-A caused a higher reduction than UV-B, but no difference among the treatments and the control was detected after 48 h.

UV-C treatments at 1.6 and 2.8 kJ/m² doses did not affect total lycopene content of fresh-cut watermelon while higher doses (4.8 and 7.2 kJ/m²) caused a slight reduction on it [10]. Although the UV-C treatments did not affect total phenolic and vitamin C contents, yet antioxidant capacity on day 0 and day 8 was increased by 4.8 and 7.2 kJ/m².

Gutiérrez et al. [34] applied 10, 20, and 30 kJ/m² UV-C to minimally processed rocket leaves, and stored them at 5°C for 12 days. No difference in total phenolic content was observed between the control and 10 kJ/m² UV-C treated samples during storage at 5°C. The total phenolic content of the samples treated at 20 and 30 kJ/m² was approximately 8% higher compared to the control after 4 days of storage, but no effects were detected afterwards. UV-C treatments caused a reduction in total carotenoid after 1 day, but no difference was reported afterwards.

Zhou et al. [105] applied a UV-C dose of 1.5 kJ/m² on peaches and stored for 35 days at 1°C. UV-C application increased the activity and gene expression of enzymes, such as, anthocyanidin synthase (ANS), chalcone synthase

(CHS), dihydroflavonol 4-reductase (DFR), flavanone 3-hydroxylase (F3H), glucose-6-phosphate dehydrogenase (G6PDH), PAL, and UDP-glucose flavonoid 3-O-glucozyltransferase (UFGT). The UV-C treated samples had higher anthocyanin accumulation compared to control. The UV-C application decreased the citric and malic acid degradation during storage.

UV-C at 5 kJ/m^2 was applied to pear and it resulted in a 176% increase in PAL activity after 48 h, 96% increase in β-1,3-glucanase activity after 24 h compared to the control [56]. Moreover, superoxide dismutase, catalase, and glutathione reductase activities were higher in the UV-C treated pears than the control fruits.

2.6.6   EFFECTS ON MICROBIAL CONTAMINATION IN FRESH PRODUCE

Contamination of fresh produce by microorganisms can cause economic losses through spoilage and, most importantly, can have negative effects on consumer health through pathogenic microorganisms. Microorganisms isolated from a variety of fresh fruits and vegetables that pose a risk to human health include: *Aeromonas hydrophila/caviae, B. cereus, E. coli* O157:H7, *Campylobacter* spp., *L. monocytogenes, Shigella* spp., *Salmonella* spp., *Y. enterocolitica,* viruses, and various molds.

In addition, microorganisms contaminating the fresh produce can cause a quality loss and shorten the storage life of especially fresh-cut produce. Thus, it is important to control these microorganisms on whole fresh or fresh-cut produce products by methods that do not adversely affect the freshness of the product. UV light has a high potential for decontamination of fresh produce products, and several research studies are summarized in Table 2.6.

Schenk et al. [82] applied UV-C at 0, 15, 31, 35, 44, 56, 66, 79 and 87 kJ/m^2 to pear slices inoculated with *L. inoccua, L. monocytogenes, E. coli,* and *Z. bailii*. The UV-C treatment at 87 kJ/m^2 caused 1.8–2.5 log and 2.6–3.4 log reductions in the microbial counts on the pear slices with and without peel, respectively. The authors found that *Listeria innocua* and *Listeria monocytogenes* were more susceptible than *E. coli* and *Z. bailii* at doses greater than 44 kJ/m^2, whereas the opposite was true at lower doses. It was stated that inactivation kinetics were successfully fitted the Weibull distribution model.

In another study, Syamaladevi et al. [94] determined that UV-C at 1.7 kJ/m^2 resulted in a 2.8 log inactivation in *Penicillium expansum* on intact pears, while 2.7 log reduction was obtained with 3.1 k /m^2 UV-C treatment

TABLE 2.6 Studies on the Effects of UV Applications on Microorganisms

Fresh Produce	Treatment Conditions	Microorganisms	Results	References
'd'Anjou' pear (intact/wounded) 'O'Henry' peach (intact)	0.59 kJ/m², 0.25 min, 1.14 kJ/m², 0.5 min, 2.16 kJ/m², 1 min, 4.00 kJ/m², 2 min, 5.71 kJ/m², 3 min, 7.56 kJ/m², 4 min 10 cm distance	*E. coli*	• 7.56 kJ/m² UV-C treatment: • reduction of 3.7 log CFU/g in intact pears; • reduction of 3.1 log CFU/g in wounded pears; • reduction of 2.9 log CFU/g in intact peaches; • The reliable lifetime (tR) from the Weibull model • 0.019 min (0.268 kJ/m²) for intact pear; • 0.062 min (0.348 kJ/m²) for wounded pears; • 0.074 min (0.371 kJ/m²) for peach; • Wounding or presence of trichomes leads to shadowing effects.	[93]
'd'Anjou' pear (*Pyrus communis L.*)	0.21, 0.43, 0.64, 1.2, 1.7, 2.1, 3.1 kJ/m² 10, 20, 30, 60, 90, 120, 180 s 10 cm distance	*P. expansum*	• At 1.7 kJ/m² UV-C dose: • 2.8 log reduction on intact pear discs; • At 3.1 kJ/m² UV-C dose: • 2.7 log reduction on wounded pear discs.	[94]
Baby spinach (*Spinacia oleracea L.*)	36 W lamp power 2.4, 7.2, 12 and 24 kJ/m² UV-C 15 cm distance	*L. monocytogenes* *S. enterica* *P. marginalis* Psychrotrophic *Enterobacteriaceae*	• 1–3 log lower *L. monocytogenes, S. enterica,* and *P. marginalis* counts compared to control on day 0, and day 14 on Petri media (*in vitro*). • 1–2 log reductions of *L. monocytogenes, S. enterica,* and *P. marginalis* compared to the control on spinach on day 0, but after 14 days 1 log lower L. monocytogenes otherwise similar counts compared to the control. • About 1.5 log reduction on psychrotrophic bacteria and *Enterobacteria* counts, and 1 log reductions on yeast and mold counts on day 0, but similar levels after 14 days of storage.	[25]

TABLE 2.6 (Continued)

Fresh Produce	Treatment Conditions	Microorganisms	Results	References
Blueberries	7.9 mW/cm² for dry UV 9.48, 23.7, 47.4 kJ/m² 4.6 mW/cm² for wet UV 5.52, 13.8, 27.6 kJ/m² 2, 5, and 10 min	E. coli O157: H7 Salmonella	• Dry UV-C treatment for 5 or 10 min • 4, 2.5 and 1.6 log reduction on E. coli O157:H7 in skin-spot, Calyx-spot, and dip innoculations • Wet UV-C treatment about 1 log more inactivations in case of the spot innoculations • No difference in the dip innoculated fruits. • Combined wet UV-C with some chemicals: • UV-C+10 ppm chlorine; • UV-C+100 ppm sodium dodecyl sulfate (SDS); • UV-C+0.5%levulinic acid+100 ppm SDS; • Not significantly affect the inactivation • About 1.5 log inactivation in both pathogens.	[60]
Fresh-cut 'Fuji' apple	15 W lamp power 5 min UV-C Storage for 15 d at 5°C	Total bacteria	• A 1.5 log reduction of total bacterial count upon 5 min exposure to UV-C. • More than 2 log lower bacterial count after 15 days of storage.	[20]
Fresh-cut 'Pink Lady' apple	15 W lamp power 1.2 kJ/m² 20 W/m² Storage for up to 14 days at 6°C	Total mesophilic bacteria Yeasts	• A 1.5 log reduction in total mesophilic bacterial counts immediately after the treatment. • 2 log and 1 log lower populations of mesophilic bacteria and yeasts at the end of 14-day storage.	[64]
Fresh-cut 'William' pear	15 W lamp power 0, 15, 31, 35, 44, 56, 66, 79, 87 kJ/m² 0–20 min 72.5 W/m² 10 cm distance	L. innocua L. monocytogenes E. coli Z. bailii	• 87 kJ/m² UV-C treatment: • 2.6 to 3.4 log reduction without peel; • 1.8 to 2.5 log reduction with peel; • L. innocua and L. monocytogenes most susceptible microorganisms at doses greater than 44 kJ/m². • At lower doses, E. coli and Z. bailii are the most susceptible microorganisms.	[82]

TABLE 2.6 *(Continued)*

Fresh Produce	Treatment Conditions	Microorganisms	Results	References
Fuji apples D'Anjou pears Cantaloupes Strawberries Raspberries	0.15–11.87 kJ/m^2	*E. coli* *L. monocytogenes*	• *L. monocytogenes* more resistant to UV-C. • UV-C at 0.15 and 0.92 kJ/m^2 caused 2.1–2.9 log and 1.2–2.1 log reductions of the *E. coli* on apples and pears, respectively. • UV-C at 7.17 kJ/m^2 caused 2 and 1-log reductions of the *E. coli* on strawberry and raspberry, respectively. • UV-C at 0.17 to 3.65 kJ/m^2 resulted in up to 1.4, 1.2, 0.7, and 0.7 log reductions of *L. monocytogenes* on apples, pears, strawberry, and cantaloupe, respectively. • More roughness of the surface lead to less inactivation due to more shadowing effects.	[2]
Iceberg Lettuce	Temperatures (4 and 25°C) Distances (10 and 50 cm) Type of exposure (one or two sides) UV intensities (1.36 to 6.80 mw/cm^2) Exposure times (0.5 to 10 min)	*E. coli* O157:H7 *Salmonella typhimurium* *Listeria monocytogenes*	• 204 mJ/cm^2 resulted in 1.45, 1.35 and 2.12 log reductions of *E. coli*, *S. typhimurium*, and *L. monocytogenes*, respectively; at 25°C while it caused 0.3, 0.57 and 1.16 log reductions at 4°C. • UV-C at 1.63 J/cm^2 applied two-sided resulted in about 8.8 log higher microbial inactivation compared to one-sided application. • As the distance from light source increased degree of inactivation decreased. • Microbial inactivation was increased with increasing UV-C intensities at 25°C and 10 cm distance from the light source.	[47]

TABLE 2.6 (Continued)

Fresh Produce	Treatment Conditions	Microorganisms	Results	References
Red Delicious apple Leaf lettuce Tomato	From 1.5 to 24 mg/cm^2	*Salmonella spp.* *E. coli* O157:H7	• About 2 log reduction of *Salmonella* and *E. coli* at 6 mJ/cm^2 on lettuce. • 2.65 and 2.79 log reduction at 24 mJ/cm^2 in *Salmonella* and *E. coli* O157:H7, respectively, on lettuce. • A 2-log average reduction of *Salmonella* on tomatoes. • A 3 log reduction of *E. coli* O157:H7 on apples.	[103]
Spinach leaves	0.16 J/cm^2 of UV-A light 0.5% acetic acid (AA) solution	*E. coli* O157: H7 *S. typhimurium* *L. monocytogenes*	• UV-A: A decrease of 0.51, 0.25, and 0.38 log in *E. coli* O157: H7, *S. typhimurium* and *L. monocytogenes*, respectively. • AA: A reduction of 0.55, 0.83, and 0.50 log for *E. coli* O157: H7, *S. typhimurium* and *L. monocytogenes*, respectively. • UV-A+AA: A reduction of 3.50, 3.29, and 4.30 log for *E. coli* O157: H7, *S. typhimurium* and *L. monocytogenes*, respectively.	[40]
Strawberries, Raspberries, Blueberries (fresh/frozen)	95 W lamp power 212 mJ/cm^2, 20 s 650 mJ/cm^2, 60 s 1331 mJ/cm^2, 120 s 4000 mJ/cm^2, 9 min	*Hepatitis A virus* (HAV) *Murine norovirus* (MNV) *E. coli* O157: H7 *S. enterica* *L. monocytogenes*	• On fresh blueberries • More than 2.5 log reduction in *MNV*; • About 0.5 log lower reduction on *HAV*; • On fresh raspberries and strawberries in the corresponding inactivation levels • A 1–1.5 log lower reduction on fresh raspberries and strawberries • On frozen berries compared to the fresh ones • A slightly lower inactivation (about 0.7 log) • On the fresh berries • Less than 1 log inactivation of *Salmonella*, *E. coli* O157:H7 and *L. monocytogenes*	[19]

on wounded pear tissues. UV-C treatment at 7.5 kJ/m^2 caused about 3.7 log reduction in *E. coli* count on intact pear surface with peel while the same dose resulted in about 3.0 log reduction on the wounded pear surface with peel [93]. The authors noted the similar level of inactivation (about 3.0 log) on *E. coli* at 7.5 kJ/m^2 on intact peach surface with peel. They also calculated the time required for a 90% reduction in the number of *E. coli* cells or the reliable lifetime (t_R) from the Weibull model and reported 0.019 min (0.268 kJ/m^2) in intact pear, 0.062 min (0.348 kJ/m^2) in wounded pears, and 0.074 min (0.371 kJ/m^2) in intact peaches. Thus, the surface characteristics of the fruit significantly affect the efficiency of the UV treatment for microbial inactivation. Wounding or presence of trichomes (as in intact peach surface) creates shadowing effects protecting the microbial cells from exposure to UV-C light [93].

UV-C light at 1.2 kJ/m^2 resulted in 1.5 log reduction in total mesophilic bacterial counts on fresh-cut apples ('Pink Lady') immediately after the treatment, and caused about 2 log and 1 log lower populations of mesophilic bacteria and yeasts, respectively, at the end of 14-day storage at 6C [64]. Similarly, Chen et al. [20] reported a 1.5 log reduction of total bacterial count on fresh-cut apples ('Fuji') on 5 min exposure to UV-C (the intensity or the dose was not reported), and this resulted in more than 2 log lower bacterial count on the samples after 15 days of storage at 5°C.

Strawberries, raspberries, and blueberries (fresh and frozen berries) were treated with UV-C for 20 s (212 mJ/cm^2), 60 s (650 mJ/cm^2), and 120 s (1331 mJ/cm^2) [19]. They found more than 2.5 log reduction in *murine norovirus* (MNV) and about 0.5 log lower reduction on *hepatitis A virus* (HAV) on fresh blueberries, while the corresponding inactivation levels were significantly lower (1–1.5 log lower) on fresh raspberries and strawberries. This seems to be due to better surface quality of blueberries (smoother) creating less shadowing effects compared to strawberry and raspberry. A slightly lower inactivation (about 0.7 log) was observed on frozen berries compared to the fresh ones. The UV-C treatments resulted in <1 log inactivation of *E. coli* O157:H7 *Salmonella*, and *L. monocytogenes* on the fresh berries.

Liu et al. [60] found that dry UV-C treatment for 5 or 10 min caused 4, 2.5 and 1.6 log reduction on *E. coli* O157:H7 inoculated onto blueberries by skin-spot, Calyx-spot, and dip inoculations, respectively. These differences were associated with different surface characteristics of the skin and calyx of the berries, whereas the latter has more uneven structures shadowing the cells from the UV-C light resulting in lower inactivation in the calyx-spot and dip-inoculated samples. The authors also applied the UV treatment to the berries in agitated water (as wet UV-C treatment) and observed about 1 log higher inactivation in the spot inoculations, but no difference in the dip inoculated fruits. It was evident that increased inactivation in the wet treatment was due to the washing effect of the agitated water. The authors also combined wet UV-C with some chemicals and found that addition of 10 ppm chlorine, 100 ppm sodium dodecyl sulfate (SDS) or 0.5% levulinic acid plus 100 ppm SDS did not significantly affect the deactivation of *S. enterica* and *E. coli* O157:H7 by the wet UV-C treatment, which caused about 1.5 log inactivation in both pathogens.

Baby spinach (*Spinacia oleracea L.*) was treated UV-C doses of 2.4, 7.2, 12, and 24 kJ/m^2 [25]. The UV-C treatments had 1–2 log lower *L. monocytogenes, S. enterica* and *P. marginalis* counts on growth media (in vitro) on day zero compared to control. The UV-C treated spinach had 1–2 log lower *L. monocytogenes, S. enteretidis* and *P. marginalis* compared to the control on day zero, but the samples had similar counts of *S. enteretidis* and *P. marginalis* while 1 log lower *L. monocytogenes* compared to the control after 14 days. The UV treatments caused about 1.5 log reduction on psychotropic bacteria and *Enterobacteria* counts, and 1 log reductions on yeast and mold counts on day zero, but the microbial populations became similar in the treated and untreated samples during the 14-day storage period.

In another study, spinach (*Spinacia oleracea L.*) leaves containing *S. typhimurium, E. coli* O157:H7, and *L. monocytogenes* were applied with 0.16 J/cm^2 of UV-A light and 0.5% acetic acid (AA) solution [40]. The UV-A treatment showed a decrease of 0.51, 0.25 and 0.38 log in *E. coli* O157:H7, *S. typhimurium,* and *L. monocytogenes*, respectively. When UV-A applied together with AA, a synergistic effect was observed with the inactivation levels of 3.50, 3.29 and 4.30 log for *E. coli* O157:H7, *S. typhimurium* and *L. monocytogenes*, respectively.

Yaun et al. [103] studied the effect of UV-C treatment at doses from 1.5 to 24 mJ/cm^2 on *Salmonella* spp. and *E. coli* O157:H7 inoculated to lettuce, apples, and tomatoes. The authors reported the dose in mW/m^2, but authors of this chapter think this is erroneous and must be mJ/cm^2. They found that both pathogens showed a similar inactivation level of about 2 log reduction at 6 mJ/cm^2 on lettuce, where the inactivation was increased to 2.65 and 2.79 log upon increasing the dose to 24 mJ/cm^2 in *Salmonella* and *E. coli* O157:H7, respectively. The authors also reported 2 log average reduction of *Salmonella* on tomatoes, and 3 log inactivation of *E. coli* O157:H7 on apples at the dose range applied.

Adhikari et al. [2] applied UV-C doses up to 11.9 kJ/m^2 to apples, canta-loupes, pears, red raspberries and strawberries and investigated the deactiva-tion of *L. monocytogenes* and *E. coli* O157:H7. They found that UV-C at 0.15 and 0.92 kJ/m^2 caused 2.1–2.9 and 1.2–2.1 log inactivation of *E. coli* O157:H7 on apples and pears surfaces, respectively. The inactivation levels of this pathogen on strawberry and raspberry were 2.0 and 1.0 log at 7.17 kJ/m^2, respectively. *L. monocytogenes* was more resistant to UV-C than *E. coli* O157:H7 in all the fruits tested. The UV-C at 0.17 to 3.65 kJ/m^2 resulted in 0.7–1.4, 0.7–1.2, 0.6–0.7 and 0.3–0.7 log reductions of *L. monocytogenes* on apple, pear, strawberry, and cantaloupe, respectively. Thus, there were obvious differences among the UV-C inactivation levels of these pathogens on different fruit surfaces, and this was associated with different surface characteristics of the fruits, where more roughness of the surface caused less inactivation due to more shadowing effects.

Kim et al. [47] found that application of 204 mJ/cm^2 (3.4 mW/cm^2 for 1 min) at 4°C to inoculated lettuce caused 0.3, 0.57 and 1.16 log inactivation of *E. coli, Salmonella,* and *L. monocytogenes*, respectively. They observed higher inactivation rates of 0.45, 1.35, and 2.12 log on *E. coli* O157:H7, *S. typhimurium*, and *L. monocytogenes*, respectively, when the same dose was applied at 25°C. The authors also compared one-sided to two-sided UV-C exposure (2.72 mW/cm^2 for up to 10 min) and found that two-sided exposure resulted in 0.8 log more inactivation of the inoculated pathogens. UV-C at 1.63 J/cm^2 resulted in 1.49, 2.05, and 1.56 log reduction upon one-sided exposure; 2, 2.9, and 2.1 log reductions on two-sided exposure on *E. coli* O157:H7, *S. typhimurium*, and *L. monocytogenes*, respectively. Since the inoculation was made only on one side of the lettuce leaves, there seems to be some additional effects of the two-sided exposure either through some means of penetration or most likely intensification of the UV-C. The authors also noted that as the distance from the light source was increased, the degree of inactivation decreased. In addition, the authors also reported that microbial inactivation was increased with increasing UV-C intensities at 25°C and 10 cm distance from the light source, but it was not clear how they achieved different intensities with the same number of lamps and distance. Their observation was mainly due to the dose-effect since the increased UV-C intensities would result in increasing dose at the same exposure time.

The effect of UV-C on microorganisms can vary depending on many factors, such as, types, strain, growth phase of microorganisms, initial micro-bial load, the nature of contamination (inoculation or natural contamination), temperature of the treatment, product characteristics (especially the surface quality), UV process parameters including wavelength, intensity, exposure time and treatment chamber design, which affects both the intensity and the exposure of samples to the UV.

2.7 CHALLENGES AND FUTURE TRENDS

Although UV technologies have been in use for many years, more and more research studies have been conducted recently. UV light has often been used for years to disinfect air, surfaces, and liquid foods. However, further research is needed to use UV light effectively in solid. Besides, the influences of UV treatment on the sensory, chemical, and physicochemical quality parameters of foods need to be better understood.

Recently, innovative technologies, such as mercury-free UV-LED lamps or pulsed UV lamps have been developed in place of the traditional UV lamps. Mercury-free UV-LEDs are expected to be used commercially more in the future because of their advantages, such as higher energy efficiency, longer use, and more stable light intensity [68]. PL lamps are expected to be used as an alternative to traditional lamps due to many advantages, such as having a wider spectrum, using short-duration pulses, high inactivation efficiency, and no repairing DNA damage [24]. Therefore, the efficacy and optimization of these new lamps need to be further explored in order to better understand and to use these in the industry.

In order to increase the efficiency of the UV light system and to design more efficient systems, it is necessary to fully understand how UV application is affected by process parameters including UV dose, UV intensity, exposure time and product parameters, such as, the surface and the optical properties of the product. One of the limitations in interpretation of the data published in the literature is that the UV treatment systems are not explicitly described. In some studies, even the UV dose applied is not correctly reported. The system design parameters include number, power, and location of UV lamps, the distance between the sample and the lamps, presence of reflectors, temperature control systems. These parameters will influence the achieved intensity, dose applied, and efficiency of homogeneous exposure of the samples to the UV.

Research on investigating the influence of dose rate on microbial inactivation and quality attributes of the samples is very limited. Since the fresh produce products are metabolically active, the dose rate, which affects the exposure time to achieve the same dose, therefore it can have major effects on the physiological response of the fresh produce tissue.

Determination of the effect of dose and dose rate on inactivation of microorganisms is important as these affect the quality and safety of foods. Comparison of UV resistance of microorganisms and kinetic modeling of microbial inactivation are important to obtain the basic data that can be used in future research. Moreover, it is quite evident from literature that inoculated microorganisms are generally inactivated by UV-C to a larger extent than the natural microflora. This is clearly related with more diverse types of microorganisms in natural flora, but it is also related with the higher degree of attachment of the natural microflora with probable existence of biofilms making them more resistant to UV-C compared to freshly inoculated samples. Modifications of inoculation procedures to reflect the conditions of natural microflora in research studies would be useful to get more realistic results. Therefore, all of these factors need to be well understood in order to use UV light technology efficiently on food applications.

The main disadvantage of UV light applications is the limited penetration depth. Knowing the depth of penetration will be useful in the selection of the UV-C dose to be applied and in predicting the direct effect of microorganisms on the surface and other parameters. In addition to the penetration of UV-C rays at the energy level in the fresh produce tissues, the determination of the penetration of the biochemical effects of UV is also important in terms of explaining the response of the fresh produce. UV-C application also leads to biochemical effects on fresh produce tissue as well as microbial inactivation. Some of these effects may be in the form of direct effect, that is, the interaction of components and energy within the rays. Some of these may be the indirect effect, where the direct effect can interact with other tissues or cells to induce further biochemical reactions. In other words, the effect of the UV can diffuse into the tissues that are not exposed to UV, causing further physiological and biochemical responses in fresh produce products. Therefore, further research is needed to determine UV-C penetration in tissue in terms of energy and biochemical effects.

2.8 SUMMARY

In this chapter, the effects of UV application on ripening and senescence, texture, storage disorder, color, bioactive compounds and microorganisms in fresh produce are evaluated extensively. While UV light applications provide effective microbial inactivation, as they should not adversely affect the physical, chemical, physicochemical, and sensory quality characteristics of the foods. UV applications can increase the shelf life and enhances the nutritional value of fruits and vegetable products. However, the beneficial effects of UV depend largely on the type of fresh produce product, and there are no standard treatment conditions to achieve positive results. Each fresh produce responds to UV treatment differently, and this must be determined to suggest the optimum treatment conditions. The combination of different applications to UV treatment can have synergistic or additive effects contributing to the quality and safety of fresh produce products. For example, the use of some chemical dips and modified atmosphere packaging commonly used in fresh produce products can be combined with UV treatment to get further benefits on product quality a shelf life; thus this needs to be investigated in future studies.

KEYWORDS

- **food quality**
- **fresh produce**
- **microbial inactivation**
- **ultraviolet dose**
- **ultraviolet intensity**
- **ultraviolet light technology**

REFERENCES

1. Abeles, F. B., Morgan, P. W., & Saltveit, M. E., (1992). *Ethylene in Plant Physiology* (2nd edn., p. 414). San Diego, CA-USA: Academic Press.
2. Adhikari, A., Syamaladevi, R. M., Killinger, K., & Sablani, S. S., (2015). Ultraviolet-C light inactivation of *Escherichia coli O157: H7* and *Listeria monocytogenes* on organic fruit surfaces. *International Journal of Food Microbiology, 210,* 136–142.
3. Ahn, S. Y., Kim, S. A., Choi, S. J., & Yun, H. K., (2015). Comparison of accumulation of stilbene compounds and stilbene related gene expression in two grape berries irradiated with different light sources. *Horticulture, Environment, and Biotechnology, 56*(1), 36–43.
4. Aiamla-or, S., Yamauchi, N., Takino, S., & Shigyo, M., (2009). Effect of UV-A and UV-B irradiation on broccoli (*Brassica oleracea L.* Italica Group) floret yellowing during storage. *Postharvest Biology and Technology, 54*(3), 177–179.

5. Alegria, C., Pinheiro, J., Duthoit, M., Gonçalves, E. M., Moldão-Martins, M., & Abreu, M., (2012). Fresh-cut carrot (cv. Nantes) quality as affected by abiotic stress (Heat Shock and UV-C Irradiation) pre-treatments. *LWT-Food Science and Technology, 48*(2), 197–203.

6. Allende, A., Tomas-Barberan, F. A., & Gil, M. I., (2006). Minimal processing for healthy traditional foods. *Trends in Food Science and Technology, 17*, 513–519.

7. Alothman, M., Bhat, R., & Karim, A. A., (2009). UV radiation-induced changes of Antioxidant capacity of fresh-cut tropical fruits. *Innovative Food Science and Emerging Technologies, 10*(4), 512–516.

8. Antonio-Gutiérrez, O. T., López-Díaz, A. S., López-Malo, A., Palou, E., & Ramírez-Corona, N., (2019). UV-C light for processing beverages: Principles, applications, and future trends: Chapter 7. In: Grumezescu, A. M., & Holban, A. M., (eds.), *Processing and Sustainability of Beverages: The Science of Beverages* (Vol. 2, pp. 205–234). UK: Woodhead Publishing.

9. Araque, L. C. O., Rodoni, L. M., Darré, M., Ortiz, C. M., Civello, P. M., & Vicente, A. R., (2018). Cyclic low dose UV-C treatments retain strawberry fruit quality more effectively than conventional pre-storage single high fluence applications. *LWT, 92*, 304–311.

10. Artés-Hernández, F., Robles, P. A., Gómez, P. A., Tomás-Callejas, A., & Artés, F., (2010). Low UV-C illumination for keeping overall quality of fresh-cut watermelon. *Postharvest Biology and Technology, 55*(2), 114–120.

11. Baldwin, E. A., & Bai, J., (2011). Physiology of fresh-cut fruits and vegetables. In: Martin-Belloso, O., & Soliva-Fortuny, R., (eds.), *Advances in Fresh-Cut Fruits and Vegetables Processing* (pp. 87–114). CRC Press; Taylor & Francis Group, NW, United States.

12. Barka, E. A., Kalantari, S., Makhlouf, J., & Arul, J., (2000). Impact of UV-C irradiation on the cell wall-degrading enzymes during ripening of tomato (*Lycopersicon esculentum L.*) fruit. *Journal of Agricultural and Food Chemistry, 48*(3), 667–671.

13. Baysal, A. H., (2018). Short-wave ultraviolet light inactivation of pathogens in fruit juices. In: Rajauria, G., & Tiwari, B. K., (eds.) *Fruit Juices* (pp. 463–510). San Diego, CA-USA: Academic Press.

14. Bintsism, T., Litopoulou-Tzanetakim, E., & Robinson R. K., (2000). Existing and potential applications of ultraviolet light in the food industry: A critical review. *Journal of the Science of Food and Agriculture, 80*(6), 637–645.

15. Bohrerova, Z., Shemer, H., Lantis, R., Impellitteri, C. A., & Linden, K. G., (2008). Comparative disinfection efficiency of pulsed and continuous-wave UV irradiation technologies. *Water Research, 42*(12), 2975–2982.

16. Braga, T. R., Silva, E. O., Rodrigues, S., & Fernandes, F. A., (2019). Drying of mangoes (*Mangifera indica L.*) with pulsed UV light as pretreatment. *Food and Bioproducts Processing, 114*, 95–102.

17. Bravo, S., Garcia-Alonso, J., Martin-Pozuelo, G., Gomez, V., Garci-Valverde, V., Navarro-Gonzalcz, I., & Periago, M. J., (2013). Effects of postharvest UV-C treatment on carotenoids and phenolic compounds of vine-ripe tomatoes. *International Journal of Food Science and Technology, 48*(8), 1744–1749.

18. Bu, J., Yu, Y., Aisikaer, G., & Ying, T., (2013). Postharvest UV-C irradiation inhibits the production of ethylene and the activity of cell wall-degrading enzymes during softening

of tomato (*Lycopersicon esculentum L.*) fruit. *Postharvest Biology and Technology, 86*, 337–345.

19. Butot, S., Cantergiani, F., Moser, M., Jean, J., Lima, A., Michot, L., Putallaza, T., Strohekera, T., & Zuber, S., (2018). UV-C inactivation of foodborne bacterial and viral pathogens and surrogates on fresh and frozen berries. *International Journal of Food Microbiology, 275*, 8–16.

20. Chen, C., Hu, W., He, Y., Jiang, A., & Zhang, R., (2016). Effect of citric acid combined with UV-C on the quality of fresh-cut apples. *Postharvest Biology and Technology, 111*, 126–131.

21. Cisneros-Zevallos, L., (2003). The use of controlled postharvest abiotic stresses as a tool for enhancing the nutraceutical content and adding value of fresh fruits and vegetables. *Journal of Food Science, 68*(5), 1560–1565.

22. Csepregi, K., Kőrösi, L., Teszlák, P., & Hideg, É., (2019). Postharvest UV-A and UV-B treatments may cause a transient decrease in grape berry skin flavonol-glycoside contents and total antioxidant capacities. *Phytochemistry Letters, 31*, 63–68.

23. Du, L., Song, J., Palmer, L. C., Fillmore, S., & Zhang, Z., (2017). Quantitative proteomic changes in development of superficial scald disorder and its response to diphenylamine and 1-MCP treatments in apple fruit. *Postharvest Biology and Technology, 123*, 33–50.

24. Elmnasser, N., Guillou, S., Leroi, F., Orange, N., Bakhrouf, A., & Federighi, M., (2007). Pulsed-light system as a novel food decontamination technology: A review. *Canadian Journal of Microbiology, 53*(7), 813–821.

25. Escalona, V. H., Aguayo, E., Martínez-Hernández, G. B., & Artés, F., (2010). UV-C doses to reduce pathogen and spoilage bacterial growth *in vitro* and in baby spinach. *Postharvest Biology and Technology, 56*(3), 223–231.

26. Faghihzadeh, F., Anaya, N. M., Hadjeres, H., Boving, T. B., & Oyanedel-Craver, V., (2019). Pulse UV light effect on microbial biomolecules and organic pollutants degradation in aqueous solutions. *Chemosphere, 216*, 677–683.

27. Fonseca, J. M., & Rushing, J., (2006). W. effect of ultraviolet-C light on quality and microbial population of fresh-cut watermelon. *Postharvest Biology and Technology, 40*(3), 256–261.

28. Food and Drug Administration, (2000). 21 CFR Part 179. Irradiation in the production, processing and handling of food. *Fed. Regist., 65*(7), 1056–71058.

29. Franck, C., Lammertyn, J., Ho, Q. T., Verboven, P., Verlinden, B., & Nicolaï, B. M., (2007). Browning disorders in pear fruit. *Postharvest Biology and Technology, 43*(1), 1–13.

30. Gómez-López, V. M., Koutchma, T., & Linden, K., (2012). Ultraviolet and pulsed light processing of fluid foods. In: Cullen, P., Tiwari, B. K., & Valdramidis, V. P., (eds.), *Novel Thermal and Non-Thermal Technologies for Fluid Foods* (pp. 185–223). New York, USA: Elsevier Inc.

31. Gonzalez-Aguilar, G., Wang, C. Y., & Buta, G. J., (2004). UV-C irradiation reduces breakdown and chilling injury of peaches during cold storage. *Journal of the Science of Food and Agriculture, 84*(5), 415–422.

32. Graça, A., Salazar, M., Quintas, C., & Nunes, C., (2013). Low dose UV-C illumination as an eco-innovative disinfection system on minimally processed apples. *Postharvest Biology and Technology, 85*, 1–7.

33. Guerrero-Beltran, J. A., & Barbosa-Canovas, G. V., (2004). Advantages and limitations on processing foods by UV light. *Food Science and Technology International, 10*(3), 137–147.

34. Gutiérrez, D. R., Char, C., Escalona, V. H., Chaves, A. R., & Rodríguez, S. D. C., (2015). Application of UV-C radiation in the conservation of minimally processed rocket (*Eruca sativa Mill.*). *Journal of Food Processing and Preservation, 39*(6), 3117–3127.

35. Hinds, L. M., Charoux, C. M., Akhter, M., O'Donnell, C. P., & Tiwari, B. K., (2019). Effectiveness of a novel UV light emitting diode based technology for the microbial inactivation of bacillus subtilis in model food systems. *Food Control*, Article ID: 106910.

36. Holck, A., Liland, K. H., Carlehög, M., & Heir, E., (2018). Reductions of *Listeria monocytogenes* on cold-smoked and raw salmon fillets by UV-C and Pulsed UV Light. *Innovative Food Science and Emerging Technologies, 50*, 1–10.

37. ISO 21348, (2007). *Space Environment (Natural and Artificial)-Process for Determining Solar Irradiances* (p. 20).

38. Jagadeesh, S. L., Charles, M. T., Gariepy, Y., Goyette, B., Raghavan, G. S. V., & Vigneault, C., (2011). Influence of postharvest UV-C hormesis on the bioactive components of tomato during post-treatment handling. *Food and Bioprocess Technology, 4*(8), 1463–1472.

39. Jean, J., Morales-Rayas, R., Anoman, M. N., & Lamhoujeb, S., (2011). Inactivation of *Hepatitis A virus* and *Norovirus* surrogate in suspension and on food-contact surfaces using pulsed UV light (pulsed light inactivation of food-borne viruses). *Food Microbiology, 28*(3), 568–572.

40. Jeong, Y. J., & Ha, J. W., (2019). Combined treatment of UV-A radiation and acetic acid to control foodborne pathogens on spinach and characterization of their synergistic bactericidal mechanisms. *Food Control, 106*, Article ID: 106698.

41. Jiang, T., Jahangir, M. M., Jiang, Z., Lu, X., & Ying, T., (2010). Influence of UV-C treatment on antioxidant capacity, antioxidant enzyme activity, and texture of postharvest shiitake (*Lentinus edodes*) Mushrooms during storage. *Postharvest Biology and Technology, 56*(3), 209–215.

42. Kader, A. A., (2013). Postharvest technology of horticultural crops: An overview from farm to fork. *Ethiop. J. Appl. Sci. Technol., 1*, 1–8.

43. Kaewsuksaeng, S., Urano, Y., Aiamla-or, S., Shigyo, M., & Yamauchi, N., (2011). Effect of UV-B irradiation on chlorophyll-degrading enzyme activities and postharvest quality in stored lime (*Citrus latifolia Tan.*). *Postharvest Biology and Technology, 61*(2/3), 124–130.

44. Kasim, R., & Kasim, M. U., (2008). The Effect of ultraviolet irradiation (UV-C) on chilling injury of cucumbers during cold storage. *Journal of Food Agriculture and Environment, 6*(1), 50–54.

45. Keklik, N. M., Krishnamurthy, K., & Demirci, A., (2012). Microbial decontamination of food by ultraviolet (UV) and pulsed UV light. In: Demirci, A., & Ngadi, M. O., (eds.), *Microbial Decontamination in the Food Industry* (pp. 344–369). Cambridge-UK: Woodhead Publishing.

46. Keshavarzfathy, M., & Taghipour, F., (2019). Radiation modeling of ultraviolet light-emitting diode (UV-LED) for water treatment. *Journal of Photochemistry and Photobiology A: Chemistry, 377*, 58–66.

47. Kim, Y. H., Jeong, S. G., Back, K. H., Park, K. H., Chung, M. S., & Kang, D. H., (2013). Effect of various conditions on inactivation of *Escherichia coli O157: H7, Salmonella typhimurium,* and *Listeria monocytogenes* in fresh-cut lettuce using ultraviolet radiation. *International Journal of Food Microbiology, 166*(3), 349–355.

48. Kim, M. J., Mikš-Krajnik, M., Kumar, A., & Yuk, H. G., (2016). Inactivation by 405 nm light emitting diode on *Escherichia coli O157: H7, Salmonella typhimurium,* and *Shigella sonnei* under refrigerated condition might be due to the loss of membrane integrity. *Food Control, 59,* 99–107.

49. Koutchma, T. N., Forney, L. J., & Moraru C. I., (2009). In: Sun, D., (ed.), *Ultraviolet Light in Food Technology: Principles and Applications* (p. 300). Contemporary Food Engineering Series; Boca Raton, USA: CRC Press, Taylor & Francis Group.

50. Koutchma, T., (2014). *Preservation and Shelf Life Extension: UV Applications for Fluid Foods* (1st edn., p. 66). New York, USA: Elsevier Inc.

51. Koutchma, T., (2014). *Food Plant Safety: UV Applications for Food and Nonfood Surfaces* (1st edn., p. 57). New York, USA: Elsevier Inc.

52. Kowalski, W., (2010). *Ultraviolet Germicidal Irradiation Handbook: UVGI for Air and Surface Disinfection* (p. 504). New York: Springer Science & Business Media.

53. Kusumler, A. S., (2011). *Ultraviyole (UV-C) Isini Uygulamasinin Patlican Ve Salataliklarda Soguk Zararlanması Uzerine Etkisi* [Effects of Ultraviolet (UV-C) Light Applications on the Chilling Injury in Eggplant and Cucumbers] (p. 210). Doctoral Dissertation, Istanbul Technical University Graduate School of Science Engineering and Technology, Istanbul, Turkey.

54. Lante, A., Tinello, F., & Nicoletto, M., (2016). UV-A light treatment for controlling enzymatic browning of fresh-cut fruits. *Innovative Food Science and Emerging Technologies, 34,* 141–147.

55. Li, D., Luo, Z., Mou, W., Wang, Y., Ying, T., & Mao, L., (2014). ABA and UV-C effects on quality, antioxidant capacity and anthocyanin contents of strawberry fruit (*Fragaria ananassa Duch.*). *Postharvest Biology and Technology, 90,* 56–62.

56. Li, J., Zhang, Q., Cui, Y., Yan, J., Cao, J., Zhao, Y., & Jiang, W., (2010). Use of UV-C treatment to inhibit the microbial growth and maintain the quality of yali pear. *Journal Food Science, 75*(7), M503–M507.

57. Liang, S., Min, J. H., Davis, M. K., Green, J. F., & Remer, D. S., (2003). Use of pulsed UV processes to destroy NDMA. *Journal of American Water Works Association, 95*(9), 121–131.

58. Liao, C., Liu, X., Gao, A., Zhao, A., Hu, J., & Li, B., (2016). Maintaining postharvest qualities of three leaf vegetables to enhance their shelf lives by multiple ultraviolet-C treatment. *LWT-Food Science and Technology, 73,* 1–5.

59. Liu, C., Jahangir, M. M., & Ying, T., (2012). Alleviation of chilling injury in postharvest tomato fruit by preconditioning with ultraviolet irradiation. *Journal of the Science of Food and Agriculture, 92*(15), 3016–3022.

60. Liu, C., Huang, Y., & Chen, H., (2015). Inactivation of *Escherichia coli O157: H7* and *Salmonella enterica* on blueberries in water using ultraviolet light. *Journal of Food Science, 80*(7), M1532–M1537.

61. López-Malo, A., & Palou, E., (2005). Ultraviolet light and food preservation. In: Barbosa-Canovas, G. V., Tapia, M. S., & Cano, M. P., (eds.), *Novel Food Processing Technologies* (pp. 405–421). Boca Raton, USA: CRC Press, Taylor & Francis Group.

62. MacGregor, S. J., Rowan, N. J., McIlvaney, L., Anderson, J. G., Fouracre, R. A., & Farish, O., (1998). Light inactivation of food related pathogenic bacteria using a pulsed power source. *Letters in Applied Microbiology, 27*(2), 67–70.

63. Maharaj, R., Arul, J., & Nadeau, P., (1999). Effect of photochemical treatment in the preservation of fresh tomato (*Lycopersicon esculentum* cv. Capello) by delaying senescence. *Postharvest Biology and Technology, 15*(1), 13–23.

64. Manzocco, L., Pievea, S. D., Bertolini, A., Bartolomeoli, I., Maifreni, M., Vianello, A., & Nicoli, M. C., (2011). Surface decontamination of fresh-cut apple by UV-C light exposure: Effects on structure, color, and sensory properties. *Postharvest Biology and Technology, 61*, 165–171.

65. Manzocco, L., & Nicoli, M. C., (2015). Surface processing existing and potential applications of ultraviolet light. *Food Science and Nutrition, 55*(4), 469–484.

66. Masschelein, W. J., & Rip, G. R., (2016). *Ultraviolet Light in Water and Wastewater Sanitation* (p. 192). Boca Raton, USA: CRC Press.

67. Michailidis, M., Karagiannis, E., Polychroniadou, C., Tanou, G., Karamanoli, K., & Molassiotis, A., (2019). Metabolic features underlying the response of sweet cherry fruit to postharvest UV-C irradiation. *Plant Physiology and Biochemistry, 144*, 49–57.

68. Muramoto, Y., Kimura, M., & Nouda, S., (2014). Development and future of ultraviolet light-emitting diodes: UV-LED will replace the UV lamp. *Semiconductor Science and Technology, 29*(8), Article ID: 084004.

69. Pan, Y., & He, Z., (2012). Effect of UV-C radiation on the quality of fresh-cut pineapples. *Procedia Engineering, 37*, 113–119.

70. Pérez-Ambrocio, A., & Guerrero-Beltrán, J. A., (2018). Effect of blue and ultraviolet-C light irradiation on bioactive compounds and antioxidant capacity of Habanero pepper (*Capsicum chinense*) during refrigeration storage. *Postharvest Biology and Technology, 135*, 19–26.

71. Pilavtepe-Çelik, M., Buzrul, S., Alpas, H., & Bozoğlu, F., (2009). Development of a new mathematical model for inactivation of *Escherichia coli O157: H7* and *Staphylococcus aureus* by high hydrostatic pressure in carrot juice and peptone water. *Journal of Food Engineering, 90*(3), 388–394.

72. Pombo, M. A., Dotto, M. C., Martínez, G. A., & Civello, P. M., (2009). UV-C irradiation delays strawberry fruit softening and modifies the expression of genes involved in cell wall degradation. *Postharvest Biology and Technology, 51*(2), 141–148.

73. Pongprasert, N., Sekozawa, Y., Sugaya, S., & Gemma, H., (2011). A novel postharvest UV-C treatment to reduce chilling injury (membrane damage, browning, and chlorophyll degradation) in banana peel. *Scientia Horticulturae, 130*(1), 73–77.

74. Promyou, S., & Supapvanich, S., (2012). Chilling injury alleviation in sweet pepper caused by UV-C treatment. *Asia Pacific Symposium II on Postharvest Research Education and Extension: APS, 1011*, 357–362.

75. Ribeiro, C., & Alvarenga, B., (2012). Prospects of UV radiation for application in postharvest technology. *Emirates Journal of Food and Agriculture, 24*(6), 586–597.

76. Rojas-Graü, M. A., Garner, E., & Martín-Belloso, O., (2010). The fresh-cut fruit and vegetables industry current situation and market trends. In: Martin-Belloso, O., & Soliva-Fortuny, R., (eds.), *Advances in Fresh-Cut Fruits and Vegetables Processing* (pp. 1–11). USA: CRC Press; Taylor & Francis Group.

77. Ruan, J., Li, M., Jin, H., Sun, L., Zhu, Y., Xu, M., & Dong, J., (2015). UV-B irradiation alleviates the deterioration of cold-stored mangoes by enhancing endogenous nitric oxide levels. *Food Chemistry, 169*, 417–423.

78. Saltveit, M., (1999). Effect of ethylene on quality of fresh fruits and vegetables. *Postharvest Biology and Technology, 15*, 279–292.

79. Sari, L. K., Setha, S., & Naradisorn, M., (2016). Effect of UV-C irradiation on postharvest quality of 'phulae' pineapple. *Scientia Horticulturae, 213*, 314–320.

80. Sastry, S. K., Datta, A. K., & Worobo, R. W., (2000). Ultraviolet light. *Journal of Food Science, 65*, 90–92.

81. Savran, H. E., & Koyuncu, M. A., (2016). The effects of superficial scald control methods having different effect mechanisms on the scald formation and α-farnesene content in apple Cv. 'granny smith'. *Scientia Horticulturae, 211*, 174–178.

82. Schenk, M., Guerrero, S., & Alzamora, S. M., (2008). Response of some microorganisms to ultraviolet treatment on fresh-cut pear. *Food Bioprocess Technology, 1*, 384–392.

83. Schaefer, R., Grapperhaus, M., Schaefer, I., & Linden, K., (2007). Pulsed UV lamp performance and comparison with UV mercury lamps. *Journal of Environmental Engineering and Science, 6*(3), 303–310.

84. Schalk, S., Adam, V., Arnold, E., Brieden, K., Voronov, A., & Witzke, H. D., (2005). UV-lamps for disinfection and advanced oxidation-lamp types, technologies and applications. *IUVA News, 8*(1), 32–37.

85. Selvarajah, S., Bauchot, A. D., & John, P., (2001). Internal browning in cold-stored pineapples is suppressed by a postharvest application of 1-methylcyclopropene. *Postharvest Biology and Technology, 23*(2), 167–170.

86. Sethi, S., Alka, J., & Bindvi, A., (2018). UV treatment of fresh fruits and vegetables. In: Siddiqui, M. W., (ed.), *Postharvest Disinfection of Fruits and Vegetables* (pp. 137–157). San Diego, USA: Academic Press.

87. Severo, J., De Oliveira, I. R., Tiecher, A., Chaves, F. C., & Rombaldi, C. V., (2015). Postharvest UV-C treatment increases bioactive, ester volatile compounds and a putative allergenic protein in strawberry. *LWT-Food Science and Technology, 64*(2), 685–692.

88. Sharma, R. R., & Demirci, A., (2003). Inactivation of *Escherichia coli* O157: H7 on inoculated alfalfa seeds with pulsed ultraviolet light and response surface modeling. *Journal of Food Science, 68*(4), 1448–1453.

89. Sheng, K., Zheng, H., Shui, S., Yan, L., Liu, C., & Zheng, L., (2018). Comparison of postharvest UV-B and UV-C treatments on table grape: Changes in phenolic compounds and their transcription of biosynthetic genes during storage. *Postharvest Biology and Technology, 138*, 74–81.

90. Song, K., Madjid, M., & Fariborz, T., (2016). Application of ultraviolet light-emitting diodes (UV-LEDs) for water disinfection: A review. *Water Research, 94*, 341–349.

91. Song, K., Taghipour, F., & Mohseni, M., (2018). Microorganisms inactivation by continuous and pulsed irradiation of ultraviolet light-emitting diodes (UV-LEDs). *Chemical Engineering Journal, 343*, 362–370.

92. Stevens, C., Khan, V. A., Lu, J. Y., Wilson, C. L., Pusey, P. L., Kabwe, M. K., Igwegbe, E. C. K., et al., (1998). The germicidal and hormetic effects of UV-C light on reducing brown rot disease and yeast microflora of peaches. *Crop Protection, 17*, 75–84.

93. Syamaladevi, R. M., Lu, X., Sablani, S. S., Insan, S. K., Adhikari, A., Killinger, K., Rasco, B., et al., (2013). Inactivation of *Escherichia coli* population on fruit surfaces

using ultraviolet-C light: Influence of fruit surface characteristics. *Food and Bioprocess Technology, 6*(11), 2959–2973.

94. Syamaladevi, R. M., Lupien, S. L., Bhuniaa, K., Sablani, S. S., Dugan, F., Rasco, B., Killinger, K., et al., (2014). UV-C light inactivation kinetics of *Penicillium expansum* on pear surfaces: Influence on physicochemical and sensory quality during storage. *Postharvest Biology and Technology, 87*, 27–32.

95. Tiecher, A., De Paula, L. A., Chaves, F. C., & Rombaldi, C. V., (2013). UV-C effect on ethylene, polyamines, and the regulation of tomato fruit ripening. *Postharvest Biology and Technology, 86*, 230–239.

96. Urban, L., Charles, F., De Miranda, M. R., & Aarrouf, J., (2016). Understanding the physiological effects of UV-C light and exploiting its agronomic potential before and after harvest. *Plant Physiology and Biochemistry Ppb, 105*, 1–11.

97. Urban, L., Sari, D. C., Orsal, B., Lopes, M., Miranda, R., & Aarrouf, J., (2018). UV-C light and pulsed light as alternatives to chemical and biological elicitors for stimulating plant natural defenses against fungal diseases. *Scientia Horticulturae, 235*, 452–459.

98. Vicente, A. R., Pineda, C., Lemoine, L., Civello, P. M., Martinez, G. A., & Chaves, A. R., (2005). UV-C treatments reduce decay, retain quality, and alleviate chilling injury in pepper. *Postharvest Biology and Technology, 35*(1), 69–78.

99. Wang, K. L. C., Li, H., & Ecker, J. R., (2002). Ethylene biosynthesis and signaling networks. *The Plant Cell, 14*(1), S131–S151.

100. Weerahewa, H. L. D., & Adikaram, N. K. B., (2005). Some biochemical factors underlying the differential susceptibility of two pineapple cultivars to internal browning disorder. *Ceylon Journal of Science (Biological Sciences), 34*, 7–20.

101. Xu, F., Wang, S., Xu, J., Liu, S., & Li, G., (2016). Effects of combined aqueous chlorine dioxide and UV-C on shelf-life quality of blueberries. *Postharvest Biology and Technology, 117*, 125–131.

102. Yang, Z., Cao, S., Su, X., & Jiang, Y., (2014). Respiratory activity and mitochondrial membrane associated with fruit senescence in postharvest peaches in response to UV-C treatment. *Food Chemistry, 161*, 16–21.

103. Yaun, B. R., Sumner, S. S., Eifert, J. D., & Marcy, J. E., (2004). Inhibition of pathogens on fresh produce by ultraviolet energy. *International Journal of Food Microbiology, 90*(1), 1–8.

104. Zhang, W., & Jiang, W., (2019). UV treatment improved the quality of postharvest fruits and vegetables by inducing resistance. *Trends in Food Science and Technology, 92*, 71–80.

105. Zhou, D., Li, R., Zhang, H., Chen, S., & Tu, K., (2020). Hot air and UV-C treatments promote anthocyanin accumulation in peach fruit through their regulations of sugars and organic acids. *Food Chemistry, 309*, 125726.

106. Zhou, D., Sun, Y., Li, M., Zhu, T., & Tu, K., (2019). Postharvest hot air and UV-C treatments enhance aroma-related volatiles by simulating the lipoxygenase pathway in peaches during cold storage. *Food Chemistry, 292*, 294–303.

107. Zhou, X., Li, Z., Lan, J., Yan, Y., & Zhu, N., (2017). Kinetics of inactivation and photoreactivation of *Escherichia coli* using ultrasound-enhanced UV-C light-emitting diodes disinfection. *Ultrasonics Sonochemistry, 35*, 471–477.

CHAPTER 3

MICROWAVE-ASSISTED EXTRACTION (MAE) TECHNOLOGY: POTENTIAL FOR EXTRACTION OF FOOD COMPONENTS

FAIZAN AHMAD, SADAF ZAIDI, and Z. R. A. A. AZAD

ABSTRACT

Various conventional methods (CMs) (such as SLE, LLE, SE) and non-conventional methods (NCMs) (such as MAE, UAE, EAE, SFE, PEF, and PLE) have been used for the separation of a compound from the plant products. However, the microwave-assisted extraction (MAE) technique is easy to implement and rapid, is less time consuming, and also provides excellent extraction efficiencies similar to or higher than those obtained with other methods. In MAE, microwave energy is applied to heat solvents in contact with samples (mainly solid samples), thus achieving the partition of the target compounds of interest from the sample into the solvent. Extractions are conducted in closed or open vessels, where the sample and solvent are combined and then exposed to microwave energy.

3.1 INTRODUCTION

Plant parts and their by-products, such as fruits, vegetables, flowers, stems, and leaves, are rich sources of naturally occurring bioactive components, minerals, and vitamins. Separation of these components can be done by various conventional methods (CMs) and non-conventional methods (NCMs) (Figure 3.1). In the past several years, various studies conducted on the separation of compounds from different plant parts and their by-products have shown that NCMs such as microwave heating, ultrasonic heating, pulse electric field, and supercritical fluid extraction (SFE) are eco-friendlier as compared to the CMs [4, 64]. NCMs are more efficient because they reduce the consumption of synthetic and organic chemicals; and reduce the extraction time and enhance the quality and yield of the extract [12, 64].

A large number of book chapters, monograms, articles, and scientific reports have already been published, where the application, importance, and various advantages of NCMs were reviewed. These reports highlight the use of extraction methods in pharmaceutical, food additives, and other sectors. In literature, very few reports are available on the extraction of bioactive compounds using microwave heating. Therefore, this chapter provides a brief review of the separation of bioactive components from plant products using MAE. Further, this chapter also highlights the primary mechanism of conventional and MAE methods, equipments used for MAE, factors affecting the extraction efficiency, food, and major food components, the importance of a balanced diet (BD), and the classification of bioactive compounds. The advantages of the microwave-assisted extraction (MAE) method compared to the traditional Soxhlet method have also been discussed at the end of the chapter.

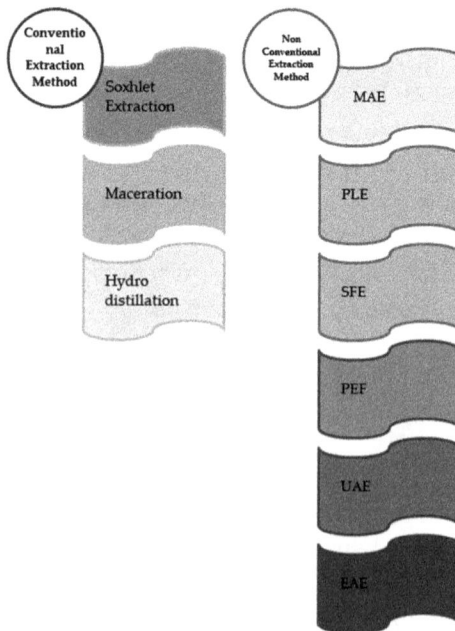

FIGURE 3.1 Conventional and non-conventional methods.

This chapter focuses on fundamentals, principles, and equipments used for MAE, and the factors affecting the extraction efficiency. Further, the chapter also includes discussions on the comparison of MAE with other food extraction techniques, and the separation of bioactive compounds from fruits and vegetables and their by-products using MAE.

3.2 FOOD AND FOOD COMPONENTS

The food that makes up our daily diet plays a vital role in maintaining our health. People these days have become conscious of what they consume as part of their daily diet and are keen to maintain a healthy lifestyle. Because of the common health issues and medical problems such as diabetes, high blood pressure, gall stones, etc., people now want to opt for a rich, healthy, and BD, which helps them to cure their respective illness and provides them with all the required balanced nutrients in ample amounts [69].

3.3 MAJOR FOOD COMPONENTS

Nutrients are essential components that our body requires in order to grow and survive. The food that we consume can mainly be classified into five major types of components (Figure 3.2), namely, vitamins, minerals, carbohydrates, protein, and fats. Other than these five basic components, food also has water, dietary fiber, and bioactive compounds in small amounts, which also hold significant importance for the body [18, 70].

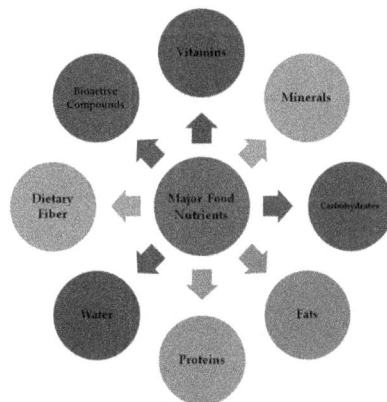

FIGURE 3.2   Major food components.

3.3.1 CARBOHYDRATES

The leading role of carbohydrates is to provide energy to the body. They can be classified as complex and simple carbohydrates:

1. **Complex Carbohydrates:** These are known as polysaccharides as they contain hundreds or thousands of monosaccharide units. They are found in wheat grain, white bread, kernels, and cakes. They are comparatively less sweet than the simple carbohydrates and are equally capable of raising the blood sugar levels, but they take time to break down.
2. **Simple Carbohydrates:** These are also called simple sugars, as they contain single monosaccharide units. They are found in natural sources of food, such as milk, fruits, and vegetables. These carbohydrates add some level of sweetness to the food and are capable of raising the blood sugar/glucose quickly, but at the same time, they are easier to break down [3].

3.3.2 PROTEINS

The primary function of proteins in the body is to help it grow and repair itself. Food items, such as milk, pulses, eggs, and meat, etc., are rich sources of proteins. Such food items that contain abundant amounts of protein are called body-building foods [32].

3.3.3 FATS

Fats provide more energy as compared to carbohydrates. Our body uses fat as a source of fuel. They play an important role because they also help in the absorption of essential vitamins like A, D, E, and K in the body. Butter, cheese, oil, etc., are some examples of foods that are rich in fat [68].

3.3.4 VITAMINS

The vitamins protect our bodies from various ailments. Apart from this, they also help in keeping our eyes, gums, bones, teeth, etc., in good shape. For example, vitamin A helps form and maintains our bones, skin, tissues, and teeth and also cures color blindness or poor visibility. Green leafy vegetables, sweet potatoes, oranges, etc., are a rich source of vitamin A. Similarly, vitamin B1 enables our body cells to turn carbohydrates into energy. It helps to cure loss of appetite or loss of weight. It is found in dried herbs, sunflower seeds, brown rice, and whole-grain cereals [17].

The foods rich in vitamin B2 are almonds, bananas, green beans, etc. The sources of vitamin B2 are foods such as mutton, fish, beef, eggs, etc. Vitamin B2 maintains our body growth and RBCs and also helps to cure skin problems. Vitamin B2 helps in the maintenance of our central nervous system and also supports RBCs in our bodies. It can be consumed to cure pale skin, replenish RBCs, and recover loss of stamina and appetite.

Vitamin C regulates healthy gums and teeth and cures gum diseases. It can be consumed through fresh herbs, cauliflower, fruits such as papaya, oranges, strawberries, and guava [55].

Vitamin D is essential for our healthy bones. Its daily consumption can help to prevent rickets and osteomalacia [42]. Its best sources are sunrays, mushrooms, fish, and eggs. It is found in soybean oil, red chili powder, apricots, green olives, and cooked spinach. It can be consumed through green leafy vegetables and soybean oil. Similarly, vitamin E1 is essential in the processing of vitamin K and in the formation of RBCs. It cures weak muscles and transmission problems in nerve impulses. Vitamin K is essential for blood coagulation and cures excessive bleeding from wounds [24].

3.3.5 MINERALS

The minerals help our body in performing vital functions, such as building strong bones, maintaining heart rate, balancing the hormones, etc.

Calcium is essential in the functioning of the nervous system and maintaining healthy bones. Its deficiency can result in weak bones and lower bone density. The primary sources of calcium are dairy products, salmon, cabbage, and broccoli [60].

Phosphorous helps in the maintenance of acid-base balance in the body. Its deficiency leads to loss of appetite, muscle weakness, and poor physique. It can be consumed from lean meats, grain, and milk [15].

Iodine helps in the formation of thyroid hormones. Therefore, its deficiency can lead to goiter, i.e., enlargement of the thyroid gland and also mental disability. It can be cured by consuming green leafy vegetables, seafood, and iodized salt on a regular basis [14].

Sodium keeps checking on blood pressure. Its deficiency can result in nausea and irritability. Table salt and celery are important sources of sodium. Iron is essential for hemoglobin formation and, therefore, in maintaining RBCs. Its deficiency can lead to anemia, weakness, and shortness of breath. Whole grains, eggs, meat, and leafy vegetables are some of the important sources of iron [46].

3.3.6   DIETARY FIBERS

Other than vitamins and minerals, dietary fibers or roughage are also essential in regulating the health of our bodies health. However, they do not provide any such specific nutrition to our body, but still, they play a significant role in the absorption of food after every meal. In other words, fibers/roughage accelerates the bowel movements and thus cures constipation. Fruits, vegetables, pulses, seeds, and nuts are some of the good sources of dietary fibers. Their consumption on a regular basis can help our bodies get rid of undigested food and problems like chronic constipation, acidity, etc.

3.4   IMPORTANCE OF A BALANCED DIET (BD)

A balanced diet (BD) is a diet that incorporates a variety of food items, providing various types of nutrients in adequate amounts that help in maintaining a healthy lifestyle. It should also include a good amount of dietary fiber and water also, along with other nutrients, vitamins, and minerals. For example, the diet that includes protein-rich pulses, various flours, cereals for carbohydrates, fats along with fruits and vegetables, meat, egg, fish, milk, etc., is a BD [29]. Other than this, it should also be kept in mind that the right amount of food, which we are consuming, is properly cooked so that it does not end up losing its nutrients. An unbalanced diet can result in major problems such as obesity and malnutrition. Therefore, a balance in the diet is essential [53].

3.5   PLANT-BASED BIOACTIVE COMPOUNDS

Bioactive compounds are naturally occurring compounds that are present in small amounts in various plants, especially in fruits, vegetables, and their by-products [37]. The consumption of these compounds provides various health benefits and therapeutic potential beyond the basic nutritional value. The bioactive compounds, such as terpenes, and terpenoids, carotenoids, anthocyanins, flavonoids, phenolic compounds, and polyphenols, are well known for their antioxidant, antimicrobial, anticancer, and many more beneficial activities [1, 59].

Numerous scientific and medical studies indicate that the consumption of a large amount of food containing bioactive compounds with antioxidant activity, including minerals, vitamins, photochemical, flavonoids, and carotenoids impart an affirmative effect on human health and reduces the effect of various diseases like Alzheimer's, cancer, diabetes, and tumor [4, 41, 49]. A large number of bioactive compounds, especially essential oils, carotenoids, and antioxidants are widely added into food and food products to increase their color, aroma, flavor, and sensory properties [50]. The direct addition of bioactive compounds in food and food products is sometimes limited due to their low solubility in the aqueous phase and their higher instability in food products during processing and preparation. In order to increase the solubility and stability of these compounds, various studies have suggested encapsulation and microencapsulation of bioactive compounds [10, 21].

The separation of compounds from plant and plant products is generally considered as a primary step, and the incorporation of bioactive compounds into the food and food product either directly or indirectly is mainly considered as a secondary step [8, 63]. A large number of studies on the extraction and isolation of these compounds have been reported in the literature using conventional and non-conventional extraction methods. Some of them, especially the MAE method, have been discussed in the later sections of this chapter.

3.6 CLASSIFICATION OF BIOACTIVE COMPOUNDS

Plants metabolism is divided into two categories, namely primary and secondary. The substances that are essential for the maintenance of cells, such as lipids, proteins, carbohydrates, and nucleic acids come in the category of primary metabolism, and the bioactive compounds are the secondary metabolites of plants. Different authors have classified the bioactive compounds into different categories. For example, Katerova et al. [35] classify the bioactive compounds of plants and plant products into three main classes, such as phenolic compounds, alkaloids, and terpenes, and terpenoids. A phenolic compound is one of the largest categories of plant bioactive compounds. On the basis of the number of carbons present in the chain, they can be classified into sixteen major groups (Table 3.1), and on the basis of their distribution, they can be classified into three categories, such as shortly distributed, widely distributed, and polymers [43, 59].

TABLE 3.1   Types of Phenolic Compounds

Structure	Class of Phenolic Compounds	Number of Carbon in the Chain
	Acetophenone	Six-Two
	Anthraquinones	Six-Two-Six
	Benzoquinones	Six
	Chromones	Six-Three
	Coumarins	Six-Three
	Flavonoids	Six-Three-Six
	Hydroxycinnamic acids	Six-Three
	Lignans	(Six-Three)$_2$

TABLE 3.1   *(Continued)*

Structure	Class of Phenolic Compounds	Number of Carbon in the Chain
	Lignins	(Six-Three)$_n$
	Naphthoquinones	Six-Four
	Phenolic acids	Six-One
	Phenylacetic acids	Six-Two
	Phenylpropenes	Six-Three
	Simple phenols	Six
	Stilbenes	Six-Two-Six
	Xanthones	Six-One-Six

Shortly distributed phenolic compounds include simple phenols, aldehydes derived from benzoic acids, resorcinol, and hydroquinone. Widely distributed ones include flavonoids and their derivatives, and polymers include tannin and lignin [25]. Further, on the basis of solubility, phenolic compounds may be classified as soluble and insoluble phenolic compounds. Due to the high amount of antiseptic, antioxidant, and anticancer properties, these phenolic compounds provide potential health benefits, and they also reduce the risk of various cardiovascular diseases. For example, such as thymol, low molecular weight phenolic compounds are used in various medicines as antiseptics [20, 75].

3.7 BASIC PRINCIPLES AND MECHANISMS OF CONVENTIONAL AND MAE METHODS

In MAE, electromagnetic waves are the main driving force to separate compounds from plant products. By this technique, extraction generally occurs due to the rupture of cells caused by electromagnetic waves [78]. The fundamentals and the working principle of the MAE process are totally different from the other general extraction methods. In conventional separation methods, heat is generally transmitted from the outside to the inside of the substrate, and mass is transmitted from the inside to the outside of it. However, in the case of MAE, both the mass and heat are transmitted in the same direction, i.e., from inside to the outside, as shown in Figure 3.3. As a result of this, the extraction process is accelerated, and a high yield of

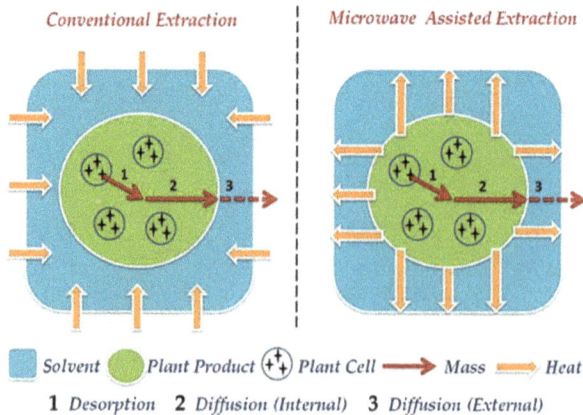

FIGURE 3.3   Mechanism of conventional and MAE method.

the extraction is obtained. In MAE, the heat is dispersed in all directions inside the solvent. In contrast to the CMs, especially in solid-liquid extraction (SLE), the heat is transmitted from the heating solvent to the inside of the sample [26, 77].

During the separation of compounds from the sample by both the MAE and the conventional method, a sample containing a compound/particle interacts with the suitable solvent, and the whole process of extraction is completed in the following six steps [52, 82]:

i.   The solvent penetrates into the sample and its matrix by diffusion.
ii. Sample components breakdown or solubilization takes place.
iii. Transportation of the solute (main compounds/particle) out from the sample matrix.
iv. Emigration of the main compound from the external surface of the sample to the solvent.
v.   Transportation of the extract with respect to the sample.
vi. Separation and isolation of extract from the solvent.

Most of the research on the MAE processing confirms that the overall process of extraction can be further classified into three major steps, such as equilibrium phase, the transition phase, and diffusion phase [19, 54], as shown in Figure 3.4:

- **Equilibrium Phase:** During an equilibrium stage, the selected solvent enters into the sample by diffusion, and the process of solubilization of solute (particles/compound) is started by the rupture of the cells. In this stage, the substrate is removed from the outer surface of the particle.
- **Transition Phase:** In the second stage of the transition phase, mass is transferred by convection, and the transfer of mass starts to appear in the solid-liquid interface.
- **Diffusion Phase:** In the last stage of diffusion, the solute containing the compounds/particles diffuses into the extracting solvent.

As compared to the other stages in the last stage of diffusion, the extraction rate is tremendously low, due to which this step is called the limiting and the irreversible step of the extraction process [77]. Figure 3.4 indicates the steps involved in the extraction of compounds/particles from various plant products using MAE. This figure also shows that the rate of separation of the compounds from the plant products is not directly dependent on time due to the various unsteady state conditions.

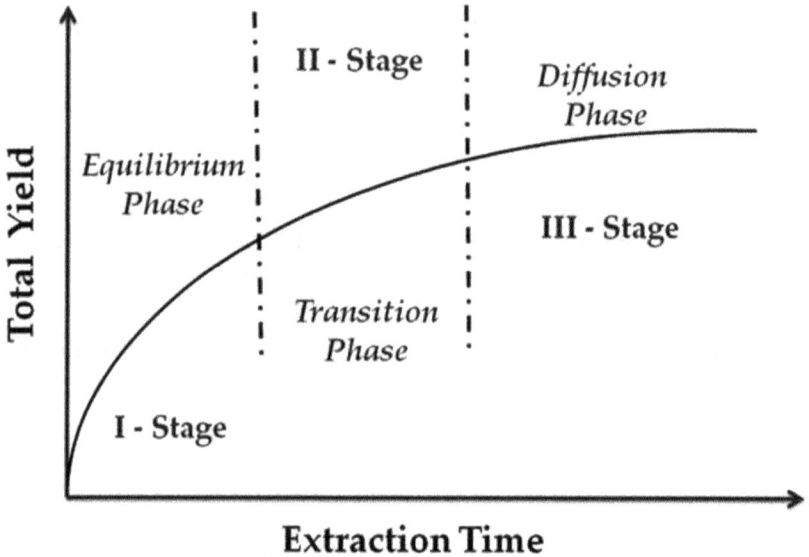

FIGURE 3.4   Major steps involved in the extraction of compounds.

3.8   FACTORS AFFECTING THE PERFORMANCE OF MAE METHOD

During the extraction, selection, and optimization of process parameters always play an important role in order to get a high amount of extraction yield. The efficacy and the performance of the MAE method widely depend on the various selected process parameters and on the other factors, such as the solvent used for the extraction, time and temperature of extraction, microwave power, the composition of sample, amount of water present in the sample and solvent to feed ratio [16, 74].

3.8.1   SOLVENT

The decision of solvent for extraction is one of the most important parameters that affect the extraction process. A good and proper choice of solvent provides a more efficient extraction process as well as a good amount of extraction yield. The decision of the solvent mainly depends on various parameters such as the solubility of the targeted bioactive compounds, which are needed to be extracted from the plant products, the interaction of selected solvent, and its penetration into the sample matrix of plant products [6, 52]. Further, it also depends on the rate of mass transfer during the separation process. In general, for the separation of compounds, the solvents should have high selectivity and solubility towards the important compounds present in the sample matrix, excluding the undesired compounds of the matrix [82]. In the case of MAE, the solvents should have a high absorbing capacity of microwave energy. Several studies conducted on MAE reported that both polar and non-polar solvents might be utilized for the extraction of compounds in MAE [4, 57].

However, the solvents that are transparent towards the microwaves cannot be used for the extraction of compounds in MAE. For example, hexane is a transparent solvent, whereas ethanol is an excellent absorbing solvent towards microwaves [11]. It has been reported that some of the solvents, such as water, ethanol, and methanol, are polar solvents that can be easily heated by microwave energy; and these solvents can be easily used in MAE.

The studies have also reported that the properties of the solvents, which are not fit for the extraction, can be modified by the addition of or combination with different solvents. For example, the addition of a small amount of water into the solvent increases the diffusion process of water into the cells of the sample matrix. This provides preferable heating and results in the enhancement of the transportation and extraction of components into the solvent at higher mass transfer rates [13]. Furthermore, the addition of salts also increases the heating rate of the solvents. To improve the extraction efficiency and heating rate of the poor microwave absorbing solvent, the solvents (such as water) can be utilized but in a smaller amount. It has been estimated that nearly 10% of water is only to be added in toluene, xylene, and hexane to improve their heating rate [11–13].

3.8.2 SOLVENT-TO-FEED RATIO

An optimal value of the solvent to feed ratio is essential to ensure effective and homogeneous heating. During the irradiation process, it must be ensured that the volume of the solvent must be adequate throughout the process to immerse the whole sample properly in the solvent [22, 78].

Both excess and deficient solvent to feed ratio create an obstruction in the extraction process. For example, excess of solvent results in poor microwave heating and additional consumption of power as the microwave radiation is absorbed by the excess solvent. However, a low quantity of solvent in the solid acts as an impediment in the mass transfer as the distribution of active compounds is concentrated in a certain region, which limits the outward movement of the compound from the cell matrix. Thus, an optimal amount of solvent to feed ratio should be used in order to achieve high efficiency in the MAE process as well as such type of solvents be used, which have high ability of providing sufficient amount of microwave heating coupled with the mobility of the separated components within the solvent itself.

3.8.3   EXTRACTION TIME

MAE method takes lesser time as compared to CMs to extract the bioactive components from the plant products. To avoid the oxidation and thermal degradation, MAE takes around half an hour to extract the compounds except in the solvent-free MAE method. It has been reported that the complete extraction of essential oil using the solvent-free MAE method takes place in one hour. In some cases of extraction, usually higher extraction time is required to achieve the complete conversion of extraction.

Therefore in such cases of MAE, the components are extracted in several steps using continuous multiple separation cycles. The multiple steps of extraction enhance the extraction yield as well as avoid the long heating of samples. During the multiple steps of extraction, an unprocessed solvent is added to the remnant, and the procedure is repeated several times until the complete depletion of the sample is achieved. The number of extraction cycles usually depends upon the size and type of the sample and on the types of solute. The results of several studies concluded that usually three cycles of 4–7 minutes each are required to extract the compounds completely from the sample matrix. For example, it has been reported that to extract the triterpenoid saponins from *Ganoderma atrum* using MAE, three cycles of five minutes each are required.

In some studies, it has been found that the increase of the extraction time increases the extraction yield up to a certain limit, but after that, the increased extraction time decreases the yield of extraction. For example, the extraction yield of astragalosides from *Radix astragali* increases with the increase of extraction time (1–5 minutes). After the achievement of maximum yield at 5 minutes, the yield decreases with the increase of extraction time [11].

3.8.4   *EXTRACTION TEMPERATURE AND MICROWAVE POWER*

Microwave power and extraction temperature affect the performance of the MAE process. Both of them are dependent on each other because the increase of microwave power increases the extraction temperature of the system. Due to this, the yield of extraction is increased [26]. Incident microwave power controls the extraction temperature as well as the quantity of energy supplied to the system. This power is converted to heat energy in the solvent. A high temperature of extraction leads to a decrease in the surface tension and viscosity of the solvent and increases the solvent power. Due to these reasons, the solubility of solutes into the solvent increases, and further, it improves the penetration power of the solvent [71].

It has been observed that firstly with the rise in temperature, the efficiency of the MAE process increases until an optimal temperature is reached, and thereafter it starts to decline with a further rise in temperature. Similarly, increasing the microwave power enhances the yield and reduces the time of extraction [7]. In the case of the MAE process, microwave power acts as a driving force to destroy the sample matrix so that the solute can easily diffuse out and dissolve in the solvent. However, sometimes a larger value of microwave power results in a low yield of extraction due to the deterioration of thermally sensitive compounds [11, 77]. Therefore, it is essential to decide the appropriate power to reduce the time requirement to achieve the required temperature as well as to avoid losses during the extraction [34].

3.9   SEPARATION OF COMPOUNDS FROM DIFFERENT PARTS OF PLANT

Product and by-products of fruits and vegetables possess potentially valuable bioactive compounds, namely, polyphenols, dietary fibers, flavonoids, and carotenoids. These bioactive compounds extracted from different sources using different extraction methods can be utilized in various industries, such as food, medicine, and cloth industries [23, 38, 39]. Apart from this, these bioactive compounds possess various types of health activities, such as: anticancer, antimicrobial, and antioxidant [4, 38]. Several reports have been published on the separation of bioactive components from the plant products in the past several years, and these are listed in Table 3.2.

TABLE 3.2   Some Compounds Separated from the Plant Product Using MAE

Fruits and Vegetables and Their by-Products	Bioactive Compounds	Extraction Method	References
Apple pomace	Pectin	MAE	[79]
Apple pomace	Polyphenols (phlorizin and quercetin)	MAE	[9]
Cranberry press cake	Isorhamnetin, Quercetin	MAE	[11]
Grape seeds	Phenolic compounds	MAE	[66]
Grape skins	Anthocyanins	MAE	[48]
Lavandula angustifolia Mill., (lavender flowers)	Linalool, Flavonoids	MSD	[30]
Orange peels	Pectin	MAE	[61]
Red raspberries	Anthocyanins	MAE	[80]
Red, yellow, white, and onion	Flavonol	MAE	[84]
Soybean germ	Lipid	MAE	[81]
Sweet potato	Polyphenols	MAE	[2, 36]
Sweetgrass leaves	5,8-Dihydroxycoumarin	MAE	[27, 62]
Tobacco leaves	Solanesol	MAE	[83]
Tomato paste	Lycopene	UMAE	[47]
Turmeric plant	Curcumin	MAE	[56]
Yellow onion	Quercetin	VMHG	[40]

3.10   EQUIPMENTS USED IN MAE

With increasing demand and numerous advantages of the MAE technique at the laboratory as well as at the industrial scale, it can be used either as MASE (microwave-assisted solvent extraction) or MSFE (microwave solvent-free extraction) [16]. In the case of the MASE method, the sample comes in contact with organic solvents and microwave energy. Initially, the polar solvents, such as water, acetic acid (AA), propanol, and butanol heat up until their boiling point and then diffuse into the sample and finally solubilize the main compounds [72]. The extraction of the compounds from the sample matrix mainly relies on the type, nature, and temperature of the solvent. To control the temperature and to apply/absorb the microwave energy to the sample properly, MASE has been performed in two types of systems: a closed vessel with controlled temperature and pressure [45], and an open vessel with atmospheric pressure [45]:

- **Closed Vessel System:** It is used in those cases where the solvent at high temperature, normally above its boiling point and at atmospheric pressure, is required to extract the compound. High temperature provides a better transfer of mass of the targeted compounds from the sample.
- **Open Vessel System:** In this, the temperature of the solvent is equal to its boiling point [72]. Through the open vessel system, a larger amount of sample can be extracted as compared to the closed vessel.

Furthermore, they also provide safety in sample handling during the separation. For the separation of components from the plant products, MASE is considered one of the most important alternatives to the solid-liquid separation method. It is used because of its several benefits, such as it reduces the extraction time and consumption of organic solvent and enhances the extraction yield. In the past several years, scientists derived a number of extraction techniques from MASE, namely, VMAE and UMAE [16]. Table 3.3 indicates the open vessel and closed vessel MAE used in different types of plant products.

TABLE 3.3 MAE Systems Performed on Different Types of Plant Products

Plant Product	Bioactive Compound	Solvent	References
Open Vessel System			
Dried apple pomace	Pectin	HCl solution	[51]
Green tea leaves	Polyphenols	Ethanol water mixture (50%)	[65]
Tobacco leaves	Solanesol	Ethanol: Hexane (3:1)	[33]
Closed Vessel System			
Dried bark of *Eucommia ulmodies*	Geniposidic and chlorogenic acid	Water methanol mixture	[51]
Dried ginger roots	Ginger	Ethanol water mixture	[72]
Olive seeds	Edible oil	Hexane	[5]
Roots of *Morinda Citrifolia*	Anthraquinones	Mixture of Ethanol and water (4:1)	[31]
Roots of *Panax Ginseng*	Saponin	60% Ethanol	[67]

3.11   LABORATORY AND INDUSTRIAL SCALE PLANTS: MAE

At the laboratory scale, MAE is one of the most widely used and in-demand extraction methods for the extraction of various major and minor food components. Most of the time, researchers are interested to analyze the components of food, waste of processed food as well as the compounds obtained from edible and non-edible plants [83]. For example, seeds, and skin of grapes from wine production, pomace from olive oil production, or olive leaves. Upadhyay et al. [76] have developed a simple MAE laboratory-scale model for extracts of green coffee beans to predict and control the industrial applications based on the evaluation of the radical scavenging activity of extracts.

From the past several years, the developments and demand of the MAE method have increased at the industrial level also. However, it has been estimated that the industrial use of MAE is much costly compared to laboratory use. However, the consumption of less solvent and the shorter extraction time reduces the cost of MAE equipment at the industrial level [74].

Compared to other extraction methods, the industrial MAE process usually takes a short extraction time with a small amount of solvent. The advantages and disadvantages of the industrial MAE method normally depend on the types of raw material (plant products, and by-products), the binding strength of components with the matrix, and their thermal stability [78]. In the past several years, many studies have been conducted in this area. For example, in 2014, Leone et al. [44] developed a microwave-assisted prototype of an industrial plant for olive oil extraction where they observed the continuous production of olive paste [73]. At an industrial scale, a company of dry wood produces a polyphenol extract during the drying of wood. Another company CRODAROM has used a continuous microwave solvent extraction (MSE) method in order to extract antioxidants from various plant products used in phytosanitary products [45].

3.12   COMPARISON OF MAE METHOD WITH SOXHLET METHOD

MAE has been considered as an alternative potential for the traditional Soxhlet method. Various studies on the comparison of MAE with Soxhlet have been conducted in the past few years. For example, Hao et al. [28] reported that 12 minutes MAE extraction gave a much higher yield compared to 12 h Soxhlet extraction (SE) of artemisinin. The results of several studies confirm that MAE took lower extraction time and a lower volume of solvent compared to the SE method. Some of the main advantages of MAE over Soxhlet are:

- A small amount of solvent (few mm) is required.
- Enhanced extraction yield.
- Mass transfer is high during the extraction because MAE provides agitation during the extraction.
- Reduces the extraction time.
- Suitable for the extraction of a small amount of useful compounds present in the sample.
- Suitable for thermolabile compounds.

3.13 SUMMARY

MAE is one of the most reliable, good, and efficient separation methods. As compared with other extraction techniques, microwave radiation in MAE destroys the extraction sample matrix leading to the quick extraction of the active compounds into the solvent after the cell is ruptured. Thus, MAE gives a higher extraction yield, provides better efficiency, and it requires a lesser extraction time. MAE has also got acceptability rapidly because most of the time, during analysis, it is the most efficient and result-oriented process in all aspects compared to the traditional extraction and analysis methods. Still, in this new era, continuous research and development are required to enhance the interpretation of extraction techniques to eliminate the practical constraints, make better and more comprehensive design, and maximization of the extraction methods for better commercial applications.

KEYWORDS

- **balanced diet**
- **bioactive compounds**
- **extraction methods**
- **food components**
- **fruits and vegetable by-products**
- **microwave-assisted extraction (MAE)**

REFERENCES

1. Ahmad, F., & Khan, S. T., (2019). Potential industrial use of compounds from by-products of fruits and vegetables. In: Malik, A., Erginkaya, Z., & Erten, H., (eds.), *Health and Safety Aspects of Food Processing Technologies* (pp. 273–307). USA: Springer.

2. Akyol, H., Riciputi, Y., Capanoglu, E., Caboni, M. F., & Verardo, V., (2016). Phenolic compounds in the potato and its by-products: An overview. *International Journal of Molecular Sciences, 17*(6), 835–840.

3. Aller, E. E. J. G., Abete, I., Astrup, A., Martinez, J. A., & Van, B. M. A., (2011). Starches, sugars, and obesity. *Nutrients, 3*(3), 341–369.

4. Altemimi, A., Lakhssassi, N., Baharlouei, A., Watson, D. G., & Lightfoot, D. A., (2017). Phytochemicals: Extraction, isolation, and identification of bioactive compounds from plant extracts. *Plants (Basel), 6*(4), 42–50.

5. Amarni, F., & Kadi, H., (2010). Kinetics study of microwave-assisted solvent extraction of oil from olive cake using hexane: Comparison with the conventional extraction. *Innovative Food Science and Emerging Technologies, 11*, 322–327.

6. Ameer, K., Shahbaz, H., & Kwon, J. H., (2017). Green extraction methods for polyphenols from plant matrices and their by-products: A review: Polyphenols extraction by green methods. *Comprehensive Reviews in Food Science and Food Safety, 16*, 295–315.

7. Ani, T. A., Calinescu, I., & Lavric, V., (2012). Microwave extraction of active principles from medicinal plants. *UPB. Sci. Bull. B, 74*(2), 129–142.

8. Azmir, J., Zaidul, I. S. M., & Rahman, M. M., (2013). Techniques for extraction of bioactive compounds from plant materials: A review. *Journal of Food Engineering, 117*(4), 426–436.

9. Bai, X. L., Yue, T. L., Yuan, Y. H., & Zhang, H. W., (2010). Optimization of microwave-assisted extraction of polyphenols from apple pomace using response surface methodology and HPLC analysis. *Journal of Separation Science, 33*(23/24), 3751–3758.

10. Bustamante, A., Hinojosa, A., Robert, P., & Escalona, V., (2017). Extraction and microencapsulation of bioactive compounds from pomegranate (*Punica granatum*) residues. *International Journal of Food Science and Technology, 52*(6), 1452–1462.

11. Veggie, P. C., Martinez, J., & Meireles, M. A. A., (2013). Fundamental of microwave extraction: Chapter 2. In: Chemat, F., & Cravotto, G., (eds.), *Microwave-Assisted Extraction for Bioactive Compounds* (pp. 15–52). New York, USA: Springer.

12. Chemat, F., Vian, M. A., & Cravotto, G., (2012). Green extraction of natural products: Concept and principles. *International Journal of Molecular Sciences, 13*(7), 8615–8627.

13. Chemat, F., Abert, V., M., Ravi, H. K., Khadhraoui, B., Hilali, S., Perino, S., & Tixier, A. S. F., (2019). Review of alternative solvents for green extraction of food and natural products: Panorama, principles, applications and prospects. *Molecules (Basel, Switzerland), 24*(16), 3007–3012.

14. Chung, H. R., (2014). Iodine and thyroid function. *Annals of Pediatric Endocrinology and Metabolism, 19*(1), 8–12.

15. Delgado, C., (2003). Rising consumption of meat and milk in developing countries has created a new food revolution. *The Journal of Nutrition, 133*, 3907–3910.

16. Destandau, E., Michel, T., & Elfakir, C., (2013). Microwave-assisted extraction. In: Rostagno, M. A., & Prado, J. M., (eds.), *Natural Product Extraction: Principles and Applications* (pp. 113–156). Cambridge, UK: RSC Publishing.

17. Devaki, S. J., & Reshma, L. R., (2017). *Vitamin C: Sources, Functions, Sensing and Analysis*. Online, doi: 10.5772/intechopen.70162.

18. Dhingra, D., Michael, M., Rajput, H., & Patil, R. T., (2012). Dietary fiber in foods: A review. *Journal of Food Science and Technology, 49*(3), 255–266.

19. Díaz, R. B., González, M. M. J., & Domínguez, G. H., (2017). Introduction. In: Dominguez G. H., & González, M. M. J., (eds.), *Water Extraction of Bioactive Compounds* (pp. 1–50). Elsevier, UK.

20. Działo, M., Mierziak, J., Korzun, U., Preisner, M., Szopa, J., & Kulma, A., (2016). The potential of plant phenolics in prevention and therapy of skin disorders. *International Journal of Molecular Sciences, 17*(2), 160–168.

21. El Abbassi, A., Sana, E. F., El Bouzidi, L., Lahrouni, M., & Khalid, N., (2016). Recent advances in microencapsulation of bioactive compounds: Chapter 8. In: Pathak, M., & Govil, J. N., (eds.), *Recent Progress in Medicinal Plants: Analytical and Processing Techniques* (pp. 129–146). New York, USA: Stadium Press LCC.

22. Eskilsson, C., & Björklund, E., (2001). Analytical-scale microwave-assisted extraction. *Journal of Chromatography A, 902*, 227–250.

23. Faizan, A., & Kumar, A., (2018). Optimization of the ultrasonic-assisted extraction process to obtain phenolic compounds from pomegranate (*Punica granatum*) peels using response surface methodology. *International Journal of Agricultural Sciences, 10*(23), 7581–7590.

24. Ferland, G., (2012). The discovery of vitamin K and its clinical applications. *Annals of Nutrition and Metabolism, 61*, 213–218.

25. Giada, M., (2013). Food phenolic compounds: Main classes, sources and their antioxidant power. In: José, A., & Moralez-González, (eds.), *Oxidative Stress and Chronic Degenerative Diseases: A Role for Antioxidants* (pp. 87–112). Intech Publisher.

26. Golmakani, M. T., & Moayyedi, M., (2015). Comparison of heat and mass transfer of different microwave-assisted extraction methods of essential oil from citrus lemon (Lisbon variety) peel. *Food Science and Nutrition, 3*(6), 506–518.

27. Grigonis, D., Venskutonis, R., Sivik, B., Sandahl, M., & Eskilsson, C. S., (2005). Comparison of different extraction techniques for isolation of antioxidants from sweetgrass (*Hierochlo odorata*). *The Journal of Supercritical Fluids, 33*, 223–233.

28. Hao, J. Y., Han, W., Huang, S. D., Xue, B. Y., & Deng, X., (2002). Microwave-assisted extraction of artemisinin from *Artemisia annua* L. *Separation and Purification Technology., 28*, 191–196.

29. Haque, Basit, Tanwani, R., & Zehra, S., (2015). Balanced diet & balanced life. *The Professional Medical Journal, 20*, 1304–1308.

30. Hassanpouraghdam, M. B., Hassani, A., Vojodi, L., & Hajisamadi, A., (2011). Essential oil constituents of *Lavandula ofcinalis* Chaix. From Northwest Iran. *Chemija, 22*, 167–171.

31. Hemwimon, S., Pavasant, P., & Shotipruk, A., (2007). Microwave-assisted extraction of antioxidative anthraquinones from roots of *Morinda citrifolia*. *Separation and Purification Technology, 54*, 44–50.

32. Henchion, M., Hayes, M., Mullen, A. M., Fenelon, M., & Tiwari, B., (2017). Future protein supply and demand: Strategies and factors influencing a sustainable equilibrium. *Foods (Basel, Switzerland), 6*(7), 53–60.

33. Hu, R. S., Wang, J., & Li, H., (2015). Simultaneous extraction of nicotine and solanesol from waste tobacco materials by the column chromatographic extraction method and their separation and purification. *Separation and Purification Technology, 146*, 1–7.

34. Hu, X., Tanaka, A., & Tanaka, R., (2013). Simple extraction methods that prevent the artifactual conversion of chlorophyll to chlorophyllide during pigment isolation from leaf samples. *Plant Methods, 9*(1), 19–25.

35. Katerova, Z., Todorova, D., Tasheva, K., & Sergiev, I., (2012). Influence of ultraviolet radiation on plant secondary metabolite production. *Genetics and Plant Physiology, 2*, 113–144.

36. Kim, M. Y., Lee, B. W., & Lee, H. U., (2019). Phenolic compounds and antioxidant activity in sweet potato after heat treatment. *Journal of the Science of Food and Agriculture, 99*(15), 6833–6840.

37. Kris-Etherton, P. M., Hecker, A., & Bonanome, A., (2002). Bioactive compounds in foods: Their role in the prevention of cardiovascular disease and cancer. *The American Journal of Medicine, 113*(9B), 71s–88s.

38. Kumar, A., Ahmad, F., & Zaidi, S., (2019). Importance of bioactive compounds present in plant products and their extraction: A review. *Agricultural Reviews, 40*(4), 249–260.

39. Kumar, K., Yadav, A. N., Kumar, V., Vyas, P., & Dhaliwal, H. S., (2017). Food waste: A potential bioresource for extraction of nutraceuticals and bioactive compounds. *BioResources and BioProcessing, 4*(1), 18–25.

40. Kwak, J. H., Seo, J. M., Kim, N. H., Arasu, M. V., Kim, S., Yoon, M. K., & Kim, S. J., (2017). Variation of quercetin glycoside derivatives in three onion (*Alliums cepa L.)* varieties. *Saudi Journal of Biological Sciences, 24*(6), 1387–1391.

41. Lako, J., Trenerry, V., Wahlqvist, M., Wattanapenpaiboon, N., Sotheeswaran, S., & Premier, R., (2007). Phytochemical flavonols, carotenoids and the antioxidant properties of a wide selection of Fijian fruit, vegetables and other readily available foods. *Food Chemistry, 101*, 1727–1741.

42. Lanham-New, S., Thompson, R., More, J., Brooke-Wavell, K., Hunking, P., & Medici, E., (2007). Importance of vitamin D, calcium, and exercise to bone health with specific reference to children and adolescents. *Nutrition Bulletin, 32*, 364–377.

43. Lattanzio, V., (2013). Phenolic compounds: Introduction: Chapter 50. In: Ramawat, K. G., & Merillon, J. M., (eds.), *Natural Products* (pp. 1543–1580). Springer-Verlag Berlin Heidelberg, New York, USA.

44. Leone, A., Tamborrino, A., Romaniello, R., Zagaria, R., & Sabella, E., (2014). Specification and implementation of a continuous microwave-assisted system for paste malaxation in an olive oil extraction plant. *Biosystems Engineering, 125*, 24–35.

45. Li, Y., Fabiano-Tixier, A. S., Vian, M., & Chemat, F., (2013). Microwave-assisted extraction of antioxidants and food colors. In: Chemat, F., & Cravotto, G., (eds.), *Microwave-Assisted Extraction of Bioactive Compounds: Theory and Practice* (pp. 103–125). New York, USA: Springer.

46. Liamis, G., Milionis, H., & Elisaf, M., (2011). Hyponatremia in patients with infectious diseases. *The Journal of Infection, 6*, 327–335.

47. Lianfu, Z., & Liu, Z., (2008). Optimization and comparison of ultrasound/microwave-assisted extraction (UMAE) and ultrasonic-assisted extraction (UAE) of lycopene from tomatoes. *Ultrasonics Sonochemistry, 15*, 731–737.

48. Liazid, A., Guerrero, R. F., Cantos, E., Palma, M., & Barroso, C. G., (2010). Microwave-assisted extraction of anthocyanins from grape skins. *Food Chemistry, 124*, 1238–1243.

49. Liu, R. H., (2013). Health-promoting components of fruits and vegetables in the diet. *Advances in Nutrition (Bethesda, Md), 4*, 384s–392s.

50. Mahfoudhi, N., Ksouri, R., & Hamdi, S., (2016). Nanoemulsions as potential delivery systems for bioactive compounds in food systems: Preparation, characterization, and applications in food industry. In: Grumezescu, A. M., (ed.), *Emulsions* (pp. 365–403). New York, USA: Academic Press.

51. Mandal, S. C., Mandal, V., & Das, A. K., (2015). Essentials of botanical extraction: Principles and applications. *Essentials of Botanical Extraction: Principles and Applications, 1*, 207–213.

52. Muhamad, I., Hassan, N., Mamat, S., Nawi, N., Rashid, W., & Tan, N., (2017). Extraction technologies and solvents of phytocompounds from plant materials: Physicochemical characterization and identification of ingredients and bioactive compounds from plant extract using various instrumentations. In: Grumezescu, A., & Holban, A. M., (eds.), *Ingredients Extraction by Physicochemical Methods in Food* (Vol. 4, pp. 523–560). New York, USA: Academic Press.

53. Müller, O., & Krawinkel, M., (2005). Malnutrition and health in developing countries. *CMAJ: Canadian Medical Association Journal (Journal de L'Association Medicale Canadienne), 173*, 279–286.

54. Mustapa, A., Martín, Á., Gallego, J. R., Mato, R., & Cocero, M., (2015). Microwave-assisted extraction of polyphenols from *Clinacanthus nutans Lindau* medicinal plant: Energy perspective and kinetics modeling. *Chemical Engineering and Processing: Process Intensification, 97, 66–74.*

55. Najeeb, S., Zafar, M. S., Khurshid, Z., Zohaib, S., & Almas, K., (2016). The role of nutrition in periodontal health: An update. *Nutrients, 8*(9), 530–537.

56. Nishiyama, T., Mae, T., & Kishida, H., (2005). Curcuminoids and sesquiterpenoids in turmeric (*Curcuma longa L.*) suppress an increase in blood glucose level in type 2 diabetic KK-Ay mice. *Journal of Agriculture and Food Chemistry, 53*(4), 959–963.

57. Özbek, H., Koçak, D., Fadıloğlu, S., Keskin, H., & Gögüs, F., (2018). Microwave-assisted extraction of non-polar compounds from pistachio hull and characterization of extracts. *Grasas y Aceites, 69*, 260–268.

58. Pagare, S., Bhatia, M., Tripathi, N., & Bansal, Y. K., (2015). Secondary metabolites of plants and their role: Overview. *Current Trends in Biotechnology and Pharmacy, 9*, 293–304.

59. Panche, A. N., Diwan, A. D., & Chandra, S. R., (2016). Flavonoids: An overview. *Journal of Nutritional Science, 5*, E47–E56.

60. Piste, P., Sayaji, D., & Avinash, M., (2012). Calcium and its role in human body. *International Journal of Research in Pharmaceutical and Biomedical Sciences (IJRPBS), 4*, 2229–3701.

61. Prakash, M. J., Sivakumar, V., Thirugnanasambandham, K., & Sridhar, R., (2013). Optimization of microwave-assisted extraction of pectin from orange peel. *Carbohydrate Polymers, 97*(2), 703–709.

62. Pukalskas, A., Van, B. T., Venskutonis, R., Linssen, J., Veldhuizen, & Groot, A., (2002). identification of radical scavengers in sweet grass (*Hierochloe odorata*). *Journal of Agricultural and Food Chemistry, 50*, 2914–2919.

63. Sasidharan, S., Chen, Y., Saravanan, D., Sundram, K. M., & Latha, L. Y., (2011). Extraction, isolation and characterization of bioactive compounds from plants' extracts. *African Journal of Traditional, Complementary, and Alternative Medicines, 8*(1), 1–10.

64. Selvamuthukumaran, M., & Shi, J., (2017). Recent advances in extraction of antioxidants from plant by-products processing industries. *Food Quality and Safety, 1*, 61–81.

65. Serdar, G., Aytaç, E., Bayrak, S., & Sökmen, M., (2016). New approaches for effective microwave-assisted extraction of caffeine and catechins from green tea. *International Journal of Secondary Metabolite, 3*, 3–13.

66. Shi, J., Yu, J., Pohorly, J., & Kakuda, Y., (2003). Polyphenolics in grape seeds - biochemistry and functionality. *Journal of Medicinal Food, 6*, 291–299.

67. Shin, B. K., Kwon, S. W., & Park, J. H., (2015). Chemical diversity of ginseng saponins from panax ginseng. *Journal of Ginseng Research, 39*(4), 287–298.

68. Singh, P., Kesharwani, R., & Keservani, R., (2017). Protein, carbohydrates, and fats. In: Bagachi, D., (ed.), *Sustained Energy for Enhanced Human Functions and Activity* (pp. 103–115). New York, USA: Academic Press.

69. Skerrett, P. J., & Willett, W. C., (2010). Essentials of healthy eating: A guide. *Journal of Midwife and Women's Health, 55*(6), 492–501.

70. Slavin, J. L., & Lloyd, B., (2012). Health benefits of fruits and vegetables. *Advances in Nutrition (Bethesda, Md), 3*(4), 506–516.

71. Soquetta, M. B., Terra, L. D. M., & Bastos, C. P., (2018). Green technologies for the extraction of bioactive compounds in fruits and vegetables. *CyTA-Journal of Food, 16*(1), 400–412.

72. Sparr, E. C., & Björklund, E., (2000). Analytical-scale microwave-assisted extraction. *Journal of Chromatography A, 902*(1), 227–250.

73. Tamborrino, A., Romaniello, R., Zagaria, R., & Leone, A., (2014). Microwave-assisted treatment for continuous olive paste conditioning: Impact on olive oil quality and yield. *Biosystems Engineering, 127*, 92–102.

74. Tatke, P., & Jaiswal, Y., (2011). An overview of microwave-assisted extraction and its applications in herbal drug research. *Research Journal of Medicinal Plant, 5*, 21–31.

75. Tungmunnithum, D., Thongboonyou, A., Pholboon, A., & Yangsabai, A., (2018). Flavonoids and other phenolic compounds from medicinal plants for pharmaceutical and medical aspects: An overview. *Medicines (Basel), 5*(3), 93–99.

76. Upadhyay, R., Ramalakshmi, K., & Rao, L. J. M., (2012). Microwave-assisted extraction of chlorogenic acids from green coffee beans. *Food Chemistry, 130*(1), 184–188.

77. Veggi, P., Martínez, J., Meireles, M., & Meireles, M. A., (2013). Microwave-assisted extraction for bioactive compounds theory and practice, fundamentals of microwave extraction. *Food Engineering Series, 4*(2), 15–53.

78. Vinatoru, M., Mason, T., & Calinescu, I., (2017). Ultrasonically assisted extraction (UAE) and microwave-assisted extraction (MAE) of functional compounds from plant materials. *TrAC Trends in Analytical Chemistry, 97*, 159–178.

79. Wang, S., Chen, F., Wu, J., Wang, Z., Liao, X., & Hu, X., (2007). Optimization of pectin extraction assisted by microwave from apple pomace using response surface methodology. *Journal of Food Engineering, 78*, 693–700.

80. Yang, Y., Yuan, X., Xu, Y., & Yu, Z., (2015). Purification of anthocyanins from extracts of red raspberry using microporous resin. *International Journal of Food Properties, 18*(5), 1046–1058.

81. Yu, D., Elfalleh, W., He, S., Ying, M., Jiang, L., Li, L., Hu, L., & Zhang, J., (2013). Physicochemical properties and minor lipid components of soybean germ. *Journal of the American Oil Chemists' Society, 90, 1551–1558.*

82. Zhang, Q. W., Lin, L. G., & Ye, W. C., (2018). Techniques for extraction and isolation of natural products: A comprehensive review. *Chinese Medicine, 13*, 20–26.
83. Zhou, H. Y., & Liu, C. Z., (2006). Microwave-assisted extraction of solanesol from tobacco leaves. *Journal of Chromatography A, 1129*, 135–139.
84. Zill, H., Vian, M., Fabiano-Tixier, A. S., Elmaataoui, M., Dangles, O., & Chemat, F., (2011). A remarkable influence of microwave extraction: Enhancement of antioxidant activity of extracted onion varieties. *Food Chemistry, 127*(4), 1472–1480.

CHAPTER 4

HIGH-PRESSURE ASSISTED FREEZING OF FOODS

PAVAN MANJUNATH GUNDU, PREETI BIRWAL,
CHAITRADEEPA G. MESTRI, and GUNJAL MAHENDRA VISHRAM

ABSTRACT

Conventional freezing leads to the formation of larger sized ice crystals as a result of the slow rate of freezing, which often causes degradation of the textural quality in the food matrix. Meanwhile, fast and rapid freezing enhances the formation of smaller and numerous ice crystals, which helps to retain the original food quality. Development of rapid assisted freezing technologies has caught the heeds of several food industrialists, and researchers, considerably due to its attributes of ice crystal formed with an improved kinetics of freezing in conjugation with the high-pressure. With a basis on qualitative, quantitative, and economical aspects of frozen food and freezing process, this chapter focuses on: (1) the basic conceptualization, principles, modern, and advanced aspects on high-pressure for assisted freezing (HPAF) process with an emphasis on ice-nucleation, crystallization, and phase transition of water with temperature and pressure as primary gradient; (2) equipments and modeling of HPAF process and thermo-physical characteristics of food components and current industrial standards; (3) the impact of HPAF on food constituents and microstructure; (4) application of this technology for enzyme and microbial inactivation, and preservation in various food industries;(5) packaging requirements, storage, and regulatory standards of HPAF food; (6) drip loss and other quality parameters HPAF food with a comparative notes on conventional and other assisted freezing techniques; and (6) future scope of HPAF with its strengths, limitations, and challenges.

4.1 INTRODUCTION

Freezing is the oldest method for food preservation, which allows retaining original and characteristic texture, taste, and nutritional value of foods better than any other existing methods of food preservation. On freezing, as residual moisture in food is converted to solid ice, suppressing the metabolism and growth/development of microbes that in turn delays the process of decomposition in food matrix. The key element of freezing process includes combined advantageous effects of both induced pressure and reduced temperature at which chemical reactions have been suppressed, growth of microorganisms is inhibited, and cellular metabolic activities are delayed [21]. Mechanical and cryogenic systems are existing methods of freezing in food chain industries.

The freezing kinetics also plays a vital role in the preservation of quality and original texture of food, as rapid freezing maintains cellular structure by forming smaller ice crystals [63]. Storage at lower temperatures has always been utilized to achieve long-term preservation of food. In relay with the development of novel technologies, industrial quick freezing has been considered as one of the efficient and satisfactory methods to preserve food quality at long-term storage [5]. Frozen foods do not require any added preservatives, as microorganisms do not grow at reduced temperature of below −9.5°C (15°F) [7]. In addition, the process of freezing requires the shortest time and less overall input of energy use and cost compared to other methods [13].

4.1.1 FROZEN FOOD MARKET

The frozen food industry has changed significantly in recent years with respect to versatile consumer demand and profile. Shelly [94] reported that the market for frozen foods in 2018 was estimated at$219.9 billion and is further aimed to achieve a net worth of$282.5 billion/year by the end of 2023. Developing countries (such as: India and China) are offering several opportunities for frozen food market with the developments and increased investments in retail landscape and cold chill market [29]. In 2018, Indian Frozen Foods Market has reached a worth of Rs. 74 billion and is further projected to reach Rs. 188 billion/year in 2024. Frozen vegetable snacks, meat products, and other fruits and vegetables set the highest market sales, capturing at 39%, 36%, and 25%, respectively, in the Indian frozen food market [1].

4.1.2   CRYSTALLIZATION AND FREEZING

Crystallization of liquid component during the process of freezing is one of the key phenomena that have to be tackled to maintain and to enhance the qualitative parameters of frozen foods. Crystallization is a phenomenon, during which the development of ice crystals will take place, and the process is driven by the rate of freezing [26]. When temperature decreases, the **density** of any liquid increases at atmospheric pressure, and its solid-state is denser than liquid state for most of the liquids. However, water behaves very differently as it reaches its highest density at 4°C. The density of water gradually decreases from 1.0 to 0.9168 g/cm^3 at 4°C and 0°C, respectively under atmospheric pressure, due to alignment of hydrogen bonds between the water molecules to give a characterized network of hexagonal structure with open spaces in the middle of formed hexagons.

The low-density and large-sized ice crystals so formed will be one of the major parameters that lead to this degradation of quality due to extensive physical and microstructural damage, protein denaturation, and degradation of other cellular constituents [95]. The location and size of these formed ice crystals in the food system largely depend on the rate of freezing and endpoint temperature. Usually, in conventional slow freezing, only a few ice nuclei are formed due to lower nucleation rate, thus inducing larger sized ice crystals. Meanwhile, with assisted freezing due to increased rate of freezing, many ice nuclei will be formed, thus inducing minute ice crystals and thus reducing qualitative damage in the food system [40, 68].

4.1.3   CONVENTIONAL AND MODERN ASSISTED FOOD FREEZING TECHNIQUES

In the conventional method of freezing, heat transfer within the entire mass of the food system will beat limited scale, and eventually resulting in a reduced rate of freezing. These critical conditions are directly responsible for the damage of food texture and quality due to the development of large-sized extracellular ice-crystals. The formation of large-sized needle-shaped ice crystals will affect food quality and its texture during the process of slow freezing. Instead, smaller and minute ice crystals can be formed when the food system is frozen at increased rate of freezing [59]. Thus, the problem of mechanical damage to cell walls, textural, and significant nutrient loss in frozen food product can be prevented with the formation of smaller, minute, and uniform ice crystals, which will get evenly distributed all over the tissue at increased freezing rate [100].

Novel and unique principle of rapid freezing and its techniques have been found successful for preservation of various food products. Rapid freezing process and its conditions rely on high capacity, high-efficiency, and improved new-age techniques, such as cryogenic, fluidized-bed, spiral, and air-blast freezing of the food system with conditions prevailing to individual quick-freezing. Meanwhile, in cryogenic, air-blast, and in other rapid or quick-freezing techniques, the formation of enlarged ice-crystals within food cells of substantial-sized food products takes place, and it will relatively have a lower rate of heat transfer due to the lesser surface to mass ratios [71]. In addition, due to the increased rate at which heat transfers in the food matrix, cryogenic freezing has been a relatively more quick method than air blast or liquid immersion freezing. However, due to thermal gradients, it can negatively affect cell components, can lead to formation of intra- and extra-cellular ice-crystals, which in turn affect the textural quality of frozen food [73]. Among these developing trends, high-pressure, and assisted freezing technique emerges as one of the promising methods of food preservation with reduced tissue damage and retention of product textural quality. This novel method helps to achieve homogenous and smaller ice-crystals in the process, reducing the freezing point of liquid and pure water at induced or assisted pressure gradient [104].

High-pressure assisted freezing (HPAF), depending on decreasing melting point of water under the assistance of pressure gradient, induces homogenous ice-nucleation within the food system [88]. HPAF and dehydro-freezing are novel techniques for preservation of perishable food commodities [51, 65]. High-pressure freezing processes are often categorized into three main classes, are distinguished based on phase diagram and how the water phase transition proceeds [101]:

- Phenomenon of freezing where transition of phase proceeds under constant high-pressure is referred as "high-pressure assisted freezing (HPAF);"
- Phase transformation on the release of applied pressure is referred as "high-pressure shift freezing (HPSF);" and

- Transformation of phase induced by a rise in the applied pressure and continued at constant high-pressure is classified as "high-pressure induced freezing (HPIF)."

Before this classification, terminologies, such as 'pressure-assisted freezing,' 'pressure-supported freezing,' and 'freezing under-pressure' were used to denote freezing processes operated or assisted with high-pressure [40, 89]. This chapter discusses HPAF of foods and food products.

4.2 HIGH-PRESSURE ASSISTED FREEZING (HPAF): THEORY AND PRINCIPLE

Phase transition and freezing rates critically account for the size, structure, and pattern of distribution of the ice crystals in any of the food matrix. For water to ice (ice-I) process during the transition of phase at atmospheric pressure, water tends to expand in its volume by 9% at 0°C, and nearly 13% at–20°C. Conventional freezing of food products significantly causes textural and structural damage in food tissues due to formation of large extracellular crystals with low-density ice-I that is much lesser than liquid water [104]. Most of the foods expand to a lesser extent compared to liquid water on freezing. Non-uniformity on volume change of the food matrix was observed due to presence of several components, which shrink to different extents when temperature is reduced. However, ice forms, like ice-II and ice-IX, have higher density than ice-I and liquid water, and can be formed under high-pressure, which contributes to effective, efficient, and qualitative freezing with very low damage to food texture and constituents without any increase in volume of the food product [65, 90].

4.2.1 PHASE TRANSITION OF WATER

4.2.1.1 ICE NUCLEATION AND CRYSTALLIZATION

In freezing, phase transformation occurs from liquid state to solid-state, where the temperature gradient is reduced significantly lower than its freezing point [49]. Almost all liquids freeze during the process of crystallization, where a uniform liquid is converted to its crystalline solids state. 'Nucleation' and followed by subsequent 'crystal growth' are two main events associated with ice crystal development. Nucleation and formation of nuclei, where the process proceeds in nanometer scale, gather to form clusters, arranging in periodic and defined pattern that interprets the crystal structure. Successive growth of formed ice-nuclei is termed as crystal growth [28]. In addition, ice-nucleation may be either homogeneous nucleation or heterogeneous [70]. Homogeneous ice-nucleation begins as there are random fluctuations of molecules in homogeneous free liquids, whereas nucleation is heterogeneous in solid foods as the presence of active cell surfaces functioning as nucleation sites [39]. The process of freezing does not begin, until temperature factor is sufficiently low enough to supply the required energy for the development of stable ice nuclei [85, 87].

4.2.1.2   SUPER-COOLING

Liquid is considered to be supercooled when it reaches its initial freezing point. Here, the liquid tends to be in metastable state, where the liquid continues to stay in that particular state, just prior development of first ice-crystal by nucleation. The process is preceded by rapid growth of crystals and it spreads throughout the volume. In addition, an impurity (such as dust and other particles) acts as a potential nucleation center. Due to high activation energy and homogenous ice nucleation, pure water freezes at lower temperatures. In the presence of nucleating substances, freezing point and melting point of water will be very near 0°C (32°F or 273.15 K) at one atmosphere of pressure [28].

Before freezing, water can supercool to–40°C (233 K or –40°F) in the absence of nucleators and under high-pressure, and water at about 2000 atmospheres of pressure can be supercooled to −70°C (−94°F or 203 K) [52]. In foods due to heterogeneous ice-nucleation, degree of supercooling will be lower than the pure water. The degree of supercooling is very crucial, as this critically influences the mechanism by which living animals and plants sustain at sub-zero temperatures and to overcome the damage due to formation of ice crystals in their tissues [70].

4.2.1.3   FORMS OF ICE

Ice is the water frozen into solid state and there exists five different crystalline structural forms of ice, such as ice-I, ice-II, ice-III, ice-IV, and ice V. Ice II to ice V aredenser than ice-I and liquid water. Meanwhile, between the range of temperature–100°C to 50°C and pressure gradient between 0 to 2.4

GPa, more than 10 ice polymorphs have been reported till date [78, 85, 87, 105]. All the ice forms seen on the earth surface are denoted as ice-I_h, which has a critical hexagonal crystalline structure and ice forms with minute traces of cubic ice is represented as ice-I_c. The solid form of ice-I persists at normal atmospheric pressure and it will have lower density than the normal water in comparison to other forms of ice. As pure water is cooled below 0°C at standard atmospheric pressure, phase transition water results in formation of ice-I_h. Meanwhile, ice-I displays decreased melting point with an increase in pressure gradient, down to 222°C at 207.5 MPa [12, 40, 90].

4.2.2 EFFECTS OF PRESSURE AND TEMPERATURE ON PHASE TRANSITION OF WATER

Many studies have been carried out to depict the exact correlation between phase transition of water and influence of temperature and pressure gradients. According to Bridgman et al. [9] and Salzmann et al. [86], ice-I remains stable at normal atmospheric pressure up to the pressure gradient of 210 MPa. In addition, above this pressure and depending on pressure/temperature gradients, different polymorphs of ice, ice-II, ice-III, ice-IV, and ice-V tend to be thermodynamically stable.

When the pressure goes up to 210 MPa, the freezing point of water is decreased, whereas above this level of pressure, increase in freezing point is observed for ice forms other than ice-I. As freezing process is an exothermic process (in which 'latent heat' is the energy released upon freezing: 'enthalpy of fusion'), it is purely equal to the amount of the energy needed to meltdown the exact same quantity of solid. Clausius-Clapeyron equation is defined as follows:

$$\frac{dT}{dP} = \frac{\Delta V . T_k}{\Delta H} \tag{1}$$

where; T as temperature and P as pressure; negative slope of equilibrium line with liquid water and ice-I interprets that latent heat (ΔH) and volume change (ΔV) in the above equation are different; and ΔH is negative irrespective of the modification in liquid-solid phase transition.

Because ice-I has lower density than the pure water due to the critical hexagonal structure in the ice with empty spaces in-between, therefore variation of volume is positive for ice-I. The phenomenon of increase in volume on freezing also leads to damage of tissues and other biological systems.

However, phase change and increment in volume is negative for liquid ice-III to V, due to positive slope of melting curve and negative phase change in enthalpy. Unlike ice-I, these ice polymers can overcome the problem of tissue damage and other quality aspects. At the same time, the supercooling phenomenon for a liquid tends to increase by pressure gradient due to range of enhanced metastable state in response to applied pressure.

Therefore more than melting point of water, temperature for homogeneous ice-nucleation decreases with increase in pressure gradient to 210 MPa. Therefore, the lowest possible temperature in order to super cool the pure liquid water is lowered from –40°C to –92°C under pressure gradient of 210 MPa [40, 78, 85, 87]. Phase diagram of water and its phase transition rates show high-pressure to be an emerging assisted technique, thus opening new scope and possibility for easy food preservation, freezing, thawing, and also for subzero temperature non-frozen storage [57, 80, 90].

4.2.3   PRINCIPLE OF HIGH-PRESSURE ASSISTED FREEZING (HPAF)

HPAF proceeds under constant pressure, when the temperature is decreased to lower than its corresponding freezing point. At atmospheric conditions, this process is much similar to conventional freezing except for pressure gradient, because HPAF is carried out under the assisted high-pressure. Typically, freezing is the phenomenon, where it starts from the surface and moves towards the center of the food matrix. As reported in the earlier studies, phenomenon of ice nucleation begins from the outermost portion of samples, which is in the undeviating touch with cooling media. As observed, ice-crystals so formed were of needle-shaped, large, and oriented radially from surface to center of the food system [34, 64, 90].

In HPSF, the product is cooled to less than 0°C under pressure and is stored in a non-frozen condition in accordance with the corresponding phase diagram and transition. Maintained pressure is released when the product reaches its desired temperature. Isostatic nature of pressure tends to induce uniform supercooling in the product on release of maintained pressure. The process is followed by homogenous development of ice-nuclei throughout the food sample due to supercooling phenomenon. On release of latent heat, it raises product temperature to its corresponding freezing point. Under environmental conditions, the process of freezing has been carried out under continuous pressure, where more number of ice crystals is formed because of low temperature and high-pressure [50, 64, 74].

The mean time for phase transformation under air-blast freezing with sizeable food samples is about 108 minutes for the formation of polygonal-shaped sizeable ice-crystals. Formed crystals were of diameter ranging from 65 up to 311 μm, with a mean value of 202 μm. Ice crystals formed in frozen samples at<50 MPa in HPAF had diameter ranging from 82 to 442 μm, with an equivalent mean value of 346 μm. This is contrast to samples at HPSF where smaller ice crystals were formed having an equivalent mean diameter of 5.6 μm, directing from 1.1 to 38.8 μm after expansion to 50 MPa. The mean value of phase transition time in HPSF was 163 minutes, notably less than in HPAF at 50 MPa. Unlike HPAF, ice nucleation occurs and is observed throughout the product in HPSF. In addition, granular shape of ice crystals with no definite orientation is formed and is found to be dispersed throughout the product [24, 68, 98, 108].

4.3   PROCESS OF OPERATION: HIGH-PRESSURE ASSISTED FREEZING (HPAF)

4.3.1   EQUIPMENT: MODELING OF HPAF PROCESS

Process of high-pressure freezing operates within pressure-resistant container, with thermostatic and thermally isolated circuits to attain the temperature <0C. It involves food and food products to pack and submerge in pressure/cooling media, followed by pressurization and freezing. To measure the operating pressure and an operating system must include a pressure gauge and thermocouples to evaluate and monitor the temperature. In addition, while selecting a potential pressurizing media, it is very critical to consider the freezing characteristics of the liquid component under pressure and its thermo-physical characteristics (such as coefficient of thermal expansion, heat capacity, and viscosity), which affects pressure influencing the temperature fluctuations.

Various pressure/cooling media either as pure substances or its mixtures (such as: ethanol/glycol ratio of 20/80 (v/v) [92], propanediol/water ratio of 55/45 (v/v) [64], ethanol/water ratio of 50/50 (v/v) [16], propylene glycol [97], castor oil/ethanol ratio of 15/85 (v/v) [55], ethylene glycol/water ratio of 75/25 (v/v) [74], glycol/water ratio of 62/38 (v/v) [56] and silicon oil [66]) have been reported in respective studies. Meanwhile, an external cooling system is usually utilized to maintain and to monitor the temperature inside the pressurization container. In a basic operational line, the pressurization container is directly submerged in thermostatic bath, where bath is set to a desired temperature [32, 91]. For industrial scaling purpose, the main vessel will be attached with cooling jacket, where the cooling media is circulated consistently [31, 40, 75].

In an operational process, a temperature-pressure gradient requires to be accurately evaluated, monitored, and regulated to ensure the rate and extent of freezing. Increase or decrease in pressure and volume (in accordance with the changes in temperature) is effectively used as indicators for phase transition. Hence, there is no other way to assure that a phase change has actually occurred without the help of these measurements [40, 69, 101].

4.3.1.1   FACTORS FOR MODELING OF HIGH-PRESSURE FREEZING

Food matrix involves complex shape, size, composition, and other physicochemical constituents. These factors are often considered as vital parameters that affect directly the effectiveness of assisted freezing. Initial models on HPAF only considered heat conduction, since combining phenomenon of phase transition and fluid motion while modeling was very difficult [20, 74, 92]. In the process of modeling of HPAF, the initial freezing point of any food sample under a particular pressure zone acts as one of the critical parameters as it influences the discontinuity in thermal properties of food sample. Composition of any food material also influences the range of initial freezing point for a particular food product. Depending on composition and thermophysical properties of food product, initial freezing point may go lower than 0°C [89].

Another considerable factor in modeling of HPAF is the temperature fluctuation due to change in pressure and other related gradients. Pressurization and depressurization always tend to increase or decrease the temperature due to expansion or compression of the food and pressurizing media. In addition, heat generated due to pressure change will exchange between the walls of the pressure medium, pressure vessel, and packaging material, thus influencing the temperature fluctuation in the corresponding sample. Delgado's group (2004) took convective phenomenon and its effects on a liquid sample into critical thinking while modeling the high-pressure freezing [61, 62, 89].

In the process of HPAF, forced convection was induced on compression when pressure media is pumped into vessel. The conductive and convective heat and mass transport conditions can affect the uniformity of high-pressure. In addition, this is influenced by the pressure chamber size, fluid viscosity, and compression rate [42–44, 84]. Convection currents within a pressurization media play a significant role in the evolution of thermal factors in the processed food samples [15, 40, 89]. Modeling for thawing process under assisted high-pressure also has been carried out in recent studies [12].

4.3.2 *FACTORS INFLUENCING THE HPAF PROCESS*

The degree of supercooling required to initialize nucleation has been strongly affected by the pressure gradient [66, 91], cooling rate, product size, and container characteristics. As nucleation phenomenon exhibits stochastic nature, therefore the result of assisted freezing can be fluctuated even under proper monitoring of all the critical parameters [14, 46]. Therefore, it is very crucial to specify the temperature range at which nucleation precedes [91]. Potatoes freezed at 70 MPa/–17°C thus showed a requirement of only negligible supercooling to form ice-I [101]. This is in contrast to a greater degree of supercooling that was found at 100 MPa/–19C and 140 MPa/–23°C. In addition, negligible supercooling was observed under 140 MPa and 209 MPa at–25°C, which was much lower than the freezing point of corresponding component [89, 91]. Need of greater degree of supercooling was observed for other kinds of ice to initiate nucleation. In order to achieve ice-V in meat and water sample, high degree of supercooling at 700 MPa with $(-10)\pm3$°C and $(-13)\pm3$°C respectively was required [69]. Therefore, higher degree supercooling is required at increased pressure gradient [89, 91]. Meanwhile, at atmospheric pressure, several solutes and macromolecules like sugars and hydrocolloids in the food system have been observed to influence the ice-nucleation and subsequent growth of crystals under the HPAF process. Sugars in the food system tend to alter glass transition temperatures (Tg) and freezing point, thus inhibiting crystal growth in the process of freezing [89].

The study to evaluate effects of hydrocolloids on freezing of sucrose solutions under 100 MPa at–22°C resulted in significant increase in degree of supercooling, which is required to induce ice-nucleation for freezing. However, the addition of hydrocolloids in the food system leads to the growth and expansion of ice crystal size distribution, and especially the addition of guar gum has significantly enhanced the increase in mean equivalent diameter of formed ice-crystals under pressure-assisted freezing [23, 40, 89].

4.4   EFFECT OF HPAF ON CONSTITUENTS AND STRUCTURES OF FOOD

The HPAF studies conducted on various food products displayed an enhancement in the microstructure of tissues compared to the traditional and other common methods, due to formation of minute sized ice-crystal in the process [99]. Meanwhile, in thermal activity during the process of phase changeover, the ice crystals have grown from outer surface to the core of the food product in HPAF; and the same has been observed with many of the HPAF products [23, 34, 50]. Due to reduction in the volume during phase changeover, polymorphs of ice except ice-I tend to cause less damage to the sample structure. Meanwhile, during the pressure release at <0°C, ice polymorphs revert to ice-I, thus reducing the possible advantages of the process [89].

4.4.1   EFFECT OF HPAF ON MICROSTRUCTURE OF FOOD

A comparative study conducted on effects of HPAF and conventional air freezing on the microstructure of eggplant tissue suggested that tissues of eggplants were observed with very low damage due to simultaneous nucleation occurred on HPAF. Whereas, damage to the microstructure was enhanced due to diminished freezing rate and increased sample volume on conventional air freezing techniques. Meanwhile, central region of eggplant in HPAF gave the same appearance as of fresh sample. In addition, all of the cells were observed to be positioned on HPAF without any cell damage. The HPAF technique was advantageous as instantaneous intracellular formation of ice with no time for water translocation [77].

The wheat starch gel frozen under 200 MPa at–25°C was most uniform and homogenous microstructure and textural appearance with reduction in freeze-induced pore size at increased pressure compared to pores formed by freezing at atmospheric pressure [100]. In comparison to gum frozen under atmospheric pressure and at 100 MPa to ice-I, no evident of enhancement in structural and textural quality was observed gellan gum gels frozen at HPAF to form ice-V between 600–686 MPa at–20°C [36]. Freezing of wheat dough at 240 MPa/–35°C and kept for one-week storage at–24°C has shown similar extensibility as of fresh dough [59, 100].

Less damage to potato tissues was observed when it was frozen to ice-III at 320 MPa, compared to freezing under 400 MPa and 200 MPa to form ice-V and ice-I, respectively [66]. Volume at 400 MPa was increased to -0.07 cm^3/g causing a little damage and when volume was increased to -0.03 cm^3/g for ice-III under 320 MPa then least damage to potato tissues was observed. In cellular membranes, phase change along with a decrease in volume of ice crystals was less destructive than phase change with increase in volume. Meanwhile, larger negative change in volume of ice crystals was more destructive to tissues than smaller negative change in volume [59, 89].

The study on the influence of high-pressure freezing on quality of sea bass muscle showed that extracellular ice was formed in HPSF, but these formed ice crystals were very minute than the crystals developed under air-blast freezing and found to have smaller spaces among the cells. Thus, neither shrinkage nor deformation was observed in the muscle fiber and no differences were observed between diverse muscle sections when compared with unfrozen sample, justifying isotropic nucleation during HPSF leading to a uniform distribution of ice crystals throughout the sample [40, 59, 89, 99].

4.4.2 EFFECT OF HPAF ON NUTRIENT PROFILE OF THE FOOD SYSTEM

Under high-pressure, protein modification can take place, when the food system is treated at 100 to 200 MPa [63]. High-pressure induces hydrophobic interaction, disintegration of salt bonds, but hydrogen bonding tends to be stronger under high-pressure. Meanwhile, covalent bonding was found to be least sensitive for pressure gradients. Oligomeric structure dissociation, aggregation, unfolding, denaturation, and gelation of proteins occur under high-pressure treatment [11]. On application of high-pressure, denaturation, and aggregation were observed in fish myofibrillar proteins [17, 72]. Significant reduction in myofibrillar protein extractability and increase in tail-muscle toughness was observed in Norway lobsters and turbot filets, when treated at 200 MPa/-18°C for 30 minutes and 140 MPa/-14°C for 12 minutes, respectively, followed by pressure release under HPSF [13]. DSC (differential scanning calorimetry) and protein extractability for high-pressure freezing resulted in partial denaturation, insolubilization, and aggregation of the myosin, along with changes in sarcoplasmic proteins [99].

4.4.3 EFFECT OF HIGH-PRESSURE FREEZING ON ENZYMES

Enzyme activity is one of the major quality indicators in a real food system. Studies on effect of HPAF at 350 MPa/-22°C and 400 MPa/-22°C, and

HPSF at 200 and 400 MPa at –22/–10°C for inactivation in enzyme solutions showed that pectin methylesterase was inactivated slightly and polyphenol oxidase (PPO) was not inactivated in the process [48]. Lipoxygenase was irreversibly inactivated and 60% of inactivation was observed at 200 MPa. Peroxidase (POD) and alpha-amylases from *Bacillus subtilis* were slightly and reversibly inactivated. In HPSF, incomplete inactivation of PPO and POD was observed in broccoli [81] and potato [82, 101].

Both the product and pressure/temperature conditions largely influence the reduction in PPO activity under HPAF and HPSF processes [5, 60, 89]. Significantly, irreversible inactivation was observed when carboxypeptidase (a serine protease) by *Saccharomyces cerevisiae* was subjected to high-pressure freezing under 200–400 MPa at 210 to 230°C [13].

Application of high-pressure freezing at 40 MPa inhibited the enzyme activity in frozen fish [89]. Further studies were conducted on characteristics of enzymes (such as: polygalacturonase, lipoxygenase, pectinmethylesterase, POD, and PPO) under high-pressure freezing, where polygalacturonase was more sensitive to HPAF treatment, as its activity was reduced by 23.4% on 100 MPa at –20°C.

In addition, lipoxygenase was found to be inactivate only at higher pressure and low temperature, whereas carrot peroxidase, potato polyphenol oxidase, tomato pectin methylesterase, and orange peel pectin methylesterase remained constant and were not affected by any temperature/pressure gradients and HPAF treatment [51]. Based on these studies, blanching treatment prior to high-pressure freezing has been highly recommended for fruits and vegetables to overcome browning after thawing [40, 51, 89, 100].

4.4.4   EFFECT OF HIGH-PRESSURE FREEZING ON MICROBIAL ACTIVITY

One of the major characteristic identities for frozen food products has been known to be microbiologically safe. High-pressure processing at or above 100 MPa effectively prevents microbial growth in food system [63]. Studies on microbial inactivation in food (both HPAF [45, 78] and HPSF [6, 18, 78–80, 83]) showed the huge potential to reduce microbial counts in an effective way.

The effectiveness and extent of microbial inactivation in HPAF strictly depends on the process conditions and applied pressure level, temperature, holding time, and product characteristics (such as its composition, water activity, pH, and type of feasible microorganisms). High extent of microbial inactivation can be achieved at increased pressure and longer holding time [18, 78, 83]. Nearly 3 log cycle of inactivation was observed when microbiologically active yeast strain in 0.85% brine solution was treated with high-pressure treatment at 180 MPa/–20°C. Meanwhile, more than 5 log cycle of inactivation ratios was noted when microbial suspensions of *Zygosaccharomyces rouxii, Aspergillus niger, Aspergillus oryzae,* and *E. coli* were treated under 140–180 MPa at –20°C [13].

High-pressure treatments at 250–350 MPa/–25°C were beneficial for inactivation of vegetative *Bacillus subtilis* cells with 2 log reduction in cells [100]. HPSF with a slow expansion rate in 18 minutes was more effective for decreasing *Listeria innocua, Pseudomonas fluorescens,* and *Micrococcus luteus* than the rapid expansion with HPSF cut salmon at 207 MPa/–21°C. Thus, it shows that expansion rate in HPSF is also an important parameter in microbial inactivation [79, 80]. Due to pressure-induced deactivation of yeast cells, inactive yeasts cells were observed in freezed dough frozen at 240 MPa/–35°C giving decreased loaf volume compared to fresh non-frozen dough on baking [100].

Subjecting frozen fish to high-pressure thawing enhanced the inhibition of microorganisms. *Staphylococcus aureus, Escherichia coli* and *Listeria monocytogenes* inoculated in milk showed reduction of 0.8, 3.5, 3.8 log cycles under HPSF of 200 MPa/–18°C for 30 minutes of holding time. Meanwhile, in different inoculated buffer solutions, these numbers escalated to 1.3, 6.1 and 4.0 log cycles, respectively [18, 100]. This clearly states that the reaction to this process of pressure treatment at low temperature can be varied by the presence of innumerable types of microorganisms and by medium characteristics [6, 18, 45, 89].

4.5 APPLICATIONS OF HIGH-PRESSURE ASSISTED FREEZING (HPAF) IN SELECTED FOOD PRODUCTS

As conventional freezing and thawing of vegetables and muscle foods can induce exudation, textural modification and degradation of qualitative parameters in food system, therefore HPAF has been considered less destructive and most effective, convenient, microbiologically-safe method to preserve the food commodity with well-maintained or enhanced qualitative characteristics [13]. High-pressure freezing can be a major part of industrial application, where ice crystallization is considered as a crucial step [18].

Many researchers have studied the application of HPAF in diversified food systems, particularly in meat systems including pork meat [22, 68, 106, 107], bovine muscle [22], and seafood products [16, 17, 105]. In addition, researchers from Unilever Group have worked and have patented the combination of assisted freezing under high-pressure processing to improve the consistency, smoothness, and to slow the meltdown phenomenon of ice creams [26, 99]. Some of the other applications and studies on high-pressure freezing for diversified food systems with its effects and results have been discussed in this section (Table 4.1).

TABLE 4.1   Applications of High-Pressure Freezing in Different Food Systems

Food System	Process Conditions	Results	References
Meat Products			
Abalone	50 MPa/−5.4°C, 100 MPa/−10.2°C, 150 MPa/−15.7°C, 200 MPa/−22.0°C	• During freezing, reduction in shear force, drip loss, lightness, and thawing was reduced when compared to conventional freezing • Decrease in aerobic colony count due to processing at increased pressure levels	[47]
Atlantic salmon	200 MPa/−20°C, 150 MPa/−14°C, 100 MPa/−8.4°C	• Well-maintained tissue structure • Formation of finer ice crystals and homogeneous distribution was observed at higher-pressure freezing	[105]
Atlantic salmon	100 MPa/−10°C, 200 MPa/−18°C	• Less drip loss after thawing and finer ice crystals were observed on high-pressure freezing • Higher mass diffusivity and less tissue damage was observed • Firmness of tissue was increased compared to sample which were subjected to air blast or those which did not undergo freezing	[2, 3]
Carp	140 MPa/−14°C	• Ice crystals of tiny size were found with intracellular space than those produced by other conventional methods of freezing • Freezing by high-pressure does not affect the texture, however, it was observed to be much better in drip loss reduction unlike freezing by air blast	[93]

TABLE 4.1   *(Continued)*

Food System	Process Conditions	Results	References
		• Free fatty acid content and thiobarbituric acid (TBA) values are relatively lower compared to samples frozen under air blast freezing	
Liver of pork; small prawns	200 MPa/−20°C, 150 MPa/−14°C, 100 MPa/−8.4°C	• Time required to change from one phase to another (phase transition) will be rapidly reduced • Formation of finer ice crystals and homogenous distribution of crystals were observed under high-pressure freezing	[96]
Muscle of beef and pork	200 MPa /−20°C	• Drip loss, tissue structure changes and protein denaturation, color change was observed on high-pressure freezing	[22]
Norway lobster	200 MPa/−18°C	• Higher toughness, lower salt soluble protein, and finer ice crystals were observed compared to samples frozen by conventional air freezing	[17]
Pork muscle	200 MPa/−17°C	• Tissue structure was greatly changed regardless of formation of finer ice crystals • Reduction in drip loss but color changes and protein denaturation were found in high-pressure frozen samples	[41]
	200 MPa/−21.5°C, 150 MPa/−16.5°C, 100 MPa/−11°C	• Small crystals of ice were observed on the surface of the sample, but comparatively large size of extracellular crystals were seen in the center location of high-pressure frozen sample • Treatment at pressures of 150 and 200 MPa lead to change in color • Reduction in drip loss but denaturation of myofibrillary protein had occurred during the process	[106, 107]

TABLE 4.1   *(Continued)*

Food System	Process Conditions	Results	References
	200 MPa/−21.2°C, 150 MPa/−14.1°C, 100 MPa/−8.9°C, 50 MPa/−4.2°C	• Color change and drip loss had occurred at pressure conditions above 150 MPa • Samples frozen at 100 MPa showed an equivalent quality as non-frozen sample	[19]
Smoked salmon mince	207 MPa/−2°C	• Freezing on high-pressure has found to be reduced in *Listeria. Innocua* and *Micrococcus luteus*, respectively, when it influenced reduction of from 1.7 and 4.6 log cycles.	[79]
	207 MPa/−22°C	• 1.4 and 1.9 cycles of log-phase that might be lowered of *Listeria innocua* on rapid and slow, and the pressure is released under high-pressure freezing	[80]
Turbot	140 MPa/−14°C	• There will be an apparent reduction in the salt soluble protein extract of viscosity and myosin heavy chain in terms of intensity was seen in the sample the frozen using the high-pressure • The harness of the tissue is increased due to the compact arrangement of the myofibrillar protein due to heavy pressure	[16]
Fruit and Vegetable Products			
Broccoli	180 MPa/−16°C, 210 MPa/−20°C	• Destruction of cellular structure after thawing and incomplete inactivation of peroxidase and polyphenol oxidase (PPO) enzymes was observed under high-pressure freezing of the sample • Texture was maintained even post thirty days of frozen storage at −20°C, but acceptability of the flavor was decreased in the sample	[36]
	210 MPa/−20.5°C	• Lower drip loss, less cell damage, and better texture in pressure-assisted treated frozen sample were observed than conventional freezing	[25]

TABLE 4.1 *(Continued)*

Food System	Process Conditions	Results	References
Carrot	100–700 MPa/−20°C	• Textural value changes, pectin release, and structural changes are inhibited in the frozen samples treated with assisted pressure of 200 to 400 MPa	[32]
	100–400 MPa/−28 to −10°C	• Expansion by volume occurs because of development of by ice I, tissues of samples frozen at 50 MPa/−15C, 100 MPa/−15°C, 150 MPa/−25°C, and 200 MPa/−28°C were extremely damaged.	[31]
	200 MPa/−15°C	• Less changes in texture and damage of cell wall damage when freezing happens and defrosting was observed, when the sample was preheated at 60°C for 30 minutes with $CaCl_2$ prior to high-pressure freezing	[102, 103]
Chinese cabbage	100–700 MPa/−20 to −18°C	• Low pectin release, reduced structural damage in tissues was observed in the samples frozen under 200 or 340 MPa than the samples frozen by conventional method • Textural changes in the frozen samples at 200, 340 or even 400 MPa were observed to be intact	[30]
Green bell pepper	200 MPa/−18°C	• Firm tissues of sample were increased in high-pressure frozen samples	[10]
Peach and mango	200 MPa/−20°C	• Nucleation of ice was quick and consistent for the sample • Loss of quality due to freeze cracking and subsequent formation of big sized ice crystal was successfully prevented	[73]
Potato	200 MPa/−15°C	• Nucleation rate of ice is increased, textural improvement and drip loss reduction was observed	[60]
	210 MPa/−20°C, 180 MPa/−16°C, 150 MPa/−12°C, 120 MPa/−8°C	• Incomplete inactivation of polyphenol oxidase and peroxidase in potato samples was observed under high-pressure freezing	[82]

TABLE 4.1 *(Continued)*

Food System	Process Conditions	Results	References
	240 MPa/−22°C	• Drip loss reduction, changes in color and effective reduction in structural damage were found in processed under high-pressure freezing treated samples and thawing which was pressure-assisted, compared to samples processed by traditional freezing method	[7]
Strawberry	200 MPa/−15°C	• Vacuum samples that is incorporated with pectin methylesterase and CaCl₂ before freezing indicates stable cell walls, contact between cells and textural alterations in texture and structure damage because of freezing under high-pressure. • Textural quality was maintained at the frozen storage for 6 weeks at−8C	[102, 103]
Tofu	200 MPa/−18°C	• Ice-crystals of smaller size were formed and quality parameters like taste, appearance, textural quality of the increased pressure-treated frozen samples post defrosting was same as those before freezing	[58]
Tofu	100–700 MPa/−20°C	• Frozen samples under high-pressure of 200 to 400 MPa showed smaller crystals of ice and slight changes in stress rupture and strain	[34]
Other Products			
Boiled egg	200 MPa/−18°C	• Size reduction of ice crystals moves rapidly and the drip loss will be lowered • All of the sudden and total failures on the strain (stress rupture) of frozen sample using high-pressure showed almost equal to non-frozen sample	[54]

TABLE 4.1 *(Continued)*

Food System	Process Conditions	Results	References
Carrageenan-gel	100–686 MPa/−20°C	• On the applied pressure levels between 200 to 400 MPa, gel was frozen only when the pressure has been released • Finer ice crystals are formed and inhibition of syneresis after thawing are observed in the frozen sample that is assisted by pressure • Addition of glucose, sucrose, and gum produced from locust bean, inhibited the ice crystals from growing	[53]
Egg custard-gel	100–686 MPa/−20°C	• On the applied pressure levels of 200 to 500 MPa for the sugar added gels and 200 to 400 for the gels in which sugar is not added the samples maintained supercooling state during pressurization and meanwhile it after freezing fine crystals has come and thawing process less changes in the textural nature has been noted • On considering sugar types, there was no notable influence, but when the quality fact was considered, 5% of sugar was added	[38]
Gelatin-gel	200 MPa/−20°C, 150 MPa/−14°C, 100 MPa/−8°C	• Crystals of ice that so fine with lower equivalent diameter was obtained under high-pressure freezing	[108]
Gels of agar	100–686 MPa/−20°C	• Gels are not frozen at the pressure levels of 200 to 400 MPa at −20°C, but during the release of pressure it will be frozen • Introducing sucrose inhibited the growth of ice crystals	[37]

4.6 QUALITY OF HIGH-PRESSURE ASSISTED FROZEN FOODS

Drip loss, water-holding capacity (WHC), microstructure, and textures are notable quality parameters for high-pressure assisted frozen foods. The extent or degree of change in these quality parameters on or after HPAF treatment was observed to be superior in comparison to conventional method of freezing.

Studies on eggplants to determine degree of drip loss and change in quality parameters under various methods of freezing showed that highest drip loss was produced in thawed eggplants due to the slow freezing rate, under conventional air freezing method. In contrast, eggplants frozen under modern HPAF technique resulted in lower drip losses and less textural damage and found effective to maintain original structural characteristics in contrast to eggplants frozen under conventional jet-air and air-blast methods. Meanwhile, in conventional air-freezing method, the rate of decrease in firmness was higher compared to samples frozen under HPAF [77]. Tofu frozen under HPSF at 200 MPa/–19°C for 45 to 90 minutes was observed to have no drip loss with well-maintained initial shape and texture [34].

Deleterious effect on hardness and drip loss of strawberry was hampered by infusion of strawberry halves with commercial fungal pectin-methylesterase and calcium chloride under vacuum prior to rapid assisted freezing. Drip loss of frozen strawberry pieces was minimized and highest original firmness value was maintained by infusion with 3% apple pectin solution and followed by immersion in PME/calcium solution prior to high-pressure freezing. PME/calcium infused and HPSF frozen strawberries firmness was well maintained during 6 weeks of frozen storage [89, 100].

A sensory study conducted on broccoli under HPSF of 180 MPa/–16°C, and 210 MPa/–20°C resulted in not acceptable flavor quality, after frozen storage of 30 days at–20°C, because of inadequate inactivation of POD and PPO. Therefore, under these conditions, blanching has been considered as the prerequisite operation to preserve the product under frozen storage [100]. In some cases, blanching also tends to cause deleterious changes in sensory parameters and quality of the product [8]. Compared to conventional freezing at 0.1 MPa/–20°C, high-pressure freezing at 200 MPa/–20°C and pressure-assisted thawing caused less damage to the rheology of cheddar or immature mozzarella [13, 55].

Heat/pressure treatments are known to increase changes in color of meat and fish products. The study to evaluate variation in color parameter of sea bass muscle under high-pressure freezing showed significant increase in its lightness, compared to no changes in the lightness under conventional air freezing technique [11]. It was also noted that deviation in WHC of muscle tissues was independent of frosting and defrosting process. Protein modification due to denaturation of myofibrillar proteins tends to be the most probable cause for decrease in WHC and for color changes in HPAF treated products [100].

The presence of solutes and hydrocolloids in the food system has also shown influence on the food quality parameters under HPAF [89]. Studies to determine the effect of zero percent, 2.5% and 5% of trehalose in HPAF chilled bean curd at 100–686 MPa/–20°C were found effective to improve the textural quality of frozen tofu by reduction in the size of the ice crystals formed under HPAF, thus preventing tofu from becoming firmer on freezing [33]. In addition, addition of sucrose @ 0%, 5%, 10% and 20% was observed to improve the quality of frozen agar under HPAF [35, 38] and gellan gum gels [36]. A substance (such as locust bean gum, alginate, gelatin, and carrageenan) that forms gel with water is generally used to prevent it from crystallizing again of ice in ice creams during frozen storage [4, 89].

4.7   PACKAGING REQUIREMENTS FOR HPAF PROCESSED FOODS

For effective packaging, food products must avoid being in contact with pressurizing and the cooling medium under HPAF and thawing. Packaging material for HPAF processed foods should possess flexibility for transmission of pressure, resistance to solvents and salts, resistance to temperature of cooling, freezing, piercing, and should not be permeable to gases and other liquids. It is preferred to pack HPAF foods under vacuum [12].

A study was conducted before and after the HPP to assess mechanical strength and migration effects of two multilayered packaging materials namely: EVOH/EVA/LDPE and PET/Al/PP, which showed no alterations in rupture module, relative deformation, and in other related mechanical strength parameters. In addition, permeability to water vapor and oxygen were found to decrease within the package after high-pressure processing [67]. Another study was conducted on different packaging materials to assess sealing methods, elastic modules, and permeability; and it showed that melted seal was one of the best methods to seal the package, which did not display any variation in migration and mechanical properties even after processing at high-pressure (400 MPa, 25°C for 30 minutes) [13, 27].

The research study used six different packaging materials and a non-uniform shaped frozen food sample to intensify the probable risk of penetrating [27]. The process of separation into constituent layers happens with all packaged materials, except when the product is packed under vacuum (where complete air was removed and the package was sealed). Packaging material, BB4L® (from Cryovac Grace, Epernon, France) containing EVA copolymer, was able to resist the pricking, and the depressurization rate did not significantly affect the package.

The thawing medium at 10°C to 20°C has been found to affect the mechanical resistance [12, 27]. The packaging material BB4L®, having copolymer Ethylene-vinyl acetate, was not able to penetrate among the various packaging materials under test. The BB4L® has four sheets, which contain Ethylene-vinyl acetate as the main copolymer constituent. Copolymer EVA constituent enhances clarity, decreases the melting temperature, extends sealing area, and tends to enhance the mechanical properties of the packaging material. The addition of ethylene butyl acrylate polymer is highly recommended during the application of sealing for frozen product because of its good stretching coefficient, which enhances the resistance to piercing [13, 27].

4.8 HPAF: ADVANTAGES, LIMITATIONS, AND CHALLENGES

The HPAF has emerged as one of the real-time techniques with its promising features. Advantageous parameters of HPAF include: depression of freezing point, reduced latent heat of phase change, inhibition, and inactivating the microorganisms as well as enzymes, short freezing time, formation of microcrystalline or vitreous ice, and structural modification with no changes in sensory and nutritional quality of the product. Other potential advantages in comparison with conventional processing and preservation include reduced time of processing, reduced loss of vitamin C and retention of flavor, texture, color, chemical integrity, and freshness of the product [12].

Although foods under HPAF produce tiny crystals of ice, yet these microcrystals may be enlarged with passing time through a phenomenon called recrystallization. Recrystallization occurs as the larger ice crystal is more heat strong with small surface energy than the ice micro-crystals in frozen foods. The process of recrystallization has been supported by fluctuation in temperature gradients of products during freezing or thawing, extended frozen storage, distribution, and at domestic storage, which critically affects the qualitative parameters of frozen foods [26, 27].

However, more research studies are needed in this specific field of HPAF to overcome few existing limitations. Inactivation of spores is a major constraint and challenge, as effective and complete inactivation of microbial spores under pressure gradients has not yet been developed. A major difficulty was identified in experimentally determining the thermophysical effects of complex food matrices under the pressure gradient [26, 27, 67]. This factor acts as a major constraint in modeling of heat transfer and phase transition in HPAF processes. The main disadvantage of all pressure-assisted frosting processes is the requirement of a high-pressure resistant (200 MPa) vessel, which is relatively expensive.

In addition, there is lack of availability of appropriate packaging system for HPAF foods and packaging of non-uniformed shaped food products, which produce air-containing pockets in package [27]. Although, HPAF is a novel technique, yet HPAF has some issues that are not easy implement in the food industry for large-scale development of equipments and other related aspects. These constraints should be highlighted in the future studies to make this technology as industrially efficient [40, 89, 90, 100].

4.9 FUTURE PERSPECTIVES

Many studies have been carried out since 1990's till-today to highlight the potential of high-pressure freezing of foods [12, 56, 74]. However, full-scale commercial application of HPAF has not been yet developed. Most of the recent studies have been conducted with lab-scale equipments. This chapter also highlights the need of much elaborated study on the basic mechanisms of phase transitions under different pressure and temperature gradients.

In addition, attention should also be given on thermophysical and characteristic studies on different ice forms, mainly, ice-III, and ice-V, with respect to various food components, which will directly overcome the constraints in the efficient modeling of HPAF processes. Impact of different concentrations of preservatives, solutes, and cryo-protectants on phenomenon of phase transformation followed by the formation of ice crystals should be evaluated in the future studies [12]. Characteristics nature of different protein components beneath subzero temperature and extreme pressure conditions should be determined in terms of compact gelation, gelation, unfolding, water retention, and solubility [56].

Upcoming studies can gradually find out the possible combination of HPAF with ultrasound because ultrasonic cavitation enhances the establishment of small and equivalent ice crystals [51]. Meanwhile, there is a need to standardize the packaging requirements and machine production cost simulations for HPAF to encourage its applications in both medium and large-scale food industries. Other packaging materials having EVA copolymer constituent, such as BB4L®, should be tested to determine its packing behavior [27].

Special emphasis should be given on pressure-assisted thawing as it can be implemented at subzero temperature with short processing time and low elimination rate. Thus, this will help to reduce threat related to microbes and other common disadvantages of defrosting at 273.15 K. Finally, future studies should focus on appropriate considerations on effects of extreme-pressure and subzero temperatures on food constituents, to study the kinetics of inactivation of microbes and enzymes, steadiness of ice crystals during storage and to assess the economic viability of HPAF method [40, 89, 90].

4.10   SUMMARY

Textural, nutritive, and qualitative degradation under conventional freezing has led to the development in rapid assisted freezing technologies. Extreme-pressure freezing as a new technique helps to achieve accelerated freezing with the formation of tiny and unvaried ice crystals. Several studies have been successfully done with the utilization of HPAF on fresh fruits, vegetables, raw meat, storage, and on the inactivation of microbes. Recent applications of HPAF have been found to preserve the microstructure of foods, with reduced drip loss and microbial reduction. More studies are required on pressure-temperature gradients and corresponding ice forms, microbial inactivation, packaging, storage stability, process modeling, and equipment development with industrial standards.

KEYWORDS

- crystallization
- food preservation
- freezing
- frozen food
- high-pressure
- ice forms
- ice-nucleation
- phase transition

REFERENCES

1. Akshay, N., (2019). *Express Food Hospitality*. https://www.foodhospitality.in/latest-updates/indian-frozen-foods-industry-will-grow-at-17-per-cent-cagr-to-be-valued-at-rs-188-bn-in-2024-cii/416937/ (accessed on 16 January 2021).
2. Alizadeh, E., Chapleau, N., De, L. M., & Le-Bail, A., (2007). Effect of different freezing processes on the microstructure of Atlantic salmon (*Salmo salar*) fillets. *Innovative Food Science and Emerging Technologies, 8*(4), 493–499.
3. Alizadeh, E., Chapleau, N., De, L. M., & Le, B. A., (2009). Impact of freezing process on salt diffusivity of seafood: Application to salmon (*Salmo salar*) using conventional and pressure shift freezing. *Food and Bioprocess Technology, 2*(3), 257–262.
4. Ando, Y., Nei, D., Kono, S., & Nabetani, H., (2017). Current state and future issues of technology development concerned with freezing and thawing of foods. *Nippon Shokuhin Kagaku Kogaku Kaishi [Journal of the Japanese Society for Food Science and Technology], 64*(8), 391–428.
5. Arthey, D., (1993). Freezing of vegetables and fruits: Chapter 9. In: Mallett, C. P., (ed.), *Frozen Food Technology* (pp. 237–240). Glasgow, UK: Blackie Academic & Professional.
6. Ballestra, P., Verret, C., Cruz, C., Largeteau, A., Demazeau, G., & El, M. A., (2010). High-pressure inactivation of *Pseudomonas* in black truffle-comparison with *Pseudomonas fluorescens* in tryptone soya broth. *High-Pressure Research, 30*(1), 104–107.
7. Benet, G. U., Chapleau, N., Lille, M., Le, B. A., Autio, K., & Knorr, D., (2006). Quality related aspects of high-pressure low temperature processed whole potatoes. *Innovative Food Science and Emerging Technologies, 7*(1, 2), 32–39.
8. Berry, M., Fletcher, J., McClure, P., & Wilkinson, J., (2008). Effects of freezing on nutritional and microbiological properties of foods. In: Evans, J. A., (ed.), *Frozen Food Science and Technology* (pp. 26–50). Oxford, UK: Blackwell Publishing Ltd.
9. Bridgman, P. W., (1912). Water, in the liquid and five solid forms, under pressure. In: *Proceedings of the American Academy of Arts and Sciences, American Academy of Arts and Sciences* (Vol. 47, pp. 441–558).
10. Castro, S. M., Van, L. A., Saraiva, J. A., Smout, C., & Hendrickx, M., (2007). Effect of temperature, pressure, and calcium soaking pre-treatments and pressure shift freezing on the texture and texture evolution of frozen green bell peppers (*Capsicum annuum*). *European Food Research and Technology, 226*(1, 2), 33–43.
11. Cheftel, J. C., & Culioli, J., (1997). Effects of high-pressure on meat: A review. *Meat Science, 46*(3), 211–236.
12. Cheftel, J. C., Levy, J., & Dumay, E., (2000). Pressure-assisted freezing and thawing: Principles and potential applications. *Food Reviews International, 16*(4), 453–483.
13. Cheftel, J. C., Thiebaud, M., & Dumay, E., (2002). Pressure-assisted freezing and thawing of foods: A review of recent studies. *International Journal of High-Pressure Research, 22*(3/4), 601–611.
14. Chen, S. L., & Lee, T. S., (1998). A study of supercooling phenomenon and freezing probability of water inside horizontal cylinders. *International Journal of Heat and Mass Transfer, 41*(4/5), 769–783.
15. Chen, C. R., Zhu, S. M., Ramaswamy, H. S., Marcotte, M., & Le, B. A., (2007). Computer simulation of high-pressure cooling of pork. *Journal of Food Engineering, 79*(2), 401–409.

16. Chevalier, D., Sequeira, M. A., Le, B. A., Simpson, B. K., & Ghoul, M., (2000). Effect of freezing conditions and storage on ice crystal and drip volume in turbot (*Scophthalmus Maximus*): Evaluation of pressure shift freezing vs. air-blast freezing. *Innovative Food Science and Emerging Technologies, 1*(3), 193–201.

17. Chevalier, D., Sequeira, M. A., Le, B. A., Simpson, B. K., & Ghoul, M., (2000). Effect of pressure shift freezing, air-blast freezing, and storage on some biochemical and physical properties of turbot (*Scophthalmus Maximus*). *LWT-Food Science and Technology, 33*(8), 570–577.

18. Choi, M. J., Min, S. G., & Hong, G. P., (2008). Effect of high-pressure shift freezing process on microbial inactivation in dairy model food system. *International Journal of Food Engineering, 4*(5), 1–7.

19. Choi, M. J., Min, S. G., & Hong, G. P., (2016). Effects of pressure-shift freezing conditions on the quality characteristics and histological changes of pork. *LWT-Food Science and Technology, 67*, 194–199.

20. Denys, S., Van-Loey, A. M., Hendrickx, M. E., & Tobback, P. P., (1997). Modeling heat transfer during highipressure freezing and thawing. *Biotechnology Progress, 13*(4), 416–423.

21. Delgado, A. E., & Sun, D. W., (2001). Heat and mass transfer models for predicting freezing processes: A review. *Journal of Food Engineering, 47*(3), 157–174.

22. Fernández, M. F., Otero, L., Solas, M. T., & Sanz, P. D., (2000). Protein denaturation and structural damage during high-pressure-shift freezing of porcine and bovine muscle. *Journal of Food Science, 65*(6), 1002–1008.

23. Fernandez, P. P., Martino, M. N., Zaritzky, N. E., Guignon, B., & Sanz, P. D., (2007). Effects of locust bean, xanthan and guar gums on the ice crystals of a sucrose solution frozen at high-pressure. *Food Hydrocolloids, 21*(4), 507–515.

24. Fernandez, P. P., Otero, L., Guignon, B., & Sanz, P. D., (2006). High-pressure shift freezing versus high-pressure assisted freezing: Effects on the microstructure of a food model. *Food Hydrocolloids, 20*(4), 510–522.

25. Fernandez, P. P., Prestamo, G., Otero, L., & Sanz, P. D., (2006). Assessment of cell damage in high-pressure-shift frozen broccoli: Comparison with market samples. *European Food Research and Technology, 224*(1), 101–107.

26. Fikiin, K., (2008). Emerging and novel freezing processes. In: Evans, J. A., (ed.), *Frozen Food Science and Technology* (pp. 101–123). Oxford, UK: Blackwell Publishing Ltd.

27. Fradin, J. F., Le, B. A., Sanz, P. D., & Molina-Garcia, A. D., (1998). Note. Behavior of packaging materials during high-pressure thawing. *Food Science and Technology International, 4*(6), 419–424.

28. Franks, F., (2003). Nucleation of ice and its management in ecosystems, philosophical transactions of the Royal Society of London. *Series A: Mathematical, Physical and Engineering Sciences, 361*(1804), 557–574.

29. Frozen Food Market, (2019). https://www.marketsandmarkets.com/Market-Reports/global-frozen-and-convenience-food-market-advanced-technologies-and-global-market-130.html (accessed on 16 January 2021).

30. Fuchigami, M., & Kato, N., (1998). High-pressure-freezing effects on textural quality of Chinese cabbage. *Journal of Food Science, 63*(1), 122–125.

31. Fuchigami, M., Kato, N., & Teramoto, A. I., (1997). High-pressure-freezing effects on textural quality of carrots. *Journal of Food Science, 62*(4), 804–808.

32. Fuchigami, M., Miyazaki, K., Kato, N., & Teramoto, A., (1997). Histological changes in high-pressure-frozen carrots. *Journal of Food Science, 62*(4), 809–812.

33. Fuchigami, M., Ogawa, N., & Teramoto, A., (2002). Trehalose and hydrostatic pressure effects on the structure and sensory properties of frozen tofu (soybean curd). *Innovative Food Science and Emerging Technologies, 3*(2), 139–147.

34. Fuchigami, M., & Teramoto, A. I., (1997). Structural and textural changes in Kinu-Tofu due to high-pressure-freezing. *Journal of Food Science, 62*(4), 828–837.

35. Fuchigami, M., & Teramoto, A., (1998). Effect of high-pressure-freezing on textural and structural quality of frozen agar gel. *The Review of High-Pressure Science and Technology, 7*, 826–828.

36. Fuchigami, M., & Teramoto, A., (2003). Texture and structure of high-pressure-frozen gellan gum gel. *Food Hydrocolloids, 17*(6), 895–899.

37. Fuchigami, M., & Teramoto, A., (2003). Changes in temperature and structure of agar gel as affected by sucrose during high-pressure freezing. *Journal of Food Science, 68*(2), 528–533.

38. Fuchigami, M., Teramoto, A., & Jibu, Y., (2006). Texture and structure of pressure-shift-frozen agar gel with high visco-elasticity. *Food Hydrocolloids, 20*(2/3), 160–169.

39. Govindarajan, A. G., & Lindow, S. E., (1988). Size of bacterial ice-nucleation sites measured in situ by radiation inactivation analysis. In: *Proceedings of the National Academy of Sciences* (Vol. 85, No, 5, pp. 1334–1338).

40. Guangming, S., Songming, Z., & Ramaswamy, H. S., (2017). High-pressure water-ice transitions in aqueous and food systems. In: Ahmed, J., Rahman, M. S., & Roos, Y. H., (eds.), *Glass Transition and Phase Transitions in Food and Biological Materials* (pp. 393–426). Oxford, UK: John Wiley & Sons Ltd.

41. Hansen, E., Trinderup, R. A., Hviid, M., Darre, M., & Skibsted, L. H., (2003). Thaw drip loss and protein characterization of drip from air-frozen, cryogen-frozen, and pressure-shift-frozen pork *Longissimus dorsi* in relation to ice crystal size. *European Food Research and Technology, 218*(1), 2–6.

42. Hartmann, C., (2002). Numerical simulation of thermodynamic and fluid-dynamic processes during the high-pressure treatment of fluid food systems. *Innovative Food Science and Emerging Technologies, 3*(1), 11–18.

43. Hartmann, C., & Delgado, A., (2002). Numerical simulation of convective and diffusive transport effects on a high-pressure-induced inactivation process. *Biotechnology and Bioengineering, 79*(1), 94–104.

44. Hartmann, C., Delgado, A., & Szymczyk, J., (2003). Convective and diffusive transport effects in a high-pressure induced inactivation process of packed food. *Journal of Food Engineering, 59*(1), 33–44.

45. Hayakawa, K., Ueno, Y., Kawamura, S., Kato, T., & Hayashi, R., (1998). Microorganism inactivation using high-pressure generation in sealed vessels under sub-zero temperature. *Applied Microbiology and Biotechnology, 50*(4), 415–418.

46. Heneghan, A. F., Wilson, P. W., & Haymet, A. D., (2002). Heterogeneous nucleation of supercooled water, and the effect of an added catalyst. In: *Proceedings of the National Academy of Sciences, 99*(15), 9631–9634.

47. Hong, G. P., & Choi, M. J., (2016). Comparison of the quality characteristics of abalone processed by high-pressure sub-zero temperature and pressure-shift freezing. *Innovative Food Science and Emerging Technologies, 33*, 19–25.

48. Indrawati, Van, L. A., Denys, S., & Hendrickx, M., (1998). Enzyme sensitivity towards high-pressure at low temperature. *Food Biotechnology, 12*(3), 263–277.
49. International Dictionary of Refrigeration, (2019). http://dictionary.iifiir.org/search.php (accessed on 16 January 2021).
50. Isaacs, N. S., (1998). *High-Pressure Food Science, Bioscience and Chemistry* (p. 510). Cambridge, UK: The Royal Society of Chemistry.
51. Islam, M. N., Zhang, M., & Adhikari, B., (2017). Ultrasound-assisted freezing of fruits and vegetables: Design, development, and applications. In: Barbosa-Canovas, G. V., Pastore, G. M., Cadogan, K., Meza, I. G., Da Silva, L. S. C., Buckle, K., Yada, R. Y., & Rosenthal, A., (eds.), *Global Food Security and Wellness* (pp. 457–487). New York, USA: Springer.
52. Jeffery, C. A., & Austin, P. H., (1997). Homogeneous nucleation of supercooled water: Results from a new equation of state. *Journal of Geophysical Research: Atmospheres, 102*(D21), 25269–25279.
53. Jibu, Y., Teramoto, A., Kuwada, H., & Fuchigami, M., (2014). Effects of high-pressure and addition of sucrose on the quality improvement of frozen-thawed carrageenan gels. Part 1: Comparison of kappa and iota carrageenan gels. *Japanese Cooking Science Journal, 47*(3), 143–154.
54. Jibu, Y., Yasukawa, K., Kuwada, H., Yokohata, N., Teramoto, A., & Fuchigami, M., (2009). Structure and texture of pressure-shift-frozen boiled egg. *Journal of Japanese Society for Cooking Science, 42*, 86–92.
55. Johnston, D. E., (2000). The effects of freezing at high-pressure on the rheology of cheddar and mozzarella cheeses. *Milchwissenschaft, 55*(10), 559–562.
56. Kalichevsky-Dong, M. T., Ablett, S., Lillford, P. J., & Knorr, D., (2000). Effects of pressure-shift freezing and conventional freezing on model food gels. *International Journal of Food Science and Technology, 35*(2), 163–172.
57. Kalichevsky, M. T., Knorr, D., & Lillford, P. J., (1995). Potential food applications of high-pressure effects on ice-water transitions. *Trends in Food Science and Technology, 6*(8), 253–259.
58. Kanda, Y., Aoki, M., & Kosugi, T., (1992). Freezing of tofu (Soybean Curd) by pressure-shift freezing and its structure. *Nippon Shokuhin Kogyo Gakkaishi , 39*(7), 608–614.
59. Kiani, H., & Sun, D. W., (2011). Water crystallization and its importance to freezing of foods: A review. *Trends in Food Science and Technology, 22*(8), 407–426.
60. Koch, H., Seyderhelm, I., Wille, P., Kalichevsky, M. T., & Knorr, D., (1996). Pressure-shift freezing and its influence on texture, color, microstructure and rehydration behavior of potato cubes. *Food/Nahrung, 40*(3), 125–131.
61. Kowalczyk, W., Hartmann, C., & Delgado, A., (2003). Freezing and thawing at the high hydrostatic pressure conditions-modeling and numerical simulation. In: *Proceedings in Applied Mathematics and Mechanics* (Vol. 3, pp. 388–389). Berlin: WILEY-VCH Verlag.
62. Kowalczyk, W., Hartmann, C., & Delgado, A., (2004). Modeling and numerical simulation of convection-driven high-pressure induced phase changes. *International Journal of Heat and Mass Transfer, 47*(5), 1079–1089.
63. LeBail, A., Chevalier, D., Mussa, D. M., & Ghoul, M., (2002). High-pressure freezing and thawing of foods: A review. *International Journal of Refrigeration, 25*(5), 504–513.

64. Levy, J., Dumay, E., Kolodziejczyk, E., & Cheftel, J. C., (1999). Freezing kinetics of a model oil-in-water emulsion under high-pressure or by pressure release. Impact on ice crystals and oil droplets. *LWT-Food Science and Technology, 32*(7), 396–405.

65. Li, B., & Sun, D. W., (2002). Novel methods for rapid freezing and thawing of foods: A review. *Journal of Food Engineering, 54*(3), 175–182.

66. Luscher, C., Schlüter, O., & Knorr, D., (2005). High-pressure-low temperature processing of foods: Impact on cell membranes, texture, color, and visual appearance of potato tissue. *Innovative Food Science and Emerging Technologies, 6*(1), 59–71.

67. Mertens, B., (1993). Packaging aspects of high-pressure food processing technology. *Packaging Technology and Science, 6*(1), 31–36.

68. Martino, M. N., Otero, L., Sanz, P. D., & Zaritzky, N. E., (1998). Size and location of ice crystals in pork frozen by high-pressure-assisted freezing as compared to classical methods. *Meat Science, 50*(3), 303–313.

69. Molina-Garcia, A. D., Otero, L., Martino, M. N., Zaritzky, N. E., Arabas, J., Szczepek, J., & Sanz, P. D., (2004). Ice VI freezing of meat: Supercooling and ultrastructural studies. *Meat Science, 66*(3), 709–718.

70. Nesvadba, P., (2008). Thermal properties and ice crystal development in frozen foods. In: Evans, J. A., (ed.), *Frozen Food Science and Technology* (pp. 1–25). Oxford – UK: Blackwell Publishing Ltd.

71. North, M. F., & Lovatt, S. J., (2006). Freezing methods and equipment. In: Hargrove, K. L., (ed.), *Food Science and Technology* (pp. 155–199). New York – USA: Marcel Dekker.

72. Ohshima, T., Ushio, H., & Koizumi, C., (1993). High-pressure processing of fish and fish products. *Trends in Food Science and Technology., 4*(11), 370–375.

73. Otero, L., Martino, M., Zaritzky, N., Solas, M., & Sanz, P. D., (2000). Preservation of microstructure in peach and mango during high-pressure-shift freezing. *Journal of Food Science, 65*(3), 466–470.

74. Otero, L., & Sanz, P. D., (2000). High-pressure shift freezing: Part 1. Amount of ice instantaneously formed in the process. *Biotechnology Progress, 16*(6), 1030–1036.

75. Otero, L., & Sanz, P. D., (2006). High-pressure-shift freezing: Main factors implied in the phase transition time. *Journal of Food Engineering, 72*(4), 354–363.

76. Otero, L., Sanz, P., Guignon, B., & Sanz, P. D., (2012). Pressure-shift nucleation: A potential tool for freeze concentration of fluid foods. *Innovative Food Science and Emerging Technologies, 13*, 86–99.

77. Otero, L., Solas, M. T., Sanz, P. D., De, E. C., & Carrasco, J. A., (1998). Contrasting effects of high-pressure-assisted freezing and conventional air freezing on eggplant tissue microstructure. *Zeitschrift Fur Lebensmitteluntersuchung Und-Forschung A [Journal of Food Inspection and Research A], 206*(5), 338–342.

78. Park, S. H., Hong, G. P., Min, S. G., & Choi, M. J., (2008). Combined high-pressure and subzero temperature phase transition on the inactivation of *Escherichia coli* ATCC 10536. *International Journal of Food Engineering, 4*(4), 01–17.

79. Picart, L., Dumay, E., Guiraud, J. P., & Cheftel, J. C., (2004). Microbial inactivation by pressure-shift freezing: Effects on smoked salmon mince inoculated with *Pseudomonas fluorescens, Micrococcus luteus* and *Listeria innocua*. *LWT-Food Science and Technology, 37*(2), 227–238.

80. Picart, L., Dumay, E., Guiraud, J. P., & Cheftel, C., (2005). Combined high-pressure-sub-zero temperature processing of smoked salmon mince: Phase transition phenomena and inactivation of *Listeria innocua*. *Journal of Food Engineering, 68*(1), 43–56.

81. Prestamo, G., Palomares, L., & Sanz, P., (2004). Broccoli (*Brassica oleracea*) treated under pressure-shift freezing process. *European Food Research and Technology, 219*(6), 598–604.

82. Prestamo, G., Palomares, L., & Sanz, P., (2005). Frozen foods treated by pressure shift freezing: Proteins and enzymes. *Journal of Food Science, 70*(1), S22–S27.

83. Prestamo, G., Pedrazuela, A., Guignon, B., & Sanz, P. D., (2007). Synergy between high-pressure, temperature and ascorbic acid on the inactivation of *Bacillus cereus*. *European Food Research and Technology, 225*(5/6), 693–698.

84. Rauh, C., Baars, A., & Delgado, A., (2009). Uniformity of enzyme inactivation in a short-time high-pressure process. *Journal of Food Engineering, 91*(1), 154–163.

85. Salzmann, C. G., (2019). Advances in the experimental exploration of water's phase diagram. *The Journal of Chemical Physics, 150*(6), 060901.1–060901.9.

86. Salzmann, C. G., Radaelli, P. G., Hallbrucker, A., Mayer, E., & Finney, J. L., (2006). The preparation and structures of hydrogen ordered phases of ice. *Science, 311*(5768), 1758–1761.

87. Salzmann, C. G., Radaelli, P. G., Mayer, E., & Finney, J. L., (2009). Ice XV: A new thermodynamically stable phase of ice. *Physical Review Letters, 103*(10), Article ID: 105701.

88. Sanz, P. D., De, E. C., Martino, M., Zaritzky, N., Otero, L., & Carrasco, J. A., (1999). Freezing rate simulation as an aid to reducing crystallization damage in foods. *Meat Science, 52*(3), 275–278.

89. Sanz, P. D., & Laura, O., (2014). High-pressure freezing. In: Sun, D. W., (ed.), *Emerging Technologies for Food Processing* (pp. 515–538). London, UK: Academic Press, Elsevier Ltd.

90. Saxena, S. R., (2013). High-pressure freezing: An emerging technology in food technology. *The Journal of Multi-Disciplinary: Knowledge Consortium of Gujarat, 7*, 01–08.

91. Schluter, O., Benet, G. U., Heinz, V., & Knorr, D., (2004). Metastable states of water and ice during pressure supported freezing of potato tissue. *Biotechnology Progress, 20*(3), 799–810.

92. Schluter, O., Heinz, V., & Knorr, D., (1998). Freezing of potato cylinders during high-pressure treatment. *Special Publication-Royal Society of Chemistry, 222*, 317–324.

93. Sequeira, M. A., Chevalier, D., Simpson, B. K., Le, B. A., & Ramaswamy, H. S., (2005). Effect of pressure shift freezing versus air blast freezing of carp (*Cyprinus carpio*) fillets: A storage study. *Journal of Food Biochemistry, 29*(5), 504–516.

94. Shelly, S., (2019). *Personal Data*. https://www.marketsandmarkets.com/PressReleases/Global-frozen-food-market.asp (accessed on 16 January 2021).

95. Shenouda, S. Y., (1980). Theories of protein denaturation during frozen storage of fish flesh. In: Chichester, C. O., Mrak, E. M., & Stewart, G. F., (eds.), *Advances in Food Research* (Vol. 26, pp. 275–311). London, UK: Academic Press.

96. Su, G., Ramaswamy, H. S., Zhu, S., Yu, Y., Hu, F., & Xu, M., (2014). Thermal characterization and ice crystal analysis in pressure shift freezing of different muscle (Shrimp and Porcine Liver) versus conventional freezing method. *Innovative Food Science and Emerging Technologies, 26*, 40–50.

97. Teramoto, A., & Fuchigami, M., (2000). Changes in temperature, texture, and structure of konnyaku (*Konjac glucomannan* Gel) during high-pressure freezing. *Journal of Food Science, 65*(3), 491–497.

98. Thiebaud, M., Dumay, E. M., & Cheftel, J. C., (2002). Pressure-shift freezing of O/W emulsions: Influence of fructose and sodium alginate on undercooling, nucleation, freezing kinetics and ice crystal size distribution. *Food Hydrocolloids, 16*(6), 527–545.

99. Tironi, V., Lebail, A., & De, L. M., (2007). Effects of pressure shift freezing and pressure-assisted thawing on sea bass (*Dicentrarchus labrax*) quality. *Journal of Food Science, 72*(7), C381–C387.

100. Urrutia, G., Arabas, J., Autio, K., Brul, S., Hendrickx, M., Kakolewski, A., Knorr, D., et al., (2007). SAFE ICE: Low-temperature pressure processing of foods: Safety and quality aspects, process parameters and consumer acceptance. *Journal of Food Engineering, 83*(2), 293–315.

101. Urrutia-Benet, G., Balogh, T., Schneider, J., & Knorr, D., (2007). Metastable phases during high-pressure-low-temperature processing of potatoes and their impact on quality-related parameters. *Journal of Food Engineering, 78*(2), 375–389.

102. Van, B. S., Grauwet, T., Van, L. A., & Hendrickx, M., (2008). Structure/processing relation of vacuum infused strawberry tissue frozen under different conditions. *European Food Research and Technology, 226*(3), 437–448.

103. Van, B. S., Messagie, I., Maes, V., Duvetter, T., Van, L. A., & Hendrickx, M., (2006). Minimizing texture loss of frozen strawberries: Effect of infusion with pectin methylesterase and calcium combined with different freezing conditions and effect of subsequent storage/thawing conditions. *European Food Research and Technology, 223*(3), 395–404.

104. You, J., Habibi, M., Rattan, N., & Ramaswamy, H. S., (2016). Pressure shift freezing and thawing. In: Balasubramaniam, V. M., Barbosa-Canovas, G. V., & Lelieveld, H. L., (eds.), *High-Pressure Processing of Food* (pp. 143–166). New York – USA: Springer.

105. Zhu, S., Bail, A. L., & Ramaswamy, H. S., (2003). Ice crystal formation in pressure shift freezing of Atlantic salmon (*Salmo salar*) as compared to classical freezing methods. *Journal of Food Processing and Preservation, 27*(6), 427–444.

106. Zhu, S., Bail, A. L., Chapleau, N., Ramaswamy, H. S., & De-Lamballerie, A. M., (2004). Pressure shift freezing of pork muscle: Effect on color, drip loss, texture, and protein stability. *Biotechnology Progress, 20*(3), 939–945.

107. Zhu, S., Bail, A. L., Ramaswamy, H. S., & Chapleau, N., (2004). Characterization of ice crystals in pork muscle formed by pressure shift freezing as compared with classical freezing methods. *Journal of Food Science, 69*(4), FEP190–FEP197.

108. Zhu, S., Ramaswamy, H. S., & Bail, A. L., (2005). Ice-crystal formation in gelatin gel during pressure shift versus conventional freezing. *Journal of Food Engineering, 66*(1), 69–76.

CHAPTER 5

MICROENCAPSULATION TECHNOLOGY: POTENTIAL IN FORMULATIONS OF PROBIOTIC FOODS

MEENATAI G. KAMBLE, AJAY CHINCHKAR, and ANURAG SINGH

ABSTRACT

Microencapsulation is a process by which active probiotic culture can be encapsulated within a polymeric material to produce particles in the micrometer to mm range. Microencapsulation provides means to protect the probiotic bacteria from the adverse acidic condition of different food matrices and storage conditions such as temperature, environmental humidity, pH, etc., Mask the aroma/flavor, improve survivability in gastrointestinal condition and transform liquids into easy to handle solid ingredients. This chapter describes the recent and advanced techniques of probiotic microencapsulation to form probiotic microcapsules and various techniques have been discussed for probiotic microencapsulation and incorporation in fermented and non-fermented foods, including microencapsulation in by using complex coacervation, extrusion, spray drying, spray chilling, prilling, emulsion, and freeze-drying.

5.1 INTRODUCTION

Probiotic bacteria help in the preparation of functional foods [54], due to their heavy load in the digestive system, which also promotes the health benefits. To act as a probiotic, it needs to survive the adverse highly acidic condition and reach to large intestine in such a number to colonize and proliferate. Most of probiotic bacteria have an inability to survive in gastric juice at a pH <2. However, most of added probiotic cells are significantly reduced during processing and storage of probiotic products. Though non-living bacteria and their metabolites contain probiotic properties [48], yet it is necessary to reach in viable condition for their characteristic property.

According to the FAO/WHO [18], probiotic food should contain $>10^7$ CFU/g load of microbes to promote positive physiological effect. *Lactobacillus* and *Bifidobacteria* genera have the status of generally recognized as safe (GRAS) microorganisms. Many experimental reports have concluded that free probiotic cells containing probiotic products have shown poor survival probiotic cells [6, 61]. Factors influencing the survival of probiotics in the digestive path are: oxygen availability, concentration of acids like lactic and acetic acid (AA) and acidity. Therefore, some methods have been adapted to increase the viability of probiotics, such as: use of resistant strains, adaption of stress, microencapsulation, and additions of micronutrients.

Protection of bacterial culture from losses is one of the most important aims of encapsulation by using the different encapsulation techniques. Microencapsulation is a technique by which probiotic cells are adhered to encapsulating matrix to survive from adverse conditions and are released at controlled conditions [15]. Microencapsulation technique keeps probiotic bacteria live at the time of processing, transportation, storage, and release them in their target organ.

Innovation in microencapsulation technologies and probiotic formulations using probiotic culture and encapsulating materials and controlled release system gives effective outcome. Microencapsulation should protect probiotics from bile salts, low pH, and other products of digestion [36]. Microencapsulation shields the microscopic organisms from outside ecological conditions, e.g., temperature, oxygen, and dampness. Different encapsulating materials are used in the encapsulation of probiotics, such as carbohydrates, proteins, fats, and waxes; and the selection of materials is based on the protection from different environmental conditions.

This chapter focuses mainly on recent information, characteristics of probiotics, challenges, and methods for microencapsulation of probiotic microbes during processing to enhance the viability.

5.2   CHARACTERISTICS OF PROBIOTIC BACTERIA FOR SURVIVAL IN GASTROINTESTINAL TRACT (GI)

Bifidobacterium and *Lactobacillus* in the intestine have exerted health beneficial effects. *Bifidobacterium* and *Lactobacillus* are inhabited in large and small intestine, respectively. They are also recognized as GRAS [35]. However, probiotics can be viable or non-viable to promote the beneficial effect [43, 52]. The selection criteria for probiotic culture depend on its production of bioactive components (Figure 5.1).

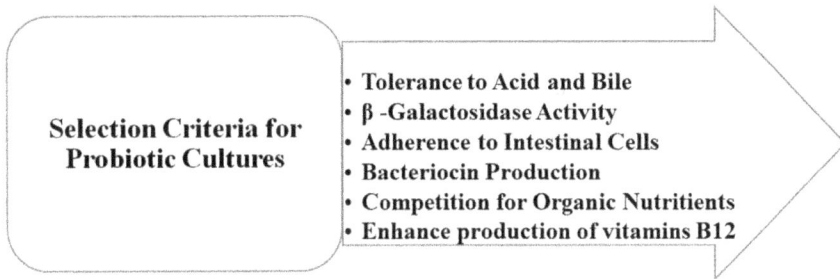

Selection Criteria for Probiotic Cultures

• Tolerance to Acid and Bile
• β -Galactosidase Activity
• Adherence to Intestinal Cells
• Bacteriocin Production
• Competition for Organic Nutritents
• Enhance production of vitamins B12

FIGURE 5.1   Selection criteria of probiotic bacteria.

Lactobacillus are Gram-positive, rod structured bacteria, and require an oxygen-free environment for their survival and get their energy from fermentative metabolism. In addition, they are strictly fermentative, aero-, and acid-tolerant, fastidious bacteria. They can ferment sugar to lactic acid fermentation. This characteristic of this genus helps it to survive at low pH.

Bifidobacteria are anaerobic, Gram-positive bacteria, which prefer pH in the range of 4.5 to 8.5. *Bifidobacteria* mainly produce acetic and lactic acid by fermenting carbohydrates. Gastric condition at pH of 2 is harmful to the free-cell survival. Microencapsulation helps in survival under these conditions. In addition, the amount of encapsulating material and structure of capsule helps in their viability.

5.3   FORMULATION CHALLENGES IN PROBIOTIC FOODS

From the innovative perspective, challenges to develop probiotic food products are:

Selection of probiotic strains, such as *Lactobacillus* or *Bifidobacterium,* etc.;

- Concentration of inoculum to be added in food;
- Rate of viability after/during processing;
- Survivability and functionality storage conditions;
- Assessment of one or multiple probiotic strain counts;
- Effect on sensory acceptability.

Figure 5.2 shows different technological challenges from processing to consumer chain. Development and endurance conditions during processing are survivability and usefulness during capacity, assessment of one or different probiotic strain checks, and sensory acceptability of probiotic-containing food products. Low pH fruit juices cannot support probiotic.

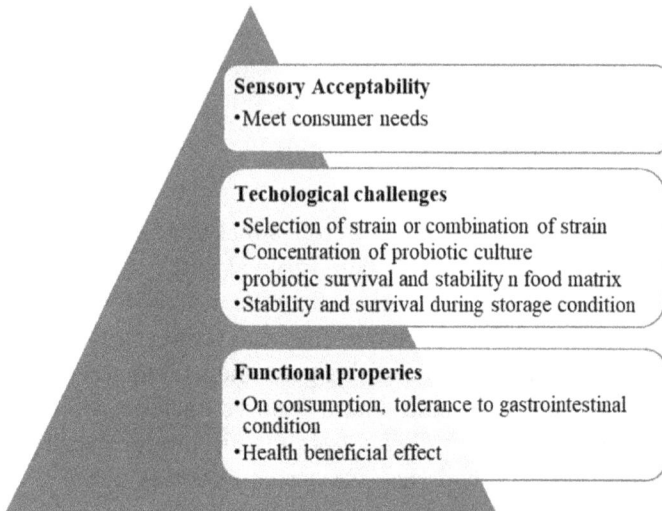

Sensory Acceptability
•Meet consumer needs

Techological challenges
•Selection of strain or combination of strain
•Concentration of probiotic culture
•probiotic survival and stability n food matrix
•Stability and survival during storage condition

Functional properies
•On consumption, tolerance to gastrointestinal condition
•Health beneficial effect

FIGURE 5.2   Technological challenges in formulation of probiotic food.

The viability of probiotic is dependent on strain employed, which should be not less than 10^7 CFU/ml at the end of finished products. Low pH of foods, higher oxygen, and processing parameters (high temperature, incubation time/temperature, rate of cooling, materials used for packaging and storage technique) are detrimental factors for the viability in juices [59].

Lactobacilli are resistant at pH of 3.7 to 4.3 in juices, whereas *Bifidobacterium* is tolerant to low acidic conditions. Nualkaekul et al. [40] suggested that dietary fibers present in fruits may shield cells from the effects of low pH. A probiotic is commercially successful only after choosing the right strain of bacteria and its viability. For long-term storage, generally dried forms of bacteria are utilized. However, such type of drying affects the viability inversely. Survival rates are generally higher in freeze-drying than spray-drying [53]. Survival rates of dried probiotic depend upon consumption and reconstitution methods, powder quality, properties of food matrix, and rehydration properties [9].

Microencapsulation size influences textural as well as sensorial properties of food. While, shape capsule affects flow properties. The size of probiotic bacteria may limit the loading capacity of microcapsule and its addition in food products and has a negative impact on barrier properties of active materials and quality of food products (in terms of sensory quality). In addition, the addition of probiotic cells in food should be free from off-flavor; and it should tolerate in simulated gastric conditions on consumption, and maintain its functionality within the host. In the case of fermented fruit juice, an increase in organic acid profile (alcohol, critic, acetic, and propionic acids, etc.), in the fermented probiotic product may reflect the metabolic activity of probiotic culture, causing reduced consumer acceptability of a food product.

Kailasapathy [24] reported that the sensory quality of probiotic can be protected through microencapsulation of probiotic cells by using suitable barrier material. Microencapsulation of probiotics masks the flavor and enhances the organoleptic characteristic to increase its acceptance by consumers. Topography of the capsule also affects the sensory quality of the final product. The dimensions of microencapsulated probiotic bacteria are given in Table 5.1. Maintaining lower investment in the cost of production is most challenging for its future success. Future advancements can address the potential difficulties for broadening the wide extent of the foods wherein probiotic cells can be successfully added.

TABLE 5.1   Survival of Some Encapsulated Probiotic Bacteria Under GI Conditions

Probiotic Microorganism	Encapsulation Method	Encapsulating Material	Size of Capsules	Survival Under GI Conditions	References
B. lactis Bb12, L. plantarum 299V, L. acidophilus La5	Fermentation	Fructo-oligosaccharide	–	Loss of viable cells near to: 0.1 log CFU/ml	[38]
B. adolescentis 15703T	Emulsion	Gelatin and Genipin; Alginate	49–53 μm	7.6 and 7.4 CFU/ml	[2]

TABLE 5.1   *(Continued)*

Probiotic Microorganism	Encapsulation Method	Encapsulating Material	Size of Capsules	Survival Under GI Conditions	References
B. animalis ssp. Lactis	Spray drying	Maltodextrin, Inulin	10.65 and 16.52 μm	–	[16]
B. longum B6	Spray drying	Gum Arabic, low fat Milk powder, Gelatin, Soluble Starch	–	93.53%, 81.26%, 87.15%, 95.47%.	[30]
B. bifidum	Extrusion	Sodium Alginate, Poly-L-Lysine	–	> 10^6 CFU/ml	[13]
Bifidobacterium	Spray drying	Waxy maize starch	5 μm	–	[41]
L. acidophilus	Extrusion	κ-carrageenan	3 mm	–	[60]
L. acidophilus and B. bifidum	Extrusion	Sodium alginate	54.25 ± 0.18 mm	–	[34]
L. acidophilus, Bifidobacterium spp.	Emulsion	Alginate and Hi-maize starch	0.5–1 mm	> 10^6 CFU/ml	[58]
L. bulgaricus	Extrusion	Sodium Alginate, Chitosan	40–80 μm	> 10^6 CFU/ml	[29]
L. casei	Extrusion	Sodium Alginate and Chitosan, Alginate, Pll-Alginate	1.89 mm	1.6×10^6, 6.7×10^3, 7.0×10^3 CFU/ml	[27]
L. casei NRRL B-442	Spray drying	Maltodextrin, Gum Arabic	< 9.5 mm, 13.3 mm	–	[1]
L. paracasei NFBC 338	Spray drying	Gum Acacia	5–15 μm	>100 folds	[15]
L. rhamnosus GG, L. acidophilus	Extrusion	Sodium Alginate	10–40 μm	–	[57]

5.4   MICROENCAPSULATION METHODS FOR PROBIOTIC ENCAPSULATION

Microencapsulation is well-organized packaging technology that entraps probiotic cells in a small capsule of protective membranes expel their contents in a controlled manner for a desirable time under given controlled conditions [50]. Microcapsule diameter ranges from few microns to 1 mm.

Microencapsulation of probiotic culture also provides protection from adverse situations [21], such as: excessive acidity, pH, oxygen, cold, and heat shock during production, storage, and gastrointestinal transit to reduce cell losses and increase viability (Table 5.1). Microencapsulation also increases the probiotic supply during food processing [17].

Kim et al. [25] have examined the impact of encapsulation on the sustainability of *L. acidophilus ATCC 43*, when exposed to stimulate the digestive system. It was observed that the encapsulated cells have more tolerance to adverse intestinal conditions than the *L. acidophilus ATCC43* cells that are non-encapsulated. The microencapsulation of probiotics decides the utilization of various innovations or distinctive exemplifying material to cover undesirable flavor, odor, to improve the survivability of the cells during production and product storage.

In dried food products, water activity is an important factor influencing the cell survival. Therefore, moisture-proof encapsulating material must be used for dried products; and non-ionizable encapsulating material protects the bacterial cells from the gastric acidity. In addition to this, suitability of encapsulation material depends on the probiotic strain (e.g., bile tolerance, acid tolerance, and oxygen tolerance). Mechanisms involved in the controlled release of material include rupturing the membrane by mechanical forces, exposure to high temperature, specific pH, and dissolution in solvent, biodegradation of polymer, microcapsules' osmotic pressure, or the above combined strategies.

Most researchers have studied the microencapsulation methods, which are mainly used for probiotic encapsulation and the outcome of research results is summarized in Table 5.2.

TABLE 5.2   List of Probiotic Cultures Encapsulated by Different Techniques

Probiotic Bacteria	Encapsulating Agent	Incorporation Technique	Major Finding	References
B. adolescentis	Chickpea protein, Alginate	Emulsification	• After 2 hr, Improved survival of encapsulated *B. adolescentis* by 5.5 times than non-encapsulated cells • The encapsulated *B. adolescentis* exposed to simulated gastric juice at pH 6.5 for 3 hr showed an increase in viability of *B. adolescentis* (~7.1 log CFU/mL)	[61]

TABLE 5.2 *(Continued)*

Probiotic Bacteria	Encapsulating Agent	Incorporation Technique	Major Finding	References
L. paracasei	Whey protein isolate, Gum Arabic	Complex coacervation	• At the end of storage, Encapsulated cells showed only ≤ 0.22 log CFU/g reduction of viable cell counts than free cell counts (≤0.64 log CFU/g) • Encapsulated culture cell, subjected to SGJ (pH 2) for 3 h showed greater survival compared to non-encapsulated cell	[6]
L. plantarum 299v; Pediococcusaci-dilactici HA-6111-2	Maltodextrin	Drying: Spray, freeze, convective hot air	• No significant decrease was observed in the bacterial count by SD and FD, while CD showed 2 log cycles reduction. During storage at 4°C, no significant difference in survivability of *Lactobacillus plantarum* 299v cells obtained by using SD, FD, and CD, respectively.	[5]
L. rhamnosus ATCC 7469	WPI, Persian Gum, Inulin	Freeze drying	• Encapsulated of *L. rhamnosus* when exposed to digestive system showed an viable *L. rhamnosus* cell • This technique reduces Aw (0.052–0.072), particle size than spray dried and electro sprayed powder Aw (0.12–0.15 and 0.15, respectively), particle size (295.7–353 nm and 341–364 nm, respectively)	[33]

TABLE 5.2   *(Continued)*

Probiotic Bacteria	Encapsulating Agent	Incorporation Technique	Major Finding	References
Lactobacillus acidophilus (JCM 1132)	Octanoic acid Triacylglycerol, hexaglyceryl, Ricinoleate, decaglycerol monolaurate	Double emulsion	• *L. acidophilus* dispersed in the juice showed a rapid decrease in cells counts and after 1.33 hr there is no live *L. acidophilus cell* • While, encapsulated *L. acidophilus* added in juice revealed 49% of cell live at 2 h	[56]
Lactobacillus plantarum ST-III	Type-A Gelatin, gum Arabic	Complex coacervation	• GE/GA/S5.5 encapsulated sample showed highest survival among others. The addition of sucrose showed increased in resistance of cell survivals against environmental stresses	[65]
Lactobacillus rhamnosus GG	Whey protein powder, casein, sunflower oil	Emulsification	• Encapsulated *L. rhamnosus GG* in combination of protein observed highest bacterial count (99%) and 97% encapsulation rate when compare to formulation made only protein	[7]
Lactobacillus rhamnosus GG (LGG)	Whey protein concentrate (WPC) Pullulan (PUL)	Gelation	• WPC-PUL gel can protect *Lactobacillus rhamnosus* LGG cells under simulated gastric juice (SGJ) (6.61 Log CFU/g) and release viable cell in simulated intestinal juice (SIJ) (9.40 Log CFU/g) J) for 4 h	[64]
Lactococcus lactis subsp. cremoris LM0230	Sodium alginate	Extrusion	• Encapsulated cell showed only <2 log reduction at room temperature • Encapsulation improved the viability of the probiotics	[63]

5.4.1 SPRAY-DRYING AND SPRAY-CHILLING METHODS

Spray drying (SD) is a basic method to remove the water molecules from the constituents. The suspension containing probiotics culture and carrier materials are injected into a HA stream inside the closed chamber to evaporate water molecules. The product feed rate, airflow rate and temperature control are important parameters of SD to reduce moisture content and develop a stable product. The large surface area may be enough to produce small droplet sizes. SD technique is based on pressurized injection of liquid material through a small aperture and the feed pump supplies the energy to reduce pressure drop in pressure due to expulsion of material through the aperture. The feed pump with nozzles produces about 3000 psi of pressure, which helps in producing desirable distribution of particle size, but it may affect the cell structure adversely. The survival rate of probiotics can be improved by the use of protectants in the media [12].

Desmond et al. [15] indicated that the SD of milk-based emulsion with the addition of the soluble fiber and gum acacia showed an increase in survivability of *L. paracasei* compared to *L. paracasei* spray-dried with only milk powder. Probiotic juices contain low atomic weight moiety that prompts lower glass progress temperature (Tg). However, when the dryer temperature reaches above the glass transition temperature (Tg), they fix it to the surface of the dryer during the operation. Therefore, there is a need of incorporation of drying-aid agent to raise the glass transition temperature of products. The maltodextrin (DE<20) and gum Arabic are the most commonly used drying-aids in food industries. Applications of SD in probiotic encapsulation or probiotic fruit juice are given in Table 5.3. This technique has been widely used on a commercial scale in industrial applications.

The one of disadvantage in spray dying technique is the higher drying temperature, which is not suitable for survival of probiotic; and addition of protectants can be done before drying to improve its survival.

Spray chilling is almost similar to SD except its operating temperature and carrier material. Spray chilling operates at low temperature conditions and it involves mixing of probiotic cell suspension and molten carrier material (melting temperature well above the operating temperature), which is sprayed through the nozzle into a cool air to allow the solidification and to form the microcapsule. Encapsulated probiotic produced are not soluble in water, and commonly carrier materials in spray cooling are: fats, waxes, fatty alcohols, fatty acids, or combinations of these materials [62]. The disadvantage of spray chilling is the active material not being consistent with its

TABLE 5.3 Applications of Encapsulated Probiotic in Fermented and Unfermented Fruit Juices

Fruit	Fermented/ Non-fermented	Probiotic Bacteria	Encapsulating Agent/Drying Aid	Incorporation Technique	Major Results	References
Acerola nectar	Non-fermented	B. animalis subsp. lactis BB-12	Cellulose Acetate Phthalate, Glycerol, Maltodextrin, Tween 80, Reconstituted Milk, Hi-maize, Trehalose	Spray drying	• After 30 days at cold storage, sample containing microencapsulated probiotic exhibited > 8 log CFU/200 ml viable cells	[3]
Apple juice	Non-fermented	L. rhamnosus GG	Chitosan-alginate, with or without inulin	Extrusion	• After 90 days, survival of encapsulated of L. rhamnosus was 4.5-folds higher, while sample containing free cells, viable count was decreased to 13.6% throughout the storage • Encapsulated cells had 27.7% survivability when exposed to GI condition	[20]
Banana juice	Fermented	L. Acidophilus	k-carrageenan	Extrusion	• L. acidophilus cells immobilized with K-carrageenan observed 3 mm in size of bead • During the fermentation, viable cell number in suspension was found to be 10^5 CFU/ml, in case of immobilized cell fermentation showed > 10^8 CFU/ml and free cell followed by fermentations showed 10^6 CFU/ml of viable counts	[60]

TABLE 5.3 *(Continued)*

Fruit	Fermented/ Non-fermented	Probiotic Bacteria	Encapsulating Agent/Drying Aid	Incorporation Technique	Major Results	References
Cashew juice	Fermented	L. casei NRRL B 442	–	Direct inoculation	• Throughout the storage period of 42 days showed 8 Log CFU/ml of L. Casei counts.	[45]
Cornelian cherry juice	Fermented / non-fermented	L. Plantarum ATCC 14917	Delignified wheat bran (DWB)	Extrusion	• After 4th week, immobilized cells retained their higher viability (9.95 log CFU/ml) when compared with immobilized cells (7.36 log CFU/ml) • Ethanol concentration was found to be in the range of (<0.3–0.9%v/v) • After fermentation observed, increase in lactic, acetic, and ethanol and decrease in sugar content • The total phenolic content was higher in samples containing immobilized cells as compared to fermented and non-fermented juice • No significant differences in sensory quality were perceived from all the samples	[31]
Fresh apple slices	Non-fermented	L. rhamnosus GG	Oligofructose, inulin, sodium alginate	Immersion	• All probiotics samples remained stable over the 14 days storage period (>10^{8} CFU/g).	[51]
Grape juice	Non-fermented	L. casei 01	Sodium alginate, chitosan	Extrusion	• Sensory analysis observed, higher acceptability of grape juice and orange containing probiotic beads (84.3 and 82.3%) • Probiotic beads increased the turbidity in the juice	[28]

TABLE 5.3 *(Continued)*

Fruit	Fermented/ Non-fermented	Probiotic Bacteria	Encapsulating Agent/Drying Aid	Incorporation Technique	Major Results	References
Grape juice	Non-fermented	*Lactobacillus acidophilus*; and *Bifidobacterium bifidum*	Sodium alginate	Extrusion	Particle size of alginate bead was found to be 54.25 ± 0.18 mm • At the end of the 60th days, the survivability of the was found to be 8.67±0.12 and 8.27±0.05 log CFU/ml, respectively for encapsulated *L. acidophilus* and *B. bifidum*, which was significantly higher than the free cells of *L. acidophilus* and *B. bifidum* containing samples (7.57±0.08 and 7.53±0.07 log CFU/ml, respectively)	[34]
Litchi juice	Non-fermented	*L. casei*	• WPC • WPC and Inulin, • WPC and gum acacia, • WPC and oligofructose • Fructo-oligosacch-aride (FOS)	Spray drying	• All samples contained >10^6 CFU/ml viable cells • Viable cells count was mostly found in WPCP + inulin coated sample • Morphology of microencapsulated samples observed circular shape of particles with hollow cavities but no fissure • PH was reduced in sample containing gum acacias (4.97) followed by WPCP + inulin, WPCP, WPCP+FOS and control • Aw was in range of 0.98–0.99 • All samples showed non-Newtonian nature • Sensory analysis observed higher acceptability in samples containing inulin	[49]

TABLE 5.3 *(Continued)*

Fruit	Fermented/ Non-fermented	Probiotic Bacteria	Encapsulating Agent/Drying Aid	Incorporation Technique	Major Results	References
Strawberry, black currant, cranberry, and pomegr-anate	Non-fermented	*L. plantarum (NCIMB 8826)*	Inulin, gum Arabic	Freeze drying	• Dried fruit powders mixed with probiotics showed that highest viable cells in blackcurrant (No decrease) juice followed by strawberry (0.3 log decrease), pomegranate (0.9 log decrease) and cranberry (~4.5 log CFU/ml survival) afterward 12 months storage • The use of inulin and gum Arabic showed no significant difference in cell existence of all samples, but they improve cell viability only in cranberry juice • On reconstitution and determination after 4 hr showed no decrease in strawberry while little decrease found in pomegranate and blackcurrant juices. While significant decrease in cranberry juice	[39]

focal point of capsulation that can have a negative impact on the protective efficiency of the capsule.

5.4.2 PRILLING

Prilling is a process, where a melted material is sprayed across the airflow in a chamber to develop capsules of solid particles. The active material containing spherical aggregates is formed by means of pipettes, syringes, or needles.

Droplets are produced at the tip continue to fall freely under the influence of gravity into the gelation bath (Figure 5.3(A)). The viscosity and nozzle diameter are important factors to produce bigger droplets of a few mm. Nozzle resonance technology is used to form microcapsules by breaking droplets through a sonically vibrating dispersion head at a specific frequency and it forms uniform droplets of 1 mm in size (Figure 5.3(B)). This is useful for low viscous solution [51].

Probiotic cell and encapsulating agents are blended altogether and the suspension is transferred at a low rate (ml/hour), where electrostatic potential on the surface of beads generates a charge that can oppose the surface tension (Figure 5.3(C)). The resulting droplet has a size up to 20 mm at 10 kV. In case of Jet-cutter technology, it is the most appropriate method for handling highly viscous liquid, and a jet of liquid is allowed through high speed series of wires, which are fixed on a turntable (Figure 5.3(D)).

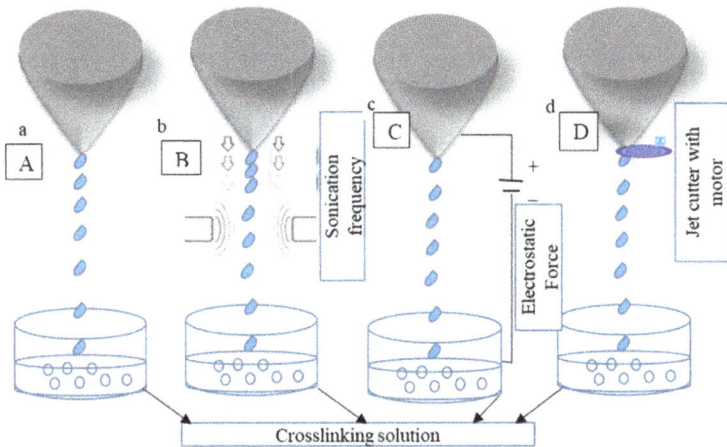

FIGURE 5.3 Different methods of prilling.

5.4.3   EMULSIFICATION

Emulsification is a process, where the dispersion of two non-miscible fluids is carried out by a stabilizer (Figures 5.4 and 5.5). The water system in oil system (W/O) and oil system in water system (O/W) are generally two types of emulsion methods. Normally in W/O system, the constant phase is oil, and the dispersed phase is water. Whereas in O/W system, the constant phase is water and the dispersed phase is oil. This process requires a considerable amount of energy to disperse one of the liquids in the form of small droplets in a continuous phase through the shearing or stirring of emulsion. The process of stirring, deformation, breakup, and droplets formation is shown in Figure 5.4. The stability is an important parameter of W/O or O/W emulsion, and typically it is stabilized with natural surfactant [12].

FIGURE 5.4   Droplet formation during emulsification process.

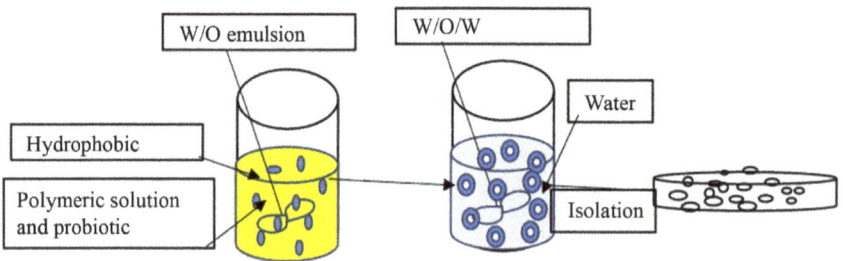

FIGURE 5.5   Double emulsion process.

L. delbrueckii ssp. Bulgaricus cells entrapped in emulsion of sesame oil showed approximately 10^4 times increase in persistence rate when it is opened to stimulate GI tract condition compared to free cells [23]. Similarly, Pimentel-González et al. [47] observed that *L. rhamnosus* cells were microencapsulated by double emulsion (Figure 5.5) and it was stabilized by using concentrated sweet whey. They observed that double emulsion stabilized with concentrated sweet whey increased *L. rhamnosus* cell viability than the free *L. rhamnosus* cells. In addition, the application of these techniques for probiotic encapsulation is reported in Tables 5.2 and 5.3.

5.4.4 EXTRUSION PROCESS

Extrusion process is a gentle encapsulation process, which minimizes cell-injuries and maintains higher viability of probiotic cells [22]. Microencapsulation by extrusion process includes suspension or emulsion of carrier material and core pass from the nozzle at high pressure. In the extrusion process, the first hydrocolloid solution is prepared, probiotic culture is added, and then the suspension is passed through the extrusion process to form the droplets into the hardening solution [26]. Details of extrusion process are shown in Figure 5.6. The diameter of the bead depends on the use of different nozzle sizes. Moreover, there is an increase in the size of microbeads due to the increase in concentration of carrier agent. This process is not suitable for large-scale production, as it requires more time for development of the microcapsule. Sodium alginate, calcium alginate, xanthan gum, gellan, locust bean gum, and κ-carrageenan are most common carrier agents that are used for microencapsulation by extrusion process (Table 5.2).

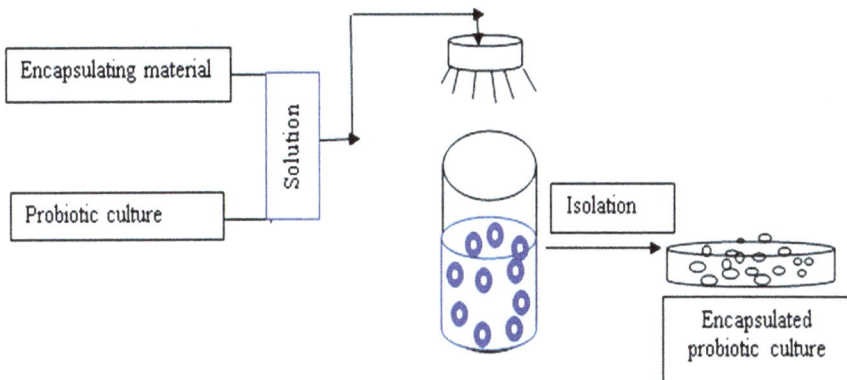

FIGURE 5.6 Extrusion process.

5.4.5   COACERVATION

Coacervation is a process of separation of colloid particles from solution that leads to the formation of microcapsules over the probiotic cells called coacervate. The process of separation of colloidal particles from the solution is achieved by changes in pH, ionic quality, and temperature of the medium [19]. In case of simple coacervation, only single macromolecules are present. Whereas in complex coacervation, more than one molecule is present having opposite charges. The complex coacervation process is mostly utilized for microencapsulation of oils and probiotic cells [22].

Few carrier materials have been studied for coacervation microencapsulation, e.g., gelatin, gum acacia, heparin, carrageenan, chitosan, soy protein, lactoglobulin, and dextran or combination of these [22]. It is a most skillful innovation due to high stacking limit >99% [22]. *B. lactis* and *L. acidophilus* cell encapsulation in pectin and casein complex by coacervation exhibited the highest acid tolerance ability of probiotic cells in GI environment [42]. Complex coacervation is the most appropriate technique for microencapsulation of probiotic bacteria.

5.4.6   FLUIDIZED BED COATING (FDC)

Fluidized bed coating (FDC) is employed for coating of probiotic bacterial cells. Probiotic bacteria are fluidized on airflow, and then coating material begins the formation of the film, followed by wetting and solidification that causes development of solid, homogenous coating on the core material [37]. Coating materials are sprayed through the nozzles to form the coating on the core matrix. Air helps the particles for uniform circulation. Coating materials, which are used for encapsulation, should be solidified at low temperatures [10].

The suspending, splashing, and cooling is constant until the capsules' walls have reached to ideal thickness. Bottom spray fluid bed process and the top spray fluid bed methods normally utilize "FDC procedure" (Figures 5.7(A) and 5.7(B)). In the case of the top spray fluidized bed process [12], nozzles are placed above the fluidizing bed for spraying of coating material onto the circling stream of particles and are normally utilized in granulation. The particle flow is counter current to spraying, involves collision in wet particles, and forms agglomeration. This is suitable for smaller particles and improves the agglomeration uniformity.

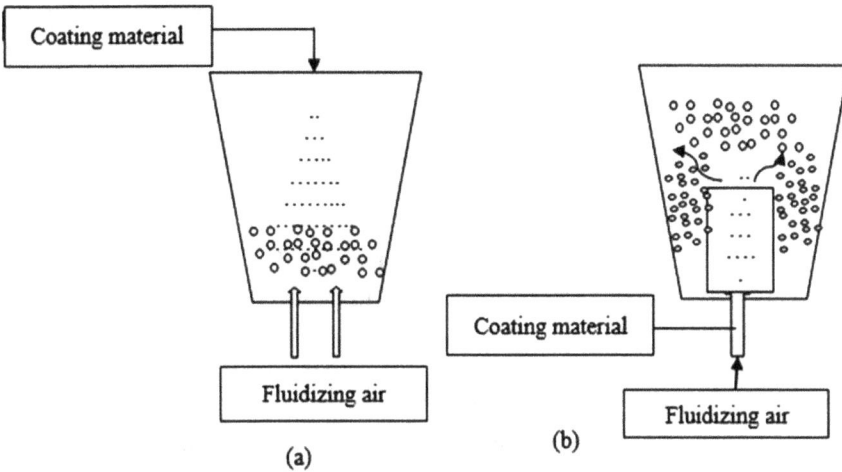

FIGURE 5.7   Fluidized bed coating process: (A) top spray fluidized bed coating; (B) bottom spray fluidized bed coating.

In bottom spray dried process, nozzles are set at the base of a bed to give the most uniform covering of streaming particles. This process can limit the agglomeration of particles during encapsulation. The uniform coating can be accomplished as particles move in, and are separated as they pass by means of the atomized spray by nozzles through the expansion region of the chamber. The fluidizing air allows solidifying of the coating material.

FDC is suitable for thermo-sensitive material and may represent major challenges. The technique consumes low energy, giving more output without affecting cell numbers [37]. The most generally coating material of probiotic is lipid-based material, such as, waxes, fatty acids, specialty oils, and other carbohydrates; and protein-based materials can be used for food application [10]. The efficiency of fluidized bed drying of *L. plantarum* in casein powder was 80% [32]. Similarly, Lee et al. [29] reported that microencapsulation bacterial culture in chitosan and alginate microparticles gave better protection and this can be an effective method for transporting viable bacterial cells in the colon.

5.4.7 FREEZE DRYING (FD)

Freeze drying (FD) process is based on the principle of sublimation to remove water from the products. The process involves freezing the product to below a critical temperature and afterward freezing is carried, where the freeze chamber works at lower pressure with expansion in-rack temperature and unbound water is expelled through sublimation. At last, auxiliary drying is done to expel water by desorption and item is gradually brought to room temperature [46, 53]. It is frequently used for encapsulation and drying of probiotic cells to produce frozen or dried-form of probiotic culture.

This technique is generally used for heat-sensitive ingredients and probiotic bacteria. The factors to be controlled during freeze-drying are: pH of medium, protective compounds, the prehistory of biomass, initial concentration of probiotic culture, freezing rate, and temperature, to sustain the viability of probiotic bacteria [8] Lyophilization of *L. delbrueckii ssp. Bulgaricus L2* showed only 4% of viable cells without any protective materials [55]. Therefore, there is a need to add protectant in the cell-suspension before drying to secure the feasibility of probiotics. Most common protectants for encapsulation of probiotic cells are low fat milk powder, milk protein, glucose, maltodextrin (DE<20) and trehalose, etc.

5.4.8 INCLUSION COMPLEXATION

Cyclodextrins are cyclic oligomers, which have the ability to form inclusion complexes with the active core. Cyclodextrins have a hydrophobic cavity inside and a hydrophilic cavity outside [25]. The size of hydrophobic inside the cavity formed is relied on the number of glucose units, with α-, β- and γ-cyclodextrins being made up of 6, 7, and 8 number of glucose units (Figure 5.8). This is mostly used for encapsulation of very sensitive substances [4]. Complex formation with hydrophobic molecules occurs due to the dislodging of water from the inside of the cyclodextrin. This may increase the dissolvability of hydrophobic-visitor particles in water and the center material is secured against deprivation [14]. One of the limitations of this technique is the high cost of material and low amount of loading capacity.

5.5 APPLICATIONS OF ENCAPSULATED PROBIOTICS IN FRUIT JUICES

Fruit and vegetables represent promising carriers for incorporation of probiotic bacteria. Non-dairy consumer is particularly interested in prebiotic-rich probiotics. However, fruit juices contain some essential nutrients, such as organic acids (citric, malic, tartaric, etc.), polyphenolic components, flavo-noids, vitamins, minerals, antioxidants, and dietary fibers.

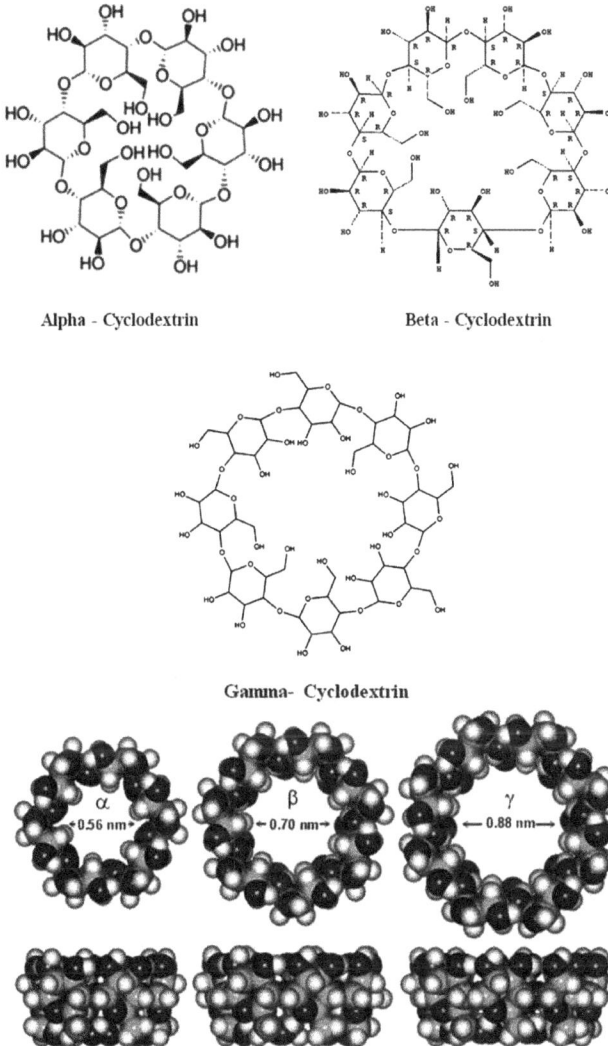

FIGURE 5.8   Chemical structure of cyclodextrin.

The vast majority of these issues may constrain the probiotic endurance in juices, such as food matrix if self, food processing, storage conditions, and GI environment, where encapsulation allows the probiotics to arrive at the digestive tract in huge numbers. Microencapsulation is similar to packing technology, which may protect the bacterial cells from low pH, acidity, oxygen, environmental conditions.

For the microencapsulation technology, many researchers have studied the wide range of materials and techniques to protect the bacterial cells that discharge their substance at controlled rates over delayed timeframes under the impact of explicit conditions. Applications of encapsulated probiotics in fermented and non-fermented fruit juice are given in Tables 5.3. The encapsulation provides more positive anaerobic conditions for sensitive bacterial cells [17]. Commercially, only few fruit-based probiotics are available in the market.

5.6 SUMMARY

Various microencapsulation techniques are available for encapsulation of probiotic to shield it from the acidic food and gastric juice of the GI tract. The advancement in microencapsulation technology can shield the probiotic culture from antagonistic ecological states of pH, acidity, temperature, so as to convey probiotic-rich microbiota in the colon.

Food industries face technological, economic challenges when the micro-encapsulated probiotic bacteria are added into the fruit juices. Different techniques to improve probiotic productivity are protection of bacteria through encapsulation, fermentation technique that is an extra cost to the food production process. The probiotic encapsulation methodology is not completely utilized at industrial levels and needs extra exploratory work to be used in food matrices. Therefore, we need to do controlled human investigations for the effectiveness of probiotics.

KEYWORDS

- fermented juice
- gastrointestinal tract
- generally recognized as safe
- microencapsulation
- non-fermented juice
- survivability

REFERENCES

1. Alves, N. N., Messaoud, G. B., Desobry, S., Costa, J. M. C., & Rodrigues, S., (2016). Effect of drying technique and feed flow rate on bacterial survival and physicochemical properties of a non-dairy fermented probiotic juice powder. *Journal of Food Engineering, 189*, 45–54.
2. Annan, N. T., Borza, A. D., & Hansen, L. T., (2008). Encapsulation of alginate-coated gelation microsphere improves the survival of probiotic *Bifidobacterium adolescentis* 15703T during exposure to simulated gastrointestinal condition. *Food Research International, 41*, 184–193.
3. Antunes, A. E. C., Liserre, A. M., Coelho, A. L. A., Menezes, C. R., Moreno, I., Yotsuyanagi, K., & Azambuja, N. C., (2013). Acerola nectar with added microencapsulated probiotic. *LWT-Food Science and Technology, 54*(1), 125–131.
4. Astray, G., Gonzalez-Barreiro, C., Mejuto, J. C., Rial-Otero, R., & Simal-Gandara, J., (2009). A review on the use of cyclodextrins in food. *Food Hydrocolloids, 3*, 1931–1940.
5. Barbosa, J., Borges, S., Amorim, M., Pereira, M. J., Oliveira, A., Pintado, M. E., & Teixeira, P., (2015). Comparison of spray drying, freeze-drying and convective hot air drying for the production of a probiotic orange powder. *Journal of Functional Foods, 17*, 340–351.
6. Bosnea, L. A., Moschakis, T., & Biliaderis, C. G., (2016). Microencapsulated cells of *Lactobacillus paracasei I* in biopolymer complex coacervates and their function in a yogurt matrix. *Food and Function, 8*(2), 554–562.
7. Burgain, J., Gaiani, C., Cailliez-Grimal, C., Jeandel, C., & Scher, J., (2013). Encapsulation of *Lactobacillus rhamnosus* GG in microparticles: Influence of casein to whey protein ratio on bacterial survival during digestion. *Innovative Food Science and Emerging Technologies, 19*, 233–242.
8. Carvalho, A. S., Silva, J., Ho, P., Teixeira, P., & Gibbs, P., (2004). Relevant factors for the preparation of freeze-dried lactic acid bacteria. *The International Dairy Journal, 14*, 835–847.
9. Carvalho, A. S., Silva, J., Ho, P., Teixeira, P., Malcata, F. X., & Gibbs, P., (2003). Effect of various growth media upon survival during storage of freeze-dried *Enterococcus faecalis* and *Enterococcus durans*. *Journal of Applied Microbiology, 94*, 947–952.
10. Champagne, C. P., & Fustier, P., (2007). Microencapsulation for the improved delivery of bioactive compounds into foods. *Current Opinion in Biotechnology, 18*, 184–190.
11. Chandramouli, V., Kailasapathy, K., Peiris, P., & Jones, M., (2004). An improved method of microencapsulation and its evaluation to protect *Lactobacillus spp.* in simulated gastric conditions. *The Journal of Microbiological Methods, 56*, 27–35.
12. Chavarri, M., Maranon, I., & Carmen, M., (2012). Encapsulation technology to protect probiotic bacteria. In: Everlon, C. R., (ed.), *Probiotics*. Intech Open; online: doi: 10.5772/50046; https://www.intechopen.com/books/probiotics/encapsulation-technology-to-protect-probiotic-bacteria (accessed on 16 January 2021).
13. Conrad, P. B., Miller, D. P., Cielenski, P. R., & De Pablo, J. J., (2000). Stabilization and preservation of *Lactobacillus acidophilus* in saccharide matrices. *Cryobiology, 41*, 17–24.
14. Cui, J., Goh, J., Kim, P., Choi, S., & Lee, B., (2000). Survival and Stability of *Bifidobacteria* loaded in alginate poly-L-lysine microparticles. *International Journal of Pharmaceutics, 210*, 51–59.

15. Del, V. E. M. M., (2004). Cyclodextrins and their uses: A review. *Process Biochemistry, 39*(9), 1033–1046.

16. Desmond, C., Ross, R. P., O'Callaghan, E., Fitzgerald, G., & Stanton, C., (2002). Improved survival of *Lactobacillus paracasei* NFBC 338 in spray-dried powders containing gum acacia. *Journal of Applied Microbiology, 93*, 1003–1011.

17. Dias, C. O., & Dos, S. O. D. A. J., (2018). Development and physicochemical characterization of microencapsulated *Bifidobacteria* in passion fruit juice: A functional non-dairy product for probiotic delivery. *Food Bioscience, 24*, 26–36.

18. Ding, W. K., & Shah, N. P., (2008). Survival of free and microencapsulated probiotic bacteria in orange and apple juices. *International Journal of Food Science and Technology, 15*, 219–232.

19. FAO/WHO, (2001). *Health and Nutritional Properties of Probiotics in Food Including Powder Milk with Live Lactic Acid Bacteria* (p. 89). Cordoba, Argentina.

20. Freitas, S., Merkle, H. P., & Gander, B., (2005). Microencapsulation by solvent extraction/evaporation: Reviewing the state of the art of microsphere preparation process technology. *Journal of Controlled Release, 102*(2), 313–332.

21. Gandomi, H., Abbaszadeh, S., Misaghi, A., Bokaie, S., & Noori, N., (2016). Effect of chitosan-alginate encapsulation with inulin on survival of *Lactobacillus rhamnosus GG* during apple juice storage and under simulated gastrointestinal conditions. *LWT-Food Science and Technology, 69*, 365–371.

22. Gibson, G. R., Rouzaud, G., Brostoff, J., & Rayment, N., (2005). An evaluation of probiotic effects in the human gut: Microbial aspects. In: *Final Technical Report for FSA Project G01022* (p. 22).

23. Gouin, S., (2004) Microencapsulation: industrial appraisal of existing technologies and trends. *Trends in Food Science and Technology, 15*, 330–347.

24. Hou, R. C. W., Lin, M. Y., Wang, M. M. C., & Tzen, J. T. C., (2003). Increase of viability of entrapped cells of *Lactobacillus delbrueckii ssp. bulgaricus* in artificial sesame oil emulsions. *Journal of Dairy Science, 86*(2), 424–428.

25. Cyclodextrin, (2019). https://en.wikipedia.org/wiki/Cyclodextrin (accessed on 16 January 2021).

26. Kailasapathy, K., (2006). Survival of free and encapsulated probiotic bacteria and their effect on the sensory properties of yoghurt. *LWT-Food Science and Technology, 39*(10), 1221–1227.

27. Kim, S. J., Cho, S. Y., Kim, S. H., Song, O. J., Shin, I. S., Cha, D. S., & Park, H. J., (2008). Effect of microencapsulation on viability and other characteristics in *Lactobacillus acidophilus ATCC 43121. LWT-Food Science and Technology, 41*, 493–500.

28. Krasaekoopt, W., Bhandari, B., & Deeth, H., (2003). Evaluation of encapsulation techniques of probiotics for yoghurt. *International Dairy Journal, 13*(1), 3–13.

29. Krasaekoopt, W., Bhandari, B., & Deeth, H., (2004). The influence of coating materials on some properties of alginate beads and survivability of microencapsulated probiotic bacteria. *International Dairy Journal, 14*, 737–743.

30. Krasaekoopt, W., & Kitsawad, K., (2010). Sensory characteristics and consumer acceptance of fruit juice containing probiotics beads in Thailand. *AU Journal of Technology, 14*, 33–38.

31. Lee, J. S., Cha, D. S., & Park, H. J., (2004). Survival of freeze-dried *Lactobacillus bulgaricus KFRI 673* in chitosan-coated calcium alginate microparticles. *Journal of Agricultural and Food Chemistry, 52*(24), 7300–7305.

32. Lian, W. C., Hsiao, H. C., & Chou, C. C., (2003). Viability of microencapsulated *Bifidobacteria* in simulated gastric juice and bile solution. *International Journal of Food Microbiology, 86,* 293–301.
33. Mantzourani, I., Nouska, C., & Terpou, A., (2018). Production of a novel functional fruit beverage consisting of cornelian cherry juice and probiotic bacteria. *Antioxidants, 7*(11), 163–169.
34. Mille, Y., Obert, J. P., Beney, L., & Gervais, P., (2004). New drying process for lactic bacteria based on their dehydration behavior in liquid medium. *Biotechnology and Bioengineering, 88,* 71–76.
35. Moayyedi, M., Eskandari, M. H., Rad, A. H. E., Ziaee, E., Khodaparast, M. H. H., & Golmakani, M. T., (2018). Effect of drying methods (electrospraying, freeze-drying, and spray drying) on survival and viability of microencapsulated *Lactobacillus rhamnosus* ATCC 7469. *Journal of Functional Foods, 40,* 391–399.
36. Mokhtari, S., Jafari, S. M., & Khomeiri, M., (2018). Survival of encapsulated probiotics in pasteurized grape juice and evaluation of their properties during storage. *Food Science and Technology International, 25*(2), 120–123.
37. Morgensen, G., Salminen, S., O'Brien, J., & Ouwehand, A. C., (2002). Inventory of microorganisms with a documented history of use in food. *Bulletin of the International Dairy Federation, 377,* 10–18.
38. Muthukumarasamy, P., Allan-Wojtas, P., & Holley, R. A., (2006). Stability of *Lactobacillus reuteri* in different types of microcapsules. *Journal of Food Science and Technology, 71,* 20–24.
39. Nag, A., & Das, S., (2013). Improving ambient temperature stability of probiotics with stress adaptation and fluidized bed drying. *Journal of Functional Foods, 5,* 170–177.
40. Nguyen, B. T., Bujna, E., Fekete, N., & Tran, A. T. M., (2019). Probiotic beverage from pineapple juice fermented with *Lactobacillus* and *Bifidobacterium* strains. *Frontiers in Nutrition, 6*(54), 1–7.
41. Nualkaekul, S., Deepika, G., & Charalampopoulos, D., (2012). Survival of freeze-dried *Lactobacillus plantarum* in instant fruit powders and reconstituted fruit juices. *Food Research International, 48*(2), 627–633.
42. Nualkaekul, S., Salmeron, I., & Charalampopoulos, D., (2001). Investigation of the factors influencing the survival of *Bifidobacterium longum* in model acidic solutions and fruit juices. *Food Chemistry, 129,* 1037–1044.
43. O'Riordan, K., Andrews, D., Buckle, K., & Conway, P., (2001). Evaluation of microencapsulation of a *Bifidobacterium* strain with starch as an approach to prolonging viability during storage. *Journal of Applied Microbiology, 91,* 1059–1066.
44. Oliveira, A. C., Moretti, T. S., Boschini, C., Baliero, J. C., Freitas, O., & Favaro-Trindade, C. S., (2007). Stability of microencapsulated *B. lactis BI 01)* and *L. acidophilus (LAC 4)* by complex coacervation followed by spray drying. *Journal of Microencapsulation, 24*(7), 685–693.
45. Ouwehand, A. C., & Salminen, S. J., (1998). The health effects of cultured milk products with viable and non-viable bacteria. *International Dairy Journal, 8,* 749–758.
46. Panghal, A., Janghub, S., Virkara, K., Gata, Y., Kumara, V., & Chhikaraa, N., (2018). Potential non-dairy probiotic products: A healthy approach. *Food Bioscience, 21,* 80–89.
47. Pereira, A. L. F., Maciel, T. C., & Rodrigues, S., (2011). Probiotic beverage from cashew apple juice fermented with *Lactobacillus casei*. *Food Research International, 44*(5), 1276–1283.

48. Pikal, M. J., (1990). Freeze-drying of protein. Part-I: Process design. *BioPharma, 3,* 18–28.
49. Pimentel-González, D. J., & Campos-Montiel, R. G., (2009). Encapsulation of *Lactobacillus rhamnosus* in double emulsions formulated with sweet whey as emulsifier and survival in simulated gastrointestinal conditions. *Food Research International, 42*(2), 292–297.
50. Plaza-Díaz, J., Ruiz-Ojeda, F., Gil-Campos, M., & Gil, A., (2018). Immune-mediated mechanisms of action of probiotics and synbiotics in treating pediatric intestinal diseases. *Nutrients, 10*(1), 42–49.
51. Poncelet, D., (2006). Microencapsulation: Fundamentals, methods and applications. In: Blitz, J. P., & Gun'Ko, V. M., (eds.), *Surface Chemistry in Biomedical and Environmental Science* (pp. 23–34). New York, USA: Springer.
52. Prakash, K. S., Bashir, K., & Mishra, V., (2017). Development of synbiotic litchi juice drink and its physiochemical, viability and sensory analysis. *Journal of Food Processing and Technology, 8*(12), 1000708.
53. Ranadheera, R. D. C. S., Baines, S. K., & Adams, M. C., (2010). Importance of food in probiotic efficacy. *Food Research International, 43,* 1–7.
54. Rossle, C., Auty, M. A. E., Brunton, N., Gormley, R. T., & Butler, F., (2010). Evaluation of fresh-cut apple slices enriched with probiotic bacteria. *Innovative Food Science and Emerging Technologies, 11*(1), 203–209.
55. Salminen, S., Ouwehand, A., Benno, Y., & Lee, Y. K., (1999). Probiotics: How should they be defined? *Trends in Food Science and Technology, 10,* 107–110.
56. Santivarangkna, C., Kulozik, U., & Foerst, P., (2007). Alternative drying processes for the industrial preservation of lactic acid starter cultures. *Biotechnology Progress, 23*(2), 302–315.
57. Shah, N., & Prajapati, J. B., (2013). Effect of carbon dioxide on sensory attributes, Physico-chemical parameters, and viability of probiotic *L. helveticus MTCC 5463* in fermented milk. *Journal of Food Science and Technology, 51*(12), 3886–3893.
58. Sheu, T. Y., & Marshall, R. T., (1993). Micro entrapment of *Lactobacilli* in calcium alginate gels. Journal *of Food Science, 58,* 557–561.
59. Shima, M., Morita, Y., Yamashita, M., & Adachi, S., (2006). Adachi protection of *Lactobacillus acidophilus* from the low pH of a model gastric juice by incorporation in a W/O/W emulsion. *Food Hydrocolloids, 20*(8), 1164–1169.
60. Sohail, A., Turner, M. S., Prabawati, E. K., Coombes, A. G. A., & Bhandari, B., (2012). Evaluation of *Lactobacillus rhamnosus GG* and *Lactobacillus acidophilus NCFM* encapsulated using a novel impinging aerosol method in fruit juice products. *International Journal of Food Microbiology, 157,* 162–166.
61. Sultana, K., Godward, G., Reynolds, N., Arumugaswamy, R., Peiris, P., & Kailasapathy, K., (2000). Encapsulation of probiotic bacteria with alginate-starch and evaluation of survival in simulated gastrointestinal conditions and in yogurt. *The International Journal of Food Microbiology, 62,* 47–55.
62. Tripathi, M. K., & Giri, S. K., (2014). Probiotic functional foods: Survival of probiotics during processing and storage. *Journal of Functional Foods, 9,* 225–241.
63. Tsen, J. H., Lin, Y. P., & King, V. A. E., (2004). Fermentation of banana media by using κ-carrageenan immobilized *Lactobacillus acidophilus. International Journal of Food Microbiology, 91,* 215–220.

64. Wang, J., Korber, D. R., Low, N. H., & Nickerson, M. T., (2014). Entrapment, survival, and release of *Bifidobacterium adolescentis* within chickpea protein-based microcapsules. *Food Research International, 55*, 20–27.

65. Yadav, V., Sharma, A., & Singh, S. K., (2015). Microencapsulation techniques applicable to food flavors research and development: Comprehensive review. *International Journal of Food and Nutritional Sciences, 4*(3), 119–124.

66. Yeung, T. W., Arroyo-Maya, I. J., McClements, D. J., & Sela, D. A., (2016). Microencapsulation of probiotics in hydrogel particles: Enhancing *Lactococcus lactis subsp. cremoris LM0230* viability using calcium alginate beads. *Food and Function, 7*(4), 1797–1804.

67. Zhang, M., Cai, D., Song, Q., Wang, Y., Sun, H., Piao, C., Yu, H., Liu, J., & Wang, Y., (2019). Effect on viability of microencapsulated *Lactobacillus rhamnosus* with the whey protein-pullulan gels in simulated gastrointestinal conditions and properties of gels. *Food Science of Animal Resources, 39*(3), 459–473.

68. Zhao, M., Wang, Y., Huang, X., Gaenzle, M., & Wu, Z., (2018). Ambient storage of microencapsulated *Lactobacillus plantarum ST-III* by complex coacervation of type-A gelatin and gum Arabic. *Food and Function, 9*(2), 1000–1008.

PART II

Prospects of Innovative
Food Preservation Strategies

CHAPTER 6

DENSE PHASE CARBON DIOXIDE (DPCD)-AIDED PRESERVATION OF FOODS

LAVANYA DEVRAJ, MOHAN G. NAIK, NIKITHA MODUPALLI,
SUKA THANGARAJU, and VENKATACHALAPATHY NATARAJAN

ABSTRACT

Dense phase carbon dioxide (DPCD) is a non-thermal food preservation technique for highly heat-sensitive, perishable food commodities like fruit juices, beverages, and other liquid foods. It is a continuous process, which utilizes pressurized CO_2 (pressure $1/10^{th}$ to that of the HHP) with a mild temperature of 30–50°C in food processing. This technology can be used in combination with other non-thermal techniques like high hydrostatic pressure (HHP), ultrasound, irradiation, and pulsed electric field (PEF). This chapter focuses on the food processing and preservation aspects of DPCD along with its advantages and disadvantages.

6.1 INTRODUCTION

The quality of food products by nullifying the activity of vegetative cells of microorganisms, yeast, molds, and enzymes can be improved through food preservation and processing using different preservation techniques [40], such as: (1) preservation by using chemicals (Class-1 preservatives: salts, sugar, oil, spices, etc., and Class-II preservatives: benzoic acid, sorbic acid, acetic acid (AA), etc.), (2) thermal processing (e.g., pasteurization, sterilization, ultra-high temperature, etc.), and (3) non-thermal processing (high-pressure processing, pulsed electric field (PEF), ultrasound technique, etc.).

Dense phase carbon dioxide (DPCD) is an emerging non-thermal method of food preservation that utilizes high pressure or pressurized carbon dioxide (CO_2: solid, liquid, and gas). This method of preservation technique has been employed not only in food processing industries to minimize the nutrient deterioration in solid foods, but also in pharmaceutical processing sectors [3]. DPCD can also be called cold pasteurization, which is mainly used to treat heat-sensitive food products, primarily liquid foods. The DPCD process involves subjecting the food products to a pressure range of 7.0 to 40 MPa at 20–60°C in a pressurized, sub- or supercritical carbon dioxide in a batch (constant), semi-batch, or continuous method for a specified period [28]. The treatment time in DPCD generally ranges from 3–9 min for constant processing and 120–140 min for semi-continuous processing.

DPCD can help to preserve the nutritional attributes of liquid foods, such as juices, squashes, and other liquid-based beverages that are sensitive to heat without giving up the organoleptic and nutritional properties [67]. The research has also affirmed positive results of DPCD against microbial activity in liquid foods. The DPCD can disrupt the microbial cells through decompression of carbon dioxide gas at 500 psi (or 3.45 MPa).

The effectiveness of DPCD against inactivating the enzymatic and microbial activity can be done by using pressurized carbon dioxide alone or in combination with other gases, e.g., nitrogen. Since carbon dioxide is an inexpensive gas with a bactericidal effect, it can be widely accommodated in wider applications in food processing industries.

There are certain advantages of DPCD compared to HHP: it is performed at 30–60°C and <30 MPa compared to HHP at 300–1200 MPa. The pressure needed during DPCD food processing is $1/10^{th}$ of the pressure utilized in the HHP method of food processing. The other significant advantages are: it is a continuous operation; it helps in the destruction of microorganisms and inactivation of enzymatic activity, and the vitamin C losses can be reduced by dissolving the CO_2 at low pH. Automatic carbonation is also possible, if necessary, through this method [3]

This chapter elaborates on the mechanism of enzyme and microbial inactivation by DPCD technique, along with its mode of action and effects on various food products. The combination of DPCD with other non-thermal preservation techniques is also discussed, elaborating on exploitation of DPCD as a highly effective residue-free, non-thermal preservation method for shelf-life extension of food products.

6.2 MECHANISM OF DPCD METHODS IN FOOD PROCESSING AND MICROBIAL INACTIVATION

DPCD utilizes anti-microbial characteristics of carbon dioxide at ambient or mild temperatures under pressure <50 MPa to extend shelf stability of sensitive foods. At elevated pressure, CO_2 can exist in gaseous, liquid (<31.1°C) or supercritical state (<31.1°C and <7.38 MPa), exhibiting physical properties characteristics of each state. At a supercritical stage, CO_2 has higher density, lower viscosity, and very low surface tension, which helps in impregnation of complex structures. Liquid and gaseous states of CO_2 come under subcritical conditions. The density of gaseous CO_2 is directly proportional to the pressure and indirectly related to temperature, according to the Ideal Gas Law. However, CO_2 density remains low in gaseous state than in the other two states [5].

In DPCD processing, the food material is treated using super- or subcritical CO_2 for a set of temperature, pressure, and time in continuous, batch or semi-batch systems. In any DPCD system, solubilization of CO_2 can be increased by using agitators and turbulent flow of liquid and pressurized gas. In stagnant systems, DPCD is given after the product has reached to a desirable temperature. Whereas in continuous operation systems, the product is initially saturated with the micro bubbling of CO_2 under high pressure, after which it is passed through a heater [61]. In semi-continuous processing, the fluid is made to flow continuously through the pressure vessel by micro-bubbling CO_2 into the product, to increase the concentration of gas pressure [66].

6.2.1 INACTIVATION OF VEGETATIVE MICROORGANISMS

The combination of time, temperature, pressure, and CO_2 concentration along with other factors (such as purity of CO_2, drop in pH, etc.), will determine the microbial inactivation impact. Researchers have shown a 7-log reduction in *Saccharomyces cerevisiae* at 40°C and 4 MPa pressure after three hours of exposure, while similar conditions did not affect the yeast [23]. Similarly, *Listeria monocytogenes* was completely incapacitated by CO_2 treatment at 35°C and 6.2 MPa pressure for 2 hours [73].

Due to the presence of peptidoglycans in the cell wall, Gram-positive bacteria are more resilient to mechanical stress than their counterparts are. This makes Gram-negative bacteria to be more intolerant to DPSD treatment than Gram-positive bacteria, though both types are susceptible to inactivation at high-pressure of CO_2 [24].

It is a fact that CO_2 can exert inhibitory effects on bacterial growth [70]. However, the actual mechanism of action for the bactericidal effect of CO_2 has not been elaborated till today. Several previous studies have reviewed some theories to explain the inhibition of microorganisms by DPCD [18, 32, 38, 45, 63]. According to Garcia-Gonzalez et al. [32], the steps involved in bacterial inactivation mechanism by DPCD are given below. These steps may occur either in singular or consecutively, but are simultaneous and interrelated in occurrences:

- Alteration of cell membranes due to pressure;
- Breakage and mechanical damage of cell wall;
- Damage to cellular organelles;
- Dissolution of DPCD in the liquid matrix of food components;
- Effect of molecular CO_2 and bicarbonate ions indirect inhibition of metabolism;
- Enzyme denaturation;
- Imbalance in electrolyte balances of viable cells;
- pH reducing activity.

The cell membrane of microorganisms mainly contains phospholipids, which are more likely to be affected by CO_2 due to its lipophilic properties [39]. Due to this, the cell membrane permeability is elevated, which lowers the extracellular pH. This causes the pressurized CO_2 gas to enter into the cell and to accumulate in the cell interior [36]. This accumulation can structurally and functionally disrupt the cell membrane permeability and can increase its fluidity. Simultaneously, both CO_2 and HCO_3^- ions cause the lowering of pH due to the chemical interactions inside the cell. The lowering of intercellular pH results in loss of cell viability, by causing denaturation of protein and enzyme and retarding the metabolic processes of cells [38, 47, 64]. Studies on the flow cytometric analysis revealed that the alteration of efflux pump occurs post-DPCD treatment due to enzymatic functions and pH changes, leading to microbial inactivation [42].

During DPCD, the CO_2 initially penetrates cells and accumulates due to its high solubility and low density. After building up to a critical level, it reacts with phospholipids and other hydrophobic compounds, thus creating disturbances in the balance of the cell biological system. The sudden depressurization leads to rapid transfer of cell constituents into the extracellular environment, leading to stimulation of cell inactivation. In addition, CO_2 at elevated pressure can modify the cell membrane by enlarging the periplasmic spaces causing leakage of cell constituents [37, 64]. One optional mechanism for microbial inactivation by DPCD is a mechanical distraction of cells caused by elevated pressure [37, 73]. Studies have demonstrated that few cells have completely been disrupted during DPCD, while some cells showed wrinkles and holes alone on the surface [23]. Research studies have shown that the cell inactivation due to DPCD is irrespective of cell rupture [17], as some cells may be completely inactivated but may remain intact or slightly deformed [37].

It has been validated that DPCD in combination with heat treatment was effective in inactivating the spores in foods. Previous studies on *Bacillus subtilis* spores under CO_2 at 5 MPa and 80°C have shown inactivation of spores effectively. It was suggested: (1) first step: the pressurized CO_2 with assistance from heat activation would penetrate into the cells; (2) second step: inactivation of previously penetrated cells, as heat activation tends to make the cells more susceptible to CO_2-assisted inactivation [6].

Another alternative theory states that DPCD results in the germination of bacterial spores at relatively low temperature due to elevated pressure. The germination of spores can result in increased sensitivity of bacteria, leading to inactivation [29]. When compared to HHP processing and heat treatment, DPCD has a higher killing effect at a comparatively low temperature, owing to its raised pressure [72].

6.2.2 INACTIVATION OF ENZYMES

DPCD treatment is effective in enzyme inactivation, especially in fruits and vegetables, where enzymes are the main reasons for quality deterioration. The enzyme inactivation is based on the effect of lowering the pH of extracellular solutions as described in this chapter. Nevertheless, the rate of enzymatic degradation does not entirely depend on pH, but is also affected by the intracellular concentration of cofactors and substrates and products that act as key elements for controlling the enzymatic activity [41]. The lowering of pH also disturbs the ionic equilibrium and volume of the cells [30]. The drop in pH causes the conversion of HCO_3^- to CO_3^{2-} ions, which can form complexes by binding to inorganic electrolytes such as calcium and magnesium. These complexes can alter several regulatory activities of the cells and can impair the protein function [36].

It has been demonstrated that CO_2 at high pressure can cause spatial and conformational alterations in secondary and tertiary protein structures [60]. DPCD can cause decomposition of α-helix structures, whereas changes in secondary structures can only cause a reduction in the activity [15, 68]. The spatial changes in protein due to DPCD can be either reversible or irreversible depending on the pressure applied. At pressure < 310 MPa, the change is reversible upon depressurization, whereas below this, the alterations are permanent and cannot be reversed [15, 35]. CO_2 in the supercritical state has been reported to alter the isoelectric profile and structure of polyphenol oxidase (PPO) protein, but no changes were reported at the atmospheric pressure [15].

6.3 APPLICATIONS OF DPCD IN FOOD PROCESSING

Studies have affirmed the bacteriostatic and inhibitory effect of CO_2 on growth and microbial multiplication in food products. *Pseudomonas* was very sensitive to CO_2 gases compared to other microorganisms like *Lactobacillus* and *Clostridium*. Other researchers conducted the sterilizing effect of CO_2 on *Escherichia coli* (both wet and dry), *Staphylococcus aureus,* and conidia of *Aspergillus niger* with the help of supercritical fluid extraction (SFE) method.

Application of DPCD in the preservation of liquid food products has been well documented, but still some research is needed in the preservation of solid foods. The difficulties in preserving the solid food product using DPCD is the slow diffusion of CO_2 in the bulk of solid foods leading to cellular damage with undesirable surface texture, which causes severe quality-related issues, limited CO_2 solubility due to decreased levels of free water on the surface of the food in connection with bacterial and enzymatic inactivation.

6.3.1 APPLICATION OF DPCD IN PRESERVATION OF LIQUID FOODS

CO_2 gas can easily be dissolved in liquids rather than in solids. The saturated solution formed after dissolving CO_2 is the analytical composition that can be expressed as an amount of pronounced solute in a solvent. Use of CO_2 in food processing is considered as "Generally regarded as safe (GRAS)" by USFDA (United Stated Food and Drug Administration).

Dense phase CO_2 in its collective form acts as an anti-microbial agent with enzyme inactivation properties at elevated pressures. The major attributes in liquid foods are physical (Color, viscosity, Brix, etc.), and chemical (pH, water activity, volatile components) properties. Elevated temperature causes change in organoleptic properties of liquid food. Table 6.1 is the compilation of different fruit beverages treated with CO_2 in response to different treatment conditions, initial microbial population, target organisms, and enzyme inactivation.

The foremost utilization of DPCD in the processing of liquid foods is microbial inactivation and apoptosis. Other uses are for enzyme inactivation, improvement of physical quality of foods and other chemical attributes. Park et al. [54] studied that treatment of DPCD on carrot juice resulted in a 60% cloud loss. Conflicting results were reported by Arreola et al. [1], Balaban et al. [4], and Yagiz et al. [75]. It was reported that the effect of DPCD had retained cloud stability in orange juice after refrigerated storage. Arreola et al. [1] observed that pectinesterase enzyme was restored during storage and the stability of the cloud is not due to the enzyme activity.

Gui et al. [33] established that there was a significant drop in browning of cloudy apple juice when treated samples were stored at 4°C compared to the untreated samples, due to the inactivation of PPO enzyme. Park et al. [54] found that carrot juice samples treated with DPCD in combination with HPP showed no viable aerobic bacterial count after a shelf life of 4 weeks at 4C refrigerated storage, while control samples showed 8.4 log cfu/ml.

Kincal et al. [43] found that DPCD without heat pasteurization at different pressures inactivated *S. typhimurium* effectively showing no viable count till 14 days of storage. Del Pozo-Insfran et al. [57] found comparable microbial viability among heat-treated and DPCD treated samples of muscadine grape juices at refrigerated storage for five weeks. They found that the yeast count in DPCD samples consistently was increased with storage time, whereas heat pasteurized samples indicated no change in the count of total aerobic microbes.

Previous research studies have also shown that DPCD treated samples did not significantly differ from non-treated samples. Several researchers evaluated the loss of sensory and nutritional properties to assess consumer perspectives using ranking tests. Balaban et al. [4], Lecky, and Balaban [44], Arreola et al. [1], Damar, and Balaban [18] indicated that DPCD treated samples showed no significant variance in flavor, aroma, and overall consumer acceptability compared with control samples. In a few studies, DPCD treated samples were rated higher than the heat pasteurized samples [18, 57].

TABLE 6.1 Use of DPCD in Microbial Inactivation in Different Beverages

Fruit Juice	Microorganisms/ Enzymes Responsible for Spoilage	Quality Parameters Before Processing	DPCD Processing Pressure	Processing Temperature and Time	Output After DPCD Processing	Shelf-Life	References
Apple juice	Enzymatic browning reaction	Browning index	30 MPa	55°C for 60 min	Degree of browning reaction was significantly reduced	—	[51]
Apple slice	Enzymatic browning Polyphenol oxidase (PPO) and pectin methylesterase (PME)	Color and textural changes	20 MPa with mild heat	25–65°C for 20 min	Complete inactivation of PPO and PME.	—	[25]
Apple juice	Alicyclobacillus acidoterrestris spores	Physicochemical properties like color, sensory, pH, and Brix	80, 100 and 120 bar for 10–40 min	65–70°C	• Surface and internal morphological changes • Complete reduction of *A. acidoterrestris* spores after DPCD	—	[2]
Apple puree	Microbial spores	pH, °Brix, color, microbial load	16.0 MPa for 40–60 min	Thermal process (35, 50, 65 and 85C) for 10–140 min	• 5-log reduction destruction in microbial spores • No significant difference in °Brix • Significant changes in pH, color	—	[33]
Coconut water (*Cocus nucifera*)	Decomposition of aroma components	Likeability and flavor	34.5 MPa, 13% CO_2	25°C for 6 min	Better retention of aroma, taste, and overall flavor compared to heat treatment	—	[19]

TABLE 6.1   *(Continued)*

Fruit Juice	Microorganisms/ Enzymes Responsible for Spoilage	Quality Parameters Before Processing	DPCD Processing Pressure	Processing Temperature and Time	Output After DPCD Processing	Shelf-Life	References
Grapefruit juice (V. Red blush) (*Citrus paradisi*)	Yeast and mold	pH, °Brix, TA, Phenolic content	13.8, 24.1 and 34.5 MPa, 5.7% CO_2	40°C for 5, 7 and 9 min	• 5-log reduction in total aerobic MO's including yeast and mold at 34.5 MPa for 7 min. • Increased cloudiness (91%) • Inactivation of pME (69.17%) • No noteworthy difference in Brix, pH, and TA • Phenolic content: unaffected • Slight difference in Ascorbic acid	6 weeks at the refrigerated condition	[26]
Guava puree (*Psidium guajava L.*)	Phytochemicals, Ascorbic acid, pH, °Brix, % TA, sensory and nutritional attributes	Undesirable changes in phytochemical composition, measurement, and comparison of chemical compounds, microbial reduction	34.5 MPa, 8% CO_2	35°C for 6.9 min	• No changes in pH, °Brix • Increase in TA and viscosity • Partial inactivation of PE • Protects polyphenolic and antioxidant levels • Delays vitamin C degradation which is highly heat sensitive	14 weeks at 4°C	[60]

TABLE 6.1 *(Continued)*

Fruit Juice	Microorganisms/ Enzymes Responsible for Spoilage	Quality Parameters Before Processing	DPCD Processing Pressure	Processing Temperature and Time	Output After DPCD Processing	Shelf-Life	References
Mandarin juice (*Citrus reticulate Blanco*) (Murcott cultivar)	Total aerobic count, Pectinesterase activity, cloud, °Brix, pH, and titratable acidity	Temperature, pressure, and residence time, pH, °Brix, color change, titratable acidity (TA)	41.1 MPa, 9 min and 7% CO_2	35°C	• Retention of cloud • Increase in yellowness and lightness • Reduction in redness • Increase in cloud • Brix, pH, and TA were the same before and after DPCD processing	–	[46]
Orange juice (OJ) (*citrus sinensis*)	Pectin methylesterase enzyme	Particle size, consistency coefficient and a* values Volatile components (ethyl butyrate, trans-hexanol, α-pinene, phellandrene, limonene, linalool, nonanal, and citronellol	40 MPa	55°C for 10–60 min 10–60 min of DPCD treatment	Reduction in particle size increased cloud stability due to homogenization during DPCD treatment • Linear decrease on ethyl butyrate and trans-2-hexanol • Increase in nonanal and citronellol compared to untreated OJ	–	[50]

TABLE 6.1 *(Continued)*

Fruit Juice	Microorganisms/ Enzymes Responsible for Spoilage	Quality Parameters Before Processing	DPCD Processing Pressure	Processing Temperature and Time	Output After DPCD Processing	Shelf-Life	References
Valencia orange juice (VOJ)	Pectin methylesterase enzyme (PME) Causes loss of cloud	Color, pH, Brix cloud stability, ascorbic acid content, and sensory properties	7–34 MPa	35–40°C for 15–18 min	• No significant difference in pH, Brix • Better retention of Vitamin C compared to control samples.	66 days under refrigeration condition	[1]
Watermelon juice (*Citrullus lanatus*)	Aerobic Microorganisms	pH, °Brix, TA, lycopene, and color	10.3, 20.6 and 34.4 MPa, CO_2: 5, 10 and 15%	30–40°C for 6 to 46 min	• 6-log reduction in native MO's at 34.4 MPa, 40°C, 10% CO_2 level for 5 minutes • 4.5 log reduction of MO's in acidified, sweetened, and carbonated products	Up to 8 weeks under refrigerated conditions	[44]

Arreola et al. [1] reported that ascorbic acid (AA) retention was elevated significantly in DPCD treated orange samples. Del-Pozo Insfran et al. [57] observed that DPCD muscadine grape juice showed no substantial changes in total anthocyanins, total phenolics, and antioxidant capacity compared to 16%, 26%, and 10% decrease in heat-treated samples, respectively. It was also stated that anthocyanin content and antioxidant capacity of DPCD samples were better than heat-treated samples even after ten weeks of storage at 4°C. Chen et al. [14] evaluated that DPCD treated melon juice retained a significant amount of volatile compounds, carotenoids, and ascorbic acid, whereas heat-treated samples showed significant losses and produced an unacceptable cooked off-flavor.

6.3.2 OTHER PROCESSED BEVERAGES THROUGH DPCD

Other than fruit-based beverages, liquid foods like beer, milk, and milk products have also been of interest for several researchers in recent times. Folkes [27] has reported that DPCD-aided pasteurization of beer samples have shown a 5-log or more reduction in yeast population. Though there was some stripping of flavor and aroma, the sensory panelists could not detect excessive changes in freshly treated and 30-day stored samples, and the aroma changes were not significant to be unacceptable by the consumers.

Similarly, Ceni et al. [12] studied the consequences of CO_2 on an endogenous milk enzyme alkaline phosphatase. The research reported that about 98.2% of alkaline phosphatase inactivation rate was observed for 0.45% of CO_2 to milk ratio, continuously treated at 8 MPa pressure and 70°C for 30 min. CO_2 treatment can also be used for the extraction of specific compounds from food. This principle was used in the extraction of cholesterol from whole milk powder by treating at 20.7 MPa and 68°C for 40 min with CO_2 flow of 6 L/min, resulting in the removal of 55.8% cholesterol with unaltered fatty acid profile, lightness, and solubility index of the sample [16]. Similarly, several other non-fruit beverages like fermented drinks and dairy products have positively benefitted by DPCD treatment (Table 6.2).

TABLE 6.2 Effects of DPCD on Non-Fruit Beverages

Beverage	Causes of Spoilage	Quality Parameters Before Processing	DPCD Pressure	Exposure Time and Temperature	Output After DPCD Processing	References
Beer	Microorganisms, yeast and molds, which causes off flavor	Physical and sensory attributes	27.6 MPa	2°C, 5% CO_2	• Better retention of aroma and flavor • Retention of freshness after storage period of 1 month at 1.7°C	[17]
Jamaica beverage (*Hibiscus sabdariffa*)	Yeasts and molds, which alters the overall acceptability of the product during storage	Anthocyanins and phenolic compounds	30.8 to 31.0 MPa	6.8 min	• Retention of overall quality of the product • Lesser loss of anthocyanin pigment • No significant changes in phenolics and antioxidants during storage interval	[58]
Kava kava (*Piper methysticum*) Traditional Polynesian beverage	Bacteria (causes alteration the organoleptic properties)	Active compounds kavalactones, δ-lactones, and 5, 6 dihydro δ-lactones	34.5 MPa	13% CO_2	• Change in initial pH from 6.3 to 5.5 • Change in mouthfeel • log microbial reduction from 4.0×10^5 to 1.0×10^2	[60]
Raw milk	Alkaline phosphatase enzyme activity, causes putridity and spoilage	Inactivation of alkaline phosphatase enzyme	8 MPa, 0.45% (w/w) concentration of CO_2	70°C for 30 min	• 98.2% inactivation rate of alkaline phosphatase	[12]
Raw skimmed milk	*Pseudomonas fluorescence* that affects the storage stability and causes spoilage	Milk fat content and freshness	20.7 MPa	35°C for 10 min	• 5.02 log reduction and bacterial inactivation	[74]

6.4   EMERGING TRENDS IN APPLICATION OF DPCD IN THE PRESERVATION OF SOLID FOODS

In liquid foods, the effectiveness of DPCD treatment greatly depends on the mode of operation, as pressure and mode of CO_2 supply will decide the level of saturation and microbial inactivation directly. In liquids, the saturation can be controlled by using atomizers or micro-bubbling systems to disperse the CO_2. This can reduce the desired exposure time significantly, thus accelerating the inactivation kinetics [71]. However, in solid foods, the CO_2 is in direct contact with the substrate. This eliminates the need for elevating the pressure or saturating the medium.

The required exposure time entirely depends on the resistance offered by the microbial strains against inactivation. In this case, the temperature is the critical process parameter, and the inactivation of microorganism or enzyme depends only on the optimum time-temperature combination [61].

Compared to liquid foods, usage of DPCD in solid food materials is not an easy task. The complex structural arrangement of solid components and low moisture content in solid foods limits the solubilization capacity of CO_2, thus affecting its bacterial and enzymatic inactivation. In addition, the complex matrix in solids slows down the diffusion of CO_2 into the bulk and may lead to quality deterioration in textural and sensorial properties of foods. However, the recent trend towards non-thermal food processing strategies has increased the importance to DPCD techniques, and several studies are being done in solid food products particularly to enhance the shelf-stability without negatively impacting on the physiochemical, nutritional, and sensory properties.

In most of previous studies, solid foods were treated using DPCD by semi-continuous or batch type system, since maintenance of high pressure in a continuous system for solid foods was not feasible. However, there is a high probability of changes in textural and sensory properties of solid foods due to excess pressure application during DPCD treatment. Hence, it is suggested to treat the food products using DPCD at low-pressures to preserve the organoleptic properties.

6.4.1   FRUITS AND VEGETABLES

One of the most common applications of DPCD in solid foods is in the inhibition of molds on whole or minimally processed fruits or vegetables. Previous studies caused a 3.5 and 1.5 log reduction for *E. coli* and *S. cerevisiae* using 5 MPa pressure, respectively at room temperature for 60 min [20].

Valverde et al. [71] reported 0.5 log reduction in inactivation of *S. cerevisiae* in minimally processed pears by using 6 MPa pressure at 25°C temperature for 10 mins; and they observed that yeast survival rates were inversely proportional to the temperature rise [71]. The effect of DPCD on yeasts and molds on sliced carrots caused significant inactivation at a rate of 1.25 log cycles at 5 MPa and 20°C for 20 mins [7].

Garcia-Gonzalez et al. [31] indicated that the yeasts have lower sensitivity towards high-pressure CO_2 than the bacterial strains in foods. They also reported that the pH of foods can directly affect the resistance of microbial strains against DPCD treatments. DPCD treatment with elevated temperature facilitates an increase in diffusivity of CO_2 and fluidity of cell membrane, thus causing metabolic reactions leading to cell apoptosis [32]. DPCD treatment on fresh spinach showed a considerable decrease in the microbial count, but the treatment adversely affected the texture and color of the leaves that became undesirable [76]. Several other authors mentioned positive results for strawberries and cucumbers, but also severe tissue destruction leading to color loss, sogginess that are critically rejected by consumers.

6.4.2   MEAT AND MARINE PRODUCTS

The use of DPCD method on meat and marine foods has recently been evaluated by several researchers, due to its ability to extract the fats at low temperatures and to remove cholesterol [13]. It has also been used to find the effectiveness of microbial inactivation to persuade pasteurization. Previous studies showed that 2 log reduction was obtained after 2 hours of treatment at 14 MPa pressure and 35°C temperature in chicken strips inoculated with microorganisms [73]. Effect of DPCD on raw pork meat treated for 60 min at 25°C and 6 MPa pressure achieved 2 log reduction of spoilage organisms. However, the treatment increased the color lightness of meat, giving a cooked appearance, which is unacceptable by consumers [10].

Other studies on surimi shrimp gel treated with DPCD at elevated temperature and pressure resulted in a reduction in α-helix of myosin of protein and increase in β-sheet, which in turn raises the gel strength of the material [34]. Similar results were reported by Rao et al. [59] in meat sausages treated with DPCD at 50 to 60°C. A most recent study on oysters treated with DPCD evaluated successfully the inhibition effects of supercritical CO_2 on the bacterial pathogens trapped in the GI system [48].

The impact of DPCD in chemical residue-free and efficient preservation for improving shelf stability of meat and meat products is an innovative and beneficial technique. However, it can give a cooked appearance to meat. It can also cause oozing out of fats from meat cuts, which can result in loss of flavor and texture, thus decreasing the consumer acceptability [62].

6.4.3 SPICES AND OTHER FOODS

The effect of DPCD as a cold or warm pasteurization method for several low-moisture foods has been successfully evaluated. Use of DPCD on paprika powder for 45 min resulted in 2 log reduction at 80°C and up to 5-log reduction at 95°C, showing effective bactericidal effect [9]. A study on combined effects on DPCD with oregano essential oil on paprika powder [11] showed the dependency of microbial inactivation due to the combined treatment, on the moisture content of the sample. In the treatment of DPCD with 2.58% Oregano essential oil at 15% moisture caused 2 log reduction in total aerobic count, whereas log reduction was 2.5 cfu/ml at 35% moisture.

Studies on the effectiveness of DPCD on *Aspergillus niger* and spores of *Aspergillus ochraceus* in a non-fermented high polyphenol cocoa powder indicated that treatment with increased pressure from 13 to 30 MPa at 65°C for 40 min resulted in the same level of inactivation of the target organisms [8]. The study also indicated that use of DPCD resulted in increased efficiency in the extraction of butter from cocoa derivatives. Another study on heavily contaminated ginseng (10^{-7} cfu/g) reported that at 10 MPa pressure and 60°C for 15 hours, <3 log reduction was obtained, which was lower than the prescribed regulations. However, addition of ethanol/water/hydrogen peroxide had resulted in complete inactivation in only 6 h of exposure time [21].

Studies also explored the use of DPCD in the extraction of essential oils and other extracts from food components. Patil et al. [56] reported extraction of oil from bioalgae was increased by 31.37% (db) at 240 bar pressure for 60 min under controlled flow of CO_2 when a mixture of 1:1 hexane and ethanol co-solvent was used at 12:1 ratio with algal solid ratio. This technique also offered an increase in total lipid yield and improved lipid purity along with a reduction in extraction time than in the conventional methods (CMs) of extraction.

6.5   EFFECTS OF DPCD IN COMBINATION WITH OTHER NON-THERMAL TECHNIQUES ON FOOD PRESERVATION

6.5.1   DPCD AND ULTRASOUND

The use of DPCD in combination with ultrasound is a relatively novel concept. Ultrasound technique further helps to reduce the time needed for CO_2 to reach saturation, especially during batch system. In addition, cavitation helps in the saturation of CO_2 in the liquid form. The high-pressure region of ultrasound increases the solubility of CO_2 in the liquid by acting as the driving force for saturation. During low-pressure cycle, the decrease in pressure makes the dissolved CO_2 to supersaturate.

Morbiato et al. [49] combined supercritical carbon dioxide with elevated power ultrasound for drying of raw chicken meat to increase the shelf-life at 40°C and 100 bars with pressurization and depressurization at a rate of 4 and 10 bar/min, respectively. The study was compared with supercritical CO_2-assisted drying and hot air (HA) oven drying method. Authors found that combination technique (supercritical CO_2 + ultrasound at 10 W) increased the drying kinetics and reduction in weight by 75.1% after 300 minutes. While in oven drying at 75C the water content was almost removed after 420 minutes which resulted a weight reduction of 74.4%.

Paniagua-Martinez et al. [53] studied the effect of supercritical CO_2 in combination with high-power ultrasound on inactivation of microbiota in pineapple juice in a continuous flow system. The experiment was carried out at 31.5°C and 100 bars for two different residence times of 3.1 and 4.6 minutes. It was found that microbiota was inactivated completely with fewer changes in quality characteristics. It was concluded that supercritical CO_2 with high power ultrasound can be an excellent substitute for cold pasteurization of pineapple juice.

Timko et al. [69] studied the ultrasound development of emulsions that are free from surfactants consisting of dense carbon dioxide and water. It was observed that emulsions with limited kinetic stability were formed at 20 kHz of pulsed ultrasound (with a critical power density of above 0.05 W cm^{-3}) after 2 minutes. Ultrasound can be used as surfactant-free technology for producing emulsions consisting of dense phase CO_2 and water, which can be further utilized for applications in chemicals and materials.

6.5.2   DPCD AND HIGH HYDROSTATIC PRESSURE (HHP)

In high hydrostatic pressure (HHP), the food products are packed before processing and thus, it eliminates post-processing contamination. Though HHP is an effective tool for removing microorganisms in liquid foods, yet it increases the enzyme activity in the food, e.g., PPO. On the other hand, DPCD is the constructive technique to inactivate the enzyme. Therefore, it is reasonable to combine DPCD and HHP to benefit from their key advantages.

Ortuño et al. [52] studied the effect of combined technology of HHP and DPCD to disable PPO, peroxidase (POD) and pectin methylesterase (PME) in feijoa (*Acca sellowiana*) puree. The treatments given to the puree were HHP, carbonation, and HHP (HHP carb), carbonation, and addition of 8.5 mL CO_2 /g puree into the headspace of the package (HHPcarb + CO_2). The samples were treated at 300, 450, and 600 MPa for 5 minutes. Authors observed that HHPcarb + CO_2 boosted the inactivation of POD, PPO, and PME and they concluded that the simultaneous application of HHP and DPCD can improve the inactivation mechanism of enzymes with carbon dioxide concentration being a key process factor.

Duong et al. [22] studied microbial and sensory effects of combined HHP and DPCD techniques and it was observed that HHPcarb+CO_2 have superior microbial inactivation. The sensory properties of the HHPcarb+CO_2 treated samples did not alter appearance and color, while it affected the texture and unsweetened feijoa samples. In addition, there was no significant difference in sweetened feijoa samples.

6.5.3   DPCD AND PULSED ELECTRIC FIELD (PFE)

The pulsed electric field (PEF) uses short electric pulses to achieve microbial inactivation of food products. These short electric pulses introduce a latent variance between the inside and outside of the membrane of microorganisms, which results in permeation of the cell membranes. PEF assists the diffusion of carbon dioxide into the microbial cell by electroporation. The protein and phospholipid bilayers of the cell membrane can be weakened due to high voltage electric field. This makes the plasma membrane to become penetrable to small particles leading to swelling and rupturing of the cell membrane.

Spilimbergo et al. [65] investigated the efficiency of the combined use of PEF and DPCD on killing the Gram-negative bacteria *S. typhimurium*. The PEF was applied at two different conditions: (1) one single pulse of 1 ms length at 30 kV/cm; and (2) 12 pulses of 4 ms length at 30 kV/cm. In both treatments, authors kept the samples at 12 MPa and 22°C/35°C for different exposure time of 0 and 45 min. Authors compared the inactivation kinetics of *S. typhimurium* under DPCD + PEF and DPCD only. They observed that PEF as an assisted technique to DPCD to enhance the inactivation kinetics and it reduced the treatment time compared to DPCD alone. This combining effect is due to the electroporation, thus causing thinning of the cell wall so that intercellular components get dispersed.

Pataro et al. [55] studied the effect of PEF as pre-treatment aid for the inactivation of *S. cerevisiae* by DPCD. They used different electric field strengths (6, 9, and 12 kV/cm) and different energy inputs (10, 20, and 40 J/ml) at 8, 11, and 14 MPa for different treatment times ranging from 3 to 30 minutes. They found that the samples treated with only PEF have shown the maximum inactivation level at 12 kV/cm and 20 J/ml with log cycle of 0.35. While the combined process showed the inactivation level of 3.13 log-cycles 6 kV/cm.

6.6   POTENTIAL FUTURE TRENDS

Though DPCD technique is highly efficient, yet there are three major issues for using this technology at large scale industrial level:

1. The optimization of energy dynamics and economics of the process. Since there is no current use of DPCD in mass production, therefore the cost of equipment and operation have not been optimized till date. In addition, purifying, and converting atmospheric CO_2 in gaseous phase into pure, dense phase supercritical fluid is an expensive and tiring operation with a high level of difficulty. Owing to the advantages of DPCD, optimization of process parameters and energy dynamics can improve its efficiency and applicability on a larger scale with minimal losses. Need for extensive research study to compare DPCD with other non-thermal techniques (such as pulsed light (PL), high-pressure processing, and cold plasma) in different food components to evaluate the suitability of DPCD for each product in terms of quality, safety, and nutritional losses, if any, sensory deterioration.
2. The lack of formulation of standard regulations for treating the food products. The research studies so far have evaluated and elaborated the safety regulations and process parameters in a product

or process-specific manner. However, if the product changes, the regulation, and process parameters need to be framed again, which is a problem in large establishments. This issue can be handled by predicting and developing models for DPCD treatments to address the solubility of CO_2 in different substrates, effects of pressure, temperature, and in combination with exposure time and CO_2 levels. This offers an advantage in predicting the solubility of CO_2 in various food components (solids or liquids) at required process conditions. Thus, this method can profoundly reduce the number of experiments required to establish a standard process protocol for each food substrate in compliance with the regulations.

3. The environmental concern for the disposal of used CO_2. The CO_2 being a greenhouse gas poses a threat to the environment and can cause global warming. In addition, CO_2 in the supercritical state under pressure can be a potential hazard in the industry to the personnel handling it. It can cause suffocation and restless when inhaled in the purest form. This problem can be combated by developing a system to trap the used CO_2, purify, and regenerate it to form DPCD again and recirculated into the system to create a cycle. This results in very fewer emissions of spent CO_2 into the atmosphere and improves the efficiency and economy of the process.

6.7   SUMMARY

DPCD preservation is a novel non-thermal food preservation technique, which utilizes pressurized carbon dioxide to inactivate the microbial and enzymatic activities. This method is used for heat-sensitive food products, especially liquid foods (e.g., fruit juices) and squashes, etc., to overcome nutrient losses, such as ascorbic acid, minerals, and to nurture the organoleptic properties, which gets quickly deteriorated during the thermal processing. Apart from liquid foods, DPCD can also be used for solid foods, such as, processing of meat products, fruits, and vegetables, and other plant materials. To enhance the processing efficiency of DPCD, it can be combined with other non-thermal processing techniques, i.e., high-pressure processing, PEF, ultrasound at mild processing conditions for effective microbial reduction and enzyme inactivation.

The benefits of DPCD are: it can be operated at 30–50°C without giving up the nutrients like anthocyanin and polyphenols; requires only 1/10th of the pressure compared to HHP; and automation of carbon dioxide can be done if needed. Thus, DPCD can also be termed as "cold pasteurization." The limitations of DPCD are the cost of pure carbon dioxide gas. Further research can be done by replacing the CO_2 gas with other gases to make the process cost-effective. DPCD is not used on a commercial scale. DPCD is an effective and highly feasible technique for improving the shelf life and storage stability of liquid and solid foods, along with other applications like DPCD aided extraction of essential oils and high-economic value components from several food ingredients.

KEYWORDS

- **dense phase carbon dioxide**
- **high hydrostatic pressure**
- **inactivation**
- **pectin methylesterase enzyme**
- **polyphenol oxidase**
- **pulsed electric field**

REFERENCES

1. Arreola, A. G., Balaban, M. O., Marshall, M. R., Peplow, A. J., Wei, C. I., & Cornell, J. A., (1991). Supercritical carbon dioxide effects on some quality attributes of single strength orange juice. *Journal of Food Science, 56*(4), 1030–1033.
2. Bae, Y. Y., Lee, H. J., Kim, S. A., & Rhee, M. S., (2009). Inactivation of *Alicyclobacillus acidoterrestris* spores in apple juice by supercritical carbon dioxide. *International Journal of Food Microbiology, 36*(1), 95–100.
3. Balaban, M. O., & Duong, T., (2014). Dense phase carbon dioxide research: Current focus and directions. *Agriculture and Agricultural Science Procedia, 2*, 2–9.
4. Balaban, M. O., Kincal, D., Hill, S., Marshall, M. R., & Wildasin, R., (2001). The synergistic use of carbon dioxide and pressure in non-thermal processing of juices. *In: IFT Annual Meeting Book of Abstracts* (pp. 3–6).
5. Ballestra, P., (2012). Effects of dense phase carbon dioxide on bacterial and fungal spores. Chapter 5. In: Balaban, M. O., & Ferrentino, G., (eds.), *Dense Phase Carbon Dioxide: Food and Pharmaceutical Applications* (Vol. 2, pp. 99–112). Wiley Online Library.
6. Ballestra, P., & Cuq, J. L., (1998). Influence of pressurized carbon dioxide on the thermal inactivation of bacterial and fungal spores. *LWT-Food Science and Technology, 31*(1), 84–88.

7. Bi, X., Wu, J., Zhang, Y., Xu, Z., & Liao, X., (2011). High pressure carbon dioxide treatment for fresh-cut carrot slices. *Innovative Food Science and Emerging Technologies, 12*(3), 298–304.

8. Calvo, L., Muguerza, B., & Cienfuegos-Jovellanos, E., (2007). Microbial inactivation and butter extraction in a cocoa derivative using high-pressure CO_2. *Journal of Supercritical Fluids, 42*(1), 80–87.

9. Calvo, L., & Torres, E., (2010). Microbial inactivation of paprika using high-pressure CO_2. *Journal of Supercritical Fluids, 52*(1), 134–141.

10. Cappelletti, M., Ferrentino, G., & Spilimbergo, S., (2015). High pressure carbon dioxide on pork raw meat: Inactivation of mesophilic bacteria and effects on color properties. *Journal of Food Engineering, 156*, 55–58.

11. Casas, J., Tello, J., Gatto, F., & Calvo, L., (2016). Microbial inactivation of paprika using oregano essential oil combined with high-pressure CO_2. *Journal of Supercritical Fluids, 116*, 57–61.

12. Ceni, G., Fernandes, S. M., Valério, C., Cansian, R. L., Oliveira, J. V., Dalla, R. C., & Mazutti, M. A., (2016). Continuous inactivation of alkaline phosphatase and *Escherichia coli* in milk using compressed carbon dioxide as inactivating agent. *Journal of CO₂ Utilization, 13*, 24–28.

13. Chao, R. R., Mulvaney, S. J., Bailey, M. E., & Fernando, L. N., (2006). Supercritical CO_2 conditions affecting extraction of lipid and cholesterol from ground beef. *Journal of Food Science, 56*(1), 183–187.

14. Chen, J., Zhang, J., Feng, Z., Song, L., Wu, J., & Hu, X., (2009). Influence of thermal and dense-phase carbon dioxide pasteurization on physicochemical properties and flavor compounds in melon juice. *Journal of Agricultural and Food Chemistry, 57*(13), 5805–5808.

15. Chen, J. S., Balaban, M., Wei, C., Marshall, M. R., & Hsu, W. Y., (1992). Inactivation of polyphenol oxidase by high-pressure carbon dioxide. *Journal of Agriculture Food Chemistry*, 2345–2349.

16. Chitra, J., Deb, S., & Mishra, H. N., (2015). Selective fractionation of cholesterol from whole milk powder: Optimization of supercritical process conditions. *International Journal of Food Science and Technology, 50*(11), 2467–2474.

17. Dagan, G. F., & Balaban, M. O., (2006). Pasteurization of beer by a continuous dense-phase CO_2 system. *Journal of Food Science, 71*(3), 164–169.

18. Damar, S., & Balaban, M. O., (2006). Review of dense phase CO_2 technology: Microbial and enzyme inactivation, and effects on food quality. *Journal of Food Science, 71*(1), 1–11.

19. Damar, S., Balaban, M. O., & Sims, C. A., (2009). Continuous dense-phase CO_2 processing of a coconut water beverage. *International Journal of Food Science and Technology, 44*(4), 666–673.

20. Debs-Louka, E., Louka, N., Abraham, G., Chabot, V., & Allaf, K., (1999). Effect of compressed carbon dioxide on microbial cell viability. *Applied and Environmental Microbiology, 65*(2), 626–631.

21. Dehghani, F., Annabi, N., Titus, M., Valtchev, P., & Tumilar, A., (2009). Sterilization of ginseng using a high-pressure CO_2 at moderate temperatures. *Biotechnology and Bioengineering, 102*(2), 569–576.

22. Duong, T., Balaban, M., Perera, C., & Bi, X., (2015). Microbial and sensory effects of combined high hydrostatic pressure and dense phase carbon dioxide process on Feijoa puree. *Journal of Food Science, 80*(11), E2478–E2485.

23. Enomoto, A., Fukushima, H., Nagai, K., Hakoda, M., & Nakamura, K., (1994). Disruption of microbial cells by the flash discharge of high-pressure carbon dioxide. *Bioscience, Biotechnology, and Biochemistry, 58*(7), 1297–1301.

24. Erkmen, O., (2012). Effects of dense phase carbon dioxide on vegetative cells: Chapter 4. In: Balaban, M. O., & Ferrentino, G., (eds.), *Dense Phase Carbon Dioxide: Food and Pharmaceutical Applications* (Vol. 2, pp. 67–97). Wiley Online Library.

25. Ferrentino, G., Barletta, D., Donsí, F., Ferrari, G., & Poletto, M., (2010). Experimental measurements and thermodynamic modeling of CO_2 solubility at high pressure in model apple juices. *Industrial and Engineering Chemistry Research, 49*(6), 2992–3000.

26. Ferrentino, G., Plaza, M. L., Ramirez-Rodrigues, M., Ferrari, G., & Balaban, M. O., (2009). Effects of dense phase carbon dioxide pasteurization on the physical and quality attributes of a red grapefruit juice. *Journal of Food Science, 74*(6), E333–E341.

27. Folkes, G., (2006). Pasteurization of beer by a continuous dense-phase CO_2 system. *Journal of Food Science, 71*(3), E165–E169.

28. Foster, J. W., Cowan, R. M., & Maag, T. A., (1962). Rupture of bacteria by explosive decompression. *Journal of Bacteriology, 83*(14), 330–334.

29. Furukawa, S., Watanabe, T., Tai, T., & Hirata, J., (2004). Effect of high-pressure gaseous carbon dioxide on the germination of bacterial spores. *International Journal of Food Microbiology, 91*(2), 209–213.

30. Gangola, P., & Rosen, B. P., (1987). Maintenance of intracellular calcium in *Escherichia coli. Journal of Biological Chemistry, 262*(26), 12570–12574.

31. Garcia-Gonzalez, L., Geeraerd, A. H., & Elst, K., (2009). Influence of type of microorganism, food ingredients, and food properties on high-pressure carbon dioxide inactivation of microorganisms. *International Journal of Food Microbiology, 129*(3), 253–263.

32. Garcia-Gonzalez, L., Geeraerd, A. H., & Spilimbergo, S., (2007). High pressure carbon dioxide inactivation of microorganisms in foods: Past, present and future. *International Journal of Food Microbiology, 117*(1), 1–28.

33. Gui, F., Wu, J., Chen, F., & Liao, X., (2006). Change of polyphenol oxidase activity, color, and browning degree during storage of cloudy apple juice treated by supercritical carbon dioxide. *European Food Research and Technology, 223*(3), 427–432.

34. Guo, M., Liu, S., Ismail, M., & Farid, M. M., (2017). Changes in the myosin secondary structure and shrimp surimi gel strength induced by dense phase carbon dioxide. *Food Chemistry, 227*, 219–226.

35. Hendrickx, M., Ludikhuyze, L., Broeck, I. V. D., & Weemaes, C., (1998). Effects of High pressure on enzymes related to food quality: Chapter 5. In: Hendrickx, M., Knorr, D., Ludikhuyze, L., Loey, A. V., & Heinz, V., (eds.), *Ultra High-Pressure Treatments in Foods* (Vol. 9, pp. 197–203). Springer.

36. Ho-mu, L., Zhiying, Y., & Li, F. C., (1993). Inactivation of *Leuconostoc dextranicum* with carbon dioxide under pressure. *The Chemical Engineering Journal, 52*(1), 29–34.

37. Hong, S. I., & Pyun, Y. R., (1999). Inactivation kinetics of *Lactobacillus plantarum* by high-pressure carbon dioxide. *Journal of Food Science, 64*(4), 728–733.

38. Hong, S. I., & Pyun, Y. R., (2001). Membrane damage and enzyme inactivation of *Lactobacillus plantarum* by high-pressure CO$_2$ treatment. *International Journal of Food Microbiology, 63*(1/2), 19–28.

39. Isenschmid, A., Marison, I. W., & Von, S. U., (1995). The influence of pressure and temperature of compressed CO$_2$ on the survival of yeast cells. *Journal of Biotechnology, 39*(3), 229–237.

40. Jan, A., Sood, M., Sofi, S. A., & Norzom, T., (2017). Non-thermal processing in food applications: A review. *International Journal of Food Science and Nutrition, 2*(6), 171–180.

41. Jones, R. P., & Greenfield, P. F., (1982). Effect of carbon dioxide on yeast growth and fermentation. *Enzyme and Microbial Technology, 4*(4), 210–223.

42. Kim, H. T., Choi, H. J., & Kim, K. H., (2009). Flow cytometric analysis of *Salmonella enterica* serotype *Typhimurium* Inactivated with supercritical carbon dioxide. *Journal of Microbiological Methods, 78*(2), 155–160.

43. Kincal, D., Hill, W. S., & Balaban, M. O., (2005). Continuous high-pressure carbon dioxide system for microbial reduction in orange juice. *Journal of Food Science, 70*(5), 249–254.

44. Lecky, M., & Balaban, M. O., (2004). Continuous high-pressure carbon dioxide processing of watermelon juice. In: *Institute of Food Technologists Annual Meeting, Las Vegas-NV, Food Science and Biotechnology* (Vol. 15, No.1, pp. 13–18).

45. Li, S., Shao, Q., Lu, Z., Duan, C., Yi, H., & Su, L., (2018). Rapid determination of crocins in saffron by near-infrared spectroscopy combined with chemometric techniques. *Spectrochimica Acta-Part A: Molecular and Biomolecular Spectroscopy, 190*, 283–289.

46. Lim, S. B., Yagiz, Y., & Balaban, M. O., (2006). Continuous high-pressure carbon dioxide processing of mandarin juice. *Food Science and Biotechnology, 15*(1), 13–18.

47. Mantoan, D., & Spilimbergo, S., (2011). Mathematical modeling of yeast inactivation of freshly squeezed apple juice under high-pressure carbon dioxide. *Critical Reviews in Food Science and Nutrition, 51*(1), 91–97.

48. Meujo, D. A. F., Kevin, D. A., Peng, J., Bowling, J. J., Liu, J., & Hamann, M. T., (2010). Reducing oyster-associated bacteria levels using supercritical fluid CO$_2$ as an agent of warm pasteurization. *International Journal of Food Microbiology, 138*(1/2), 63–70.

49. Morbiato, G., Zambon, A., & Toffoletto, M., (2019). Supercritical carbon dioxide combined with high power ultrasound as innovative drying process for chicken breast. *Journal of Supercritical Fluids, 147*, 24–32.

50. Niu, L., Hu, X., Wu, J., Liao, X., Chen, F., Zhao, G., & Wang, Z., (2010). Effect of dense phase carbon dioxide process on physicochemical properties and flavor compounds of orange juice. *Journal of Food Processing and Preservation, 34*(2), 530–548.

51. Niu, S., Xu, Z., Fang, Y., Zhang, L., Yang, Y., Liao, X., & Hu, X., (2010). Comparative study on cloudy apple juice qualities from apple slices treated by high-pressure carbon dioxide and mild heat. *Innovative Food Science and Emerging Technologies, 11*(1), 91–97.

52. Ortuño, C., Duong, T., Balaban, M. O., & Benedito, J., (2013). Combined high hydrostatic pressure and carbon dioxide inactivation of pectin methylesterase, polyphenol oxidase and peroxidase in Feijoa puree. *Journal of Supercritical Fluids, 82*, 56–62.

53. Paniagua-Martínez, I., Mulet, A., García-Alvarado, M. A., & Benedito, J., (2018). Inactivation of the microbiota and effect on the quality attributes of pineapple juice

using a continuous flow ultrasound-assisted supercritical carbon dioxide system. *Food Science and Technology International, 24*(7), 547–554.

54. Park, S. J., Lee, J. I., & Park, J., (2002). Effects of a combined process of high-pressure carbon dioxide and high hydrostatic pressure on the quality of carrot juice. *Journal of Food Science, 67*(5), 1827–1834.
55. Pataro, G., Ferrentino, G., Ricciardi, C., & Ferrari, G., (2010). Pulsed electric fields assisted microbial inactivation of *s. cerevisiae* cells by high-pressure carbon dioxide. *Journal of Supercritical Fluids, 54*(1), 120–128.
56. Patil, P. D., Dandamudi, K. P. R., Wang, J., Deng, Q., & Deng, S., (2018). Extraction of bio-oils from algae with supercritical carbon dioxide and co-solvents. *Journal of Supercritical Fluids, 135*, 60–68.
57. Pozo-Insfran, D. D., Balaban, M. O., & Talcott, S. T., (2006). Microbial stability, phytochemical retention, and organoleptic attributes of dense phase CO_2 processed muscadine grape juice. *Journal of Agricultural and Food Chemistry, 54*(15), 5468–5473.
58. Ramírez-Rodrigues, M. M., Plaza, M. L., & Ferrentino, G., (2013). Effect of dense phase carbon dioxide processing on microbial stability and physicochemical attributes of *Hibiscus sabdariffa* beverage. *Journal of Food Process Engineering, 36*(1), 125–133.
59. Rao, W., Li, X., Wang, Z., Yang, Y., & Qu, Y., (2017). Dense phase carbon dioxide combined with mild heating induced myosin denaturation, texture improvement, and gel properties of sausage. *Journal of Food Process Engineering, 40*(2), 1–10.
60. Rawson, A., Tiwari, B. K., Brunton, N., & Brennan, C., (2012). Application of supercritical carbon dioxide to fruit and vegetables: Extraction, processing, and preservation. *Food Reviews International, 28*, 253–276.
61. Shimoda, M., Cocunubo-Castellanos, J., & Kago, H., (2001). The influence of dissolved CO_2 concentration on the death kinetics of *Saccharomyces cerevisiae*. *Journal of Applied Microbiology, 91*(2), 306–311.
62. Sirisee, U., Hsieh, F., & Huff, H. E., (1998). Microbial safety of supercritical carbon dioxide processes. *Journal of Food Processing and Preservation, 22*(5), 387–403.
63. Spilimbergo, S., & Bertucco, A., (2003). Non-thermal bacteria inactivation with dense CO_2. *Biotechnology and Bioengineering, 84*(6), 627–638.
64. Spilimbergo, S., Bertucco, A., Basso, G., & Bertoloni, G., (2005). Determination of extracellular and intracellular pH of *Bacillus subtilis* suspension under CO_2 treatment. *Biotechnology and Bioengineering, 92*(4), 447–451.
65. Spilimbergo, S., Cappelletti, M., & Tamburini, S., (2014). Partial permeabilization and depolarization of *Salmonella enterica typhimurium* Cells after treatment with pulsed electric fields and high-pressure carbon dioxide. *Process Biochemistry, 49*(12), 2055–2062.
66. Spilimbergo, S., Dehghani, F., Bertucco, A., & Foster, N. R., (2003). Inactivation of bacteria and spores by pulsed electric field and high pressure CO_2 at low temperature. *Biotechnology Bioengineering, 82*(1), 118–125.
67. Spilimbergo, S., Elvassore, N., & Bertucco, A., (2002). Microbial inactivation by high-pressure. *Journal of Supercritical Fluids, 22*(1), 55–63.
68. Taylor, P., Ishikawa, H., Shimoda, M., Tamaya, K., Yonekura, A., Kawano, T., Ishikawa, H., Shimoda, M., Amaya, K. T., & Yonekura, A., (1997). Inactivation of *Bacillus* spores by the supercritical carbon dioxide micro-bubble method. *Bioscience, Biotechnology, and Biochemistry, 61*(6), 1022, 1023.

69. Timko, M. T., Marre, S., & Maag, A. R., (2016). Formation and characterization of emulsions consisting of dense carbon dioxide and water: Ultrasound. *Journal of Supercritical Fluids, 109*, 51–60.

70. Valley, G., & Rettger, L. F., (1927). The influence of carbon dioxide on bacteria. *Journal of Bacteriology, 14*(2), 101.

71. Valverde, M. T., Marín-Iniesta, F., & Calvo, L., (2010). Inactivation of *Saccharomyces cerevisiae* in conference pear with high-pressure carbon dioxide and effects on pear quality. *Journal of Food Engineering, 98*(4), 421–428.

72. Watanabe, T., Furukawa, S., Tai, T., & Hirata, J., (2003). High-pressure carbon dioxide decreases the heat tolerance of the bacterial spores. *Food Science and Technology Research, 9*(4), 342–344.

73. Wei, C. I., Balaban, M. O., Fernando, S. Y., & Peplow, A. J., (1991). Bacterial effect of high-pressure CO_2 treatment on foods spiked with *Listeria* or *Salmonella. Journal of Food Protection, 54*(3), 189–193.

74. Werner, B. G., & Hotchkiss, J. H., (2006). Continuous flow non-thermal CO_2 processing: The lethal effects of subcritical and supercritical CO_2 on total microbial populations and bacterial spores in raw milk. *Journal of Dairy Science, 89*(3), 872–881.

75. Yagiz, Y., Lim, S. L., & Balaban, M. O., (2005). Continuous high pressure CO_2 processing of mandarin juice. In*: IFT Annual Meeting Book of Abstracts* (pp. 15–20). Chicago, IL-USA: Institute of Food Technologists (IFT).

76. Zhong, Q., Black, D. G., Davidson, P. M., & Golden, D. A., (2008). Non-thermal inactivation of *Escherichia coli* K-12 on spinach leaves, using dense phase carbon dioxide. *Journal of Food Protection, 71*(5), 1015–1017.

CHAPTER 7

IONIC LIQUIDS: APPLICATIONS IN FOOD SCIENCE AND FOOD PROCESSING

CAROLINA ELISA DEMAMAN ORO, VICTOR DE AGUIAR PEDOTT, MARCELO LUIS MIGNONI, ROGÉRIO MARCOS DALLAGO, MARCUS VINÍCIUS TRES, and GIOVANI LEONE ZABOT

ABSTRACT

Ionic liquids (ILs) can be used in chemical synthesis, biocatalytic transformations, electrochemical, and analytical device designs, biosensors constructions, and food-related separation processes. Imidazole-based ILs are generally used as extraction solvents and in food analysis. Based on this context, this book chapter focuses on the characteristics and main applications of ILs in food-related areas. The chapter also includes physicochemical, toxicological, and thermodynamic parameters; and discusses ILs composed of ions based on natural products. Future outlooks on the subject and the major technological challenges and advantages of using these attractive liquids are also discussed.

7.1 INTRODUCTION

The composition of ionic liquids (ILs) is an organic or inorganic anion and a large organic cation, which exists in a liquid state at 25°C and are more inert electrochemically than the standard electrolytes [25, 62]. They can be classified according to their anionic or cationic counterparts [72]. ILs are non-flammable substances and have particular properties, such as strong solubilization power, small vapor pressure, low volatility, and high solvating capacity, polarity, and chemical and thermal stability [25]. ILs can be used in biocatalytic transformations, chemical synthesis, electrochemical device designs, biosensors, analytic devices, and separation processes [69, 78].

The physicochemical and specific characteristics of ILs can be easily tuned for specific necessities after considerable blends of cations and anions. For instance, they can be applied as the extraction solvent and co-extractant agent for lithium isotope separation [65], for cellulose recovery from industrial paper dried sludge [27], and for simultaneous saccharification and fermentation of sugarcane bagasse by co-fermenting xylose and glucose [2]. In addition, hydrophobic ILs may be used for the treatment of extracting pollutants from liquid solutions [34]. For applications in the food industry, physicochemical, toxicological, and thermodynamic parameters have great importance, mainly the toxicological parameters. Ionic liquid toxicity is influenced by its components (anions and cations) and hydrophobic or hydrophilic properties [61].

Imidazole-based ILs are generally used as extractions solvents and in food analysis [69], but these ILs are toxic and poorly biodegradable. The high solubility and stability of imidazole-based ILs in water must be considered for industrial processes since such properties make ILs persistent pollutants and breakthrough classical water treatments [60]. ILs constituted of ions based on natural products can overcome the toxicity issues, forming biocompatible and biodegradable compounds [58]. Choline-based ILs are biocompatible compounds with properties similar to quaternary ammonium compounds, which are cationic surfactants with biocide activity [61].

This chapter presents an overview and use of ILs in food-related areas; and characteristics of ILs.

7.2   GENERAL CHARACTERISTICS OF IONIC LIQUIDS (ILS)

Association of different anions with organic cations constitutes ILs. Properties like miscibility and thermal stability are dominated by the anion, while the density, viscosity, shape, surface tension, symmetry, and alkyl length are influenced by the cation [6]. The originated properties allow the use of ILs as solvents [14], catalysts [76], structure-directing agents [40], lubricants [13], and surfactants [38], etc.

In ILs synthesis, commonly cations and anions are used. Some cations are imidazolium, pyridinium, pyrroridinium, and their poly-alkyl derivatives. Some anions are bis(trifluoromethylsulfonyl)imide (NTf_2), trifluoromethylsulfate (TfO), dicyanamide ($N(CN)_2$), tetrafluoroborate (BF_4), hexafluorophosphate (PF_6), chloride, bromide, iodide, nitrate, perchlorate, formate, and acetate [6]. Simple anions compose less stable ILs since they are not liquid at temperatures near to 25°C [53].

The adaptive power of ILs ions to the media provides valuable physicochemical properties, such as low vapor pressure and melting point, high thermal stability, high electrical conductivity, adjustable viscosity, and polarity [6]. These characteristics can be changed depending on how much impurities are present in the medium. Consequently, purification of ILs is important to avoid such changes in their characteristics.

7.2.1   MELTING POINT AND VAPOR PRESSURE

The cation-anion interaction of ILs is attributed to the influence of anions on the melting point. The increase in alkyl chains can increase the melting point [12]. Generally, size, and symmetry of cation [51], charge and charge distribution have influence on the melting point since van der Waals and columbic interactions are dominant forces in ILs [42]. Strong interactions (such as hydrogen bonding) involve charge delocalization and symmetry, thus can cause high melting points [16].

ILs are commonly known as non-volatile solvents. Some ILs may undergo distillation process, enabling their purification [17]. Generally, ILs present low vapor pressure at room temperature (approximately 25°C), but the measurement of vapor pressures of ILs is possible at high temperatures [59].

The vapor pressure varies depending on the ionic composition of ILs. The alkyl length can influence vapor pressure, but it has a high dependence on the ions that are used. For Ntf_2-based ILs, an increment of alkyl length reduces the vapor pressure by changes in cohesive forces, assuming that the melting point has no influence on vapor pressure [6, 59].

7.2.2   THERMAL STABILITY AND CONDUCTIVITY

ILs present high thermal stability that can be checked by thermogravimetric analysis under vacuum [6]. It is possible to observe the estimated onset temperature (temperature of rapid mass change) and, likewise, the decomposition of ILs occurs at temperatures lower than the onset temperature through thermo-gravimetric analysis (TGA) [75].

Thermal stability is also dependent on the interactions of the ions, mainly by the anion. Cation effect on thermal stability is attributed to alkyl length and reactivity. Overall, an increment in alkyl length causes a reduction in the thermal stability. Based on cation reactivity, pyrrolidinium ILs present high thermochemical stability than their non-cyclic tetraalkylammonium, imidazolium, and pyridinium [43]. The anion's influence is related by its hydrophobicity and nucleophilicity [33] and an increase in these properties can decrease the thermal stability values. The thermal stability of most used anions is expressed as $[PF_6] >[Beti] > [NTf_2] > [BF_4] > [Me]$ [6, 12, 43].

The conductivity is due to the presence of ions in ILs are constituted of ions, and it varies for different ILs. Commonly, imidazolium-based ILs exhibit high conductivity values (10 mS/cm) compared to lower values of conductivity for tetraalkylammonium, piperidinium, pyrrolidinium, and pyridinium-based ILs [35]. In pyrrolidinium-based ILs, the conductivity levels are influenced by temperature, viscosity, and diffusion [21]. If such parameters are satisfactorily controlled, it is possible to use ILs as supercapacitors [9], electrolytes [21], and as substitute for ion-lithium batteries [54].

7.2.3   VISCOSITY AND POLARITY

Viscosity is a barrier to applications of ILs in several areas since a high value reduces organic reaction rates and diffusion rates of redox species [12]. At 25°C, ILs show high viscosities, whereas some moderate viscosities are attributed to Emim $C(CN)_3$ (15cP), Emim $N(CN)_2$ (16 cP), and allylmim $N(CN)_2$ (20 cP) [6]. The viscosity of ILs is mainly influenced by cation-anion interactions and alkyl length [30]. For lower viscosity values, van der Waals interactions prevail. When H-bonding interaction is predominant, the viscosity tends to increase [6, 30]. An increment of cation alkyl length increases the viscosity [12]. In the liquid fluids, the viscosity of common ILs decreases at high temperatures.

The polarity is defined by the sum of all intermolecular interactions between solute and solvents, excluding interactions that lead to chemical transformations [57]. Several interactions as Columbic, dipole, permanent, and induced H-bonding, and electron pair donor-acceptor can influence on ILs polarity. The polarity depends on the technique used for determination. However, there is no right polarity scale and different scales give different polarities. Indeed, the chosen scale depends on the specific application [55].

For polarity determination, the most used scale is Reichardt's polarity scale ($E_T(30)$), where high values of $E_T(30)$ correspond to high solvent polarities [56]. Reichardt's parameter favors H-bonding, which is not suitable for other contributions. For instance, for more precise polarity determination, multi-parameter scales are used as Kamlet-Taft parameter, which associates the dipolarity/polarizability effects (π^*) with hydrogen bond basicity (β) and acidity (α) effects, thus providing a precise description of solvents polarities [55, 56].

The cation-anion combinations can influence the ILs polarity, whereas the anion's influence on polarity is ranked as follows [19]:

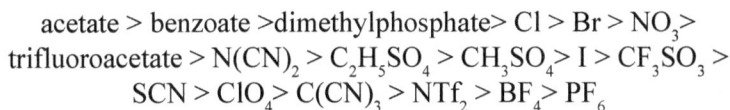

$$\text{acetate} > \text{benzoate} > \text{dimethylphosphate} > Cl > Br > NO_3 >$$
$$\text{trifluoroacetate} > N(CN)_2 > C_2H_5SO_4 > CH_3SO_4 > I > CF_3SO_3 >$$
$$SCN > ClO_4 > C(CN)_3 > NTf_2 > BF_4 > PF_6$$

The cation's influence is dependent on the alkyl chain attached to the charged center. For an alkyl length, long alkyl chains decrease the polarity, for which the order for cations is ranked as follows [41]: ammonium > imidazolium > pyridinium > pyrrolidinium [41].

7.3 COMPOSITION OF IONS IN IONIC LIQUIDS (ILS) BASED ON NATURAL PRODUCTS

Although ILs are considered as green solvents, such general description seems unsuitable because the toxicity levels of some ILs caused by the anions and cations used for their synthesis have hazard problems for humans and the environment [79]. Aqueous effluents containing ILs should be analyzed in terms of toxicity, due to its high solubility levels and stability on water. In fact, it can become a pollutant on wastewater, whereas thiol-based ILs are most harmful [60, 68].

Toxicity is a major drawback of applying ILs in biotechnological and food industries [64]. The use of nature-based anions and cations has demonstrated to reduce the toxicity [29, 69]. Low environmental hazard cations and anions for ILs synthesis include: amino acids, carboxylic acids, and non-heterocyclic ammonium cations as cholinium chloride, thus showing low environmental impacts.

Biodegradability is another important parameter when analyzing the green approach of ILs. Conventional ILs based on pyridinium and imidazolium present low values of biodegradability [24]. The drawback of biodegradability could be minimized by using long alkyl chains, aromatic rings, and hydrolyzable groups [37]. ILs synthesized with cations and anions from natural products show higher biodegradability compared with conventional ILs [32, 60], which reinforces the use of this new group of biocompatible ILs.

For applying ILs based on natural products in the food industry, it is needed to assess the biodegradability and toxicity levels. The biodegradability has been evaluated by some studies under aerobic and anaerobic conditions, showing high values [28, 32, 37]. It is also necessary to perform genetic toxicity tests, studies with rodents and nonrodents, reproduction toxicity evaluations, metabolism, and pharmacokinetic studies, and human tests. Some substances used for ILs synthesis are recognized as nontoxic for application in the food industry, such as cholinium chloride, carboxylic acids, saccharin, and acesulfame potassium [69]. The use of cations and anions from natural sources for ILs synthesis has advantages compared with conventional ILs, since their precursors, like vegetable oils and animal foods, have large-scale production. Likewise, the production cost of ILs can be reduced when produced from natural sources [49].

7.3.1 PHYSICOCHEMICAL, TOXICOLOGICAL, AND THERMODYNAMIC PARAMETERS

The use of anions and cations from natural products for ILs synthesis makes them a different group of ILs with high values of biodegradability and low toxicity, thus enabling the "green" concept. The change in ILs constituents could cause changes in their general properties since the cation-anion interactions originate particular properties.

The new group of ILs is based on cholinium cations and amino acids and carboxylic acids as anions. Many varieties of carboxylic acids and amino acids could be used in the synthesis, whereas the anions structures affect the final properties of the synthesized ILs [15, 66]. In order to compare some properties between biocompatible and commonly used ILs, Table 7.1 presents main and general properties of ILs based on cholinium and amino/carboxylic acids. Properties of some conventional ILs are:

- 1-Allyl-1-methylpyrrolidinium   bis(trifluoromethane-sulfonyl)imide [AMPyrr][TFSI],
- 1-butyl-1-methylpyrrolidinium-bis(trifluoromethylsulfonyl)imide [bmpyrr][NTf$_2$],

- 1-Butyl-3-methylimidazolium Trifluoromethanesulfonate[bmin] [TfO],
- 1-Butyl-3-methylimidazolium-tetrafluoroborate [bmin][BF$_4$],
- 1-ethyl-2,3-dimethylimidazolium-bis(trifluoromethylsulfonyl)imide [em$_2$im][NTf$_2$],
- 1-ethyl-3-methylimidazolium-bis(trifluoromethylsulfonyl)imide [emin][NTf$_2$],
- 1-hexyl-3-methylimidazolium chloride [hmin][Cl],
- 1-Methyl-1-propylpiperidinium bis(trifluoromethane-sulfonyl)imide [PMPip][TFSI],
- 1-Methyl-1-propylpyrrolidinium bis(trifluorometh-anesulfonyl) imide [PMPyrr][TFSI],
- Pyrrolidinium-octanoate [pyrr][C$_7$CO$_2$].

Cation-anion interactions have a great influence on ILs properties (Table 7.1). The properties of the ILs differ from one to another depending on the ions in use. In general, ILs based on natural products show higher viscosity and low density compared with conventional ILs.

Changes in ion structures of biocompatible ILs have the same patterns of interaction as conventional ILs. For amino acid anions, an increase in anion size causes increases in viscosity, refractive index, and temperatures of decomposition. It also causes reductions in density and conductivity [66], whereas similar aspects are observed for ILs synthesized with carboxylic acids as anions [45].

TABLE 7.1   Physical Properties of Biocompatible and Commonly Used Ionic Liquids

Ionic Liquid	ρ (g/cm³)	η (mPa.s)	η_D (-)	σ (µs/cm)	References
Bioactive Compounds					
Choline propanoate	1.071	290.200	1.468	–	[45]
Choline butanoate	1.046	630.600	1.470	–	[45]
Choline hexanoate	1.015	710.700	1.471	–	[45]
Cholinium glycinate	1.145	182.300	1.523	67.700	[66]
Cholinium L-alaninate	1.113	385.600	1.546	21.300	[66]
Cholinium β-alaninate	1.120	5092.100	1.532	7.100	[66]
Cholinium prolinate	1.121	10643.800	1.542	0.300	[66]
Cholinium serinate	1.191	11543.700	1.545	9.300	[66]

TABLE 7.1 *(Continued)*

Ionic Liquid	ρ (g/cm³)	η (mPa.s)	η_D (-)	σ (μs/cm)	References
Cholinium lysinate	1.152	187.000	1.515	0.840	[15]
Cholinium leucinate	1.052	7.980	1.488	12.360	[15]
Common Ionic Liquids					
[bmin][TfO]	1.290	90.000	1.438	3.700	[12]
[bmin][BF₄]	1.120	233.000	1.429	1.700	[12]
[bmpyrr][NTf₂]	1.410	85.000	–	2.200	[12]
[emin][NTf₂]	1.519	28.000	1.423	8.800	[12]
[em₂im][NTf₂]	1.510	88.000	1.430	3.200	[12]
[hmin][Cl]	1.030	716.000	1.515	–	[12]
[pyrr][C₇CO₂]	0.945	62.700	1.458	8.520	[3]
[PMpip][TFSI]	1.320	141.000	–	2.900	[74]
[PMpyrr][TFSI]	1.270	59.000	–	4.900	[74]
[AMpyrr][TFSI]	1.430	52.000	–	5.700	[74]

Legend: ρ: density; η: absolute viscosity; η_D: refractive index; σ: conductivity.

The new group of ILs, based on anions and cations from natural products, demonstrated to have low hazard potential for the environment and humans [29, 32]. Toxicity parameters depend on cation/anion interactions, but probably the anion has a major effect on toxicity values [46]. The main effect of anions on toxicity is due to its structure, especially anions containing F (e.g., NTF_2^- and BF_4^-), which are more toxic as a consequence of hydrolysis of fluorinated compounds forming hydrofluoric acid [44]. The use of natural anions as amino acids based on imidazolium, pyridinium, and pyrrolidinium could minimize the toxicity to acceptable levels. Anions in the form of BF_4 increase the toxicity levels even for natural cations as cholinium [18].

The cations influence the toxicity due to increases in hydrophobicity caused by long alkyl chain length and aromatic constituents, leading to strong interactions with cell walls and increasing the toxicity potential [44]. The choline cation is a good alternative for the synthesis of ILs due to high melting point, solubility on water, and capacity of becoming deep eutectic solvents when metal salts as zinc (II) chloride and tin (II) chloride are added. However, low values of toxicity and high capacity of biodegradability can be highlighted [20].

7.3.2   BIOCOMPATIBLE AND BIODEGRADABLE COMPOUNDS

Biodegradability has high importance when using ILs in the food industry since they have high solubility and stability on water, which could cause water pollution. Generally, conventional ILs (based on pyridinium, imidazolium, and pyrrolidinium cations) present low values of biodegradability. The alkyl chain length of cations can influence biodegradability parameters. Overall, the anions do not present a high influence on ILs biodegradability [23, 60].

For ILs based on natural products, high values of biodegradability have been mainly caused by the cations and anions in use, such as amino acids, carboxylic acids, and cholinium cation [56, 57, 66]. Biocompatible ILs show low values of toxicity and high levels of biodegradability, thus enabling their use in the food industry. The versatility of ILs is important in their success due to the possibility of preparing ILs with defined characteristics for a particular application. Furthermore, the preparation of biodegradable ILs using ions obtained from natural sources should be explored for the development of sample preparation methods with truly eco-friendly solvents [48].

7.4   APPLICATIONS OF IONIC LIQUIDS (ILS) IN FOOD-RELATED AREAS

ILs have been investigated to replace hazardous, volatile, and flammable organic solvents in different applications in the industry. They are miniaturized methods in which a small amount of ILs is used in the extraction process, thus enhancing the environmental sustainability of these methods [48].

Volatile organic compounds, including halogenated hydrocarbons, aromatic, and aliphatic hydrocarbons, some esters, alcohols, aldehydes, ethers, and ketones, are mainly used as organic phase solvents. Developing novel green solvents is one of the topics at the forefront of innovation in green chemistry. Besides being eco-friendly, natural ILs provide shorter extraction time, simplicity, low cost and better selectivity [52]. Therefore, the incorporation of ILs produced from natural products in extraction procedure allows complying most of these requirements [48].

ILs can be custom-made with different functions to meet a diverse range of requirements for widespread applications [68]. The main attraction of ILs stands out in their structure, which can be controlled to produce desired chemical properties. In particular, as extraction phase, the selectivity of ILs can be increased by introducing functional groups that impart specific chemical functionalities, thus enhancing specific extraction capabilities [26]. After proper selection of cations and anions, an ionic liquid that exhibits exceptional selectivity toward a group of target compounds can be created [52].

Solvent-based extraction methods are: single-drop microextraction (SDME), liquid-liquid extraction (LLE), dispersive liquid-liquid micro-extraction (DLLME), and hollow fiber liquid-phase microextraction [52]. Some other applications involve the immobilization of ILs as a liquid phase in membranes. One of the main benefits of ILs membrane-based separation processes is that they allow continuous operation with a minimal amount of ionic liquid (active phase). There are different methodologies for ionic liquid immobilization. In order to exploit the synthesized ionic liquid as a carrier in the membrane, ionic liquid must be soluble in an organic solvent and insoluble in water for the membrane impregnation. Therefore, the choice of the chain length needs to find the true balance between hydrophobic and solvent solubility properties [36].

In recent years, ILs have been applied as surface modifiers to increase adsorption sites of mesoporous silica [63]. Chen et al. [11] studied zeolite to replace silica gel to immobilize ILs because silica gel exhibited a slightly lower loading capacity for ILs. Thus, highly viscous and poorly water-soluble ionic liquid [C4MIM] PF6 was first immobilized on NaY zeolite using the simple and time-saving impregnation method with the assistance of dichloromethane.

The main research approaches and applications for food-related areas using ILs are biotechnology, analysis, and chemical engineering. ILs can be used for biocatalysis, biotransformation, lipase production and activation, lipids processing, synthesis of antioxidants, synthesis of sugar/carbohydrates, matrices for mass spectrometry, gas chromatography columns, extraction, and recovery, separation, and coating, etc., [50]. The choice of the technique and the ILs to be applied is determinant for reaching success.

7.4.1   SURVEY OF PATENTS: USE OF IONIC LIQUIDS (ILS) IN FOOD PROCESSING

The patents' survey was done by searching for information in the last few years in the main international patent repositories like the United States Patent and Trademark Office (USPTO), World Intellectual Property Organization (WIPO), and Brazilian National Institute of Industrial Property (INPI), among others.

Patents about ILs are widely available in the database. According to WIPO (2019), after using the "PATENTSCOPE-Search International and National Patent Collections" with the criteria for search as 'ILs,' more than 4155 granted and deposited patents are available for consultation for a period ranging from 1937 from 2019. By analyzing patents from 2010 to 2019, it is possible to observe that the largest number of records occurred during this period (2217 deposited patents, approximately 53.4% of all), as shown in Figure 7.1. When the terms 'ILs' and 'food' are searched, it is observed that ILs in food-related areas contribute with approximately 1% of total patents related to ILs in the main international patent repositories. The number of granted and deposited patents worldwide on the subject dealt in this book chapter is presented in Figure 7.2.

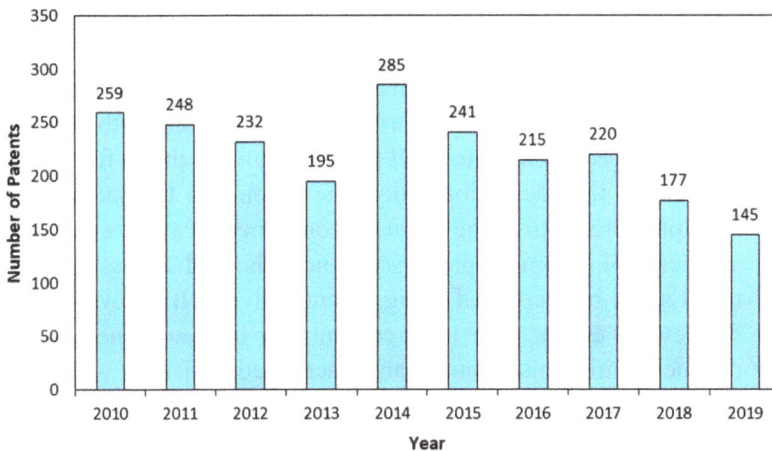

FIGURE 7.1   Number of patents dealing with ionic liquids recorded with the period ranging from 2010 to 2019.
Source: Data from: World Intellectual Property Organization (WIPO).

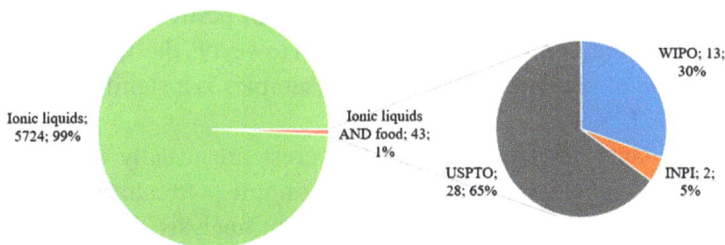

FIGURE 7.2   Number of patents on ionic liquids and those applied in food-related areas recorded worldwide.
Source: Data from: WIPO, USPTO, and INPI.

The main international patent repositories feature several areas, where the topics on ILs applied in food-related areas are covered in patents. Therefore, a brief list of the main subjects dealt with in different patents is presented below:

- **Fungal Biomass Extract Patent:** It presents the application of a fungal extract. It comprises one or more non-ionic polysaccharides, which represent technological additives for manufacturing a food-grade liquid. The invention can be used in processing of some fermented beverages, especially beers and wines [67].
- **Enhanced Food Soil Removal Patent:** It relates to the cleaning composition including an ionic liquid and preferably no co-surfactants are provided. The compositions are substantially free of alkylphenol ethoxylates, including nonylphenol ethoxylates. The cleaning compositions provide superior cleaning efficacy for various materials, including proteins and food materials, providing an effective, biorenewable, and eco-friendly alternative to nonylphenol ethoxylates [8].
- **Process Feedstock Patent:** It claims some methods for producing lipids with the use of ionic liquid solutions. In the sequence, after the lipids production, the biomass components can be applied in the synthesis of chemical precursors, biofuels, and foods. In addition, the ILs can be recovered using an anti-solvent. It allows further use for several cycles. The lipid content can be determined and lipid-producing microbial groups can be screened [39].

7.5 ADVANTAGES OF USING IONIC LIQUIDS (ILS) IN FOOD PROCESSING

For food quality and safety analysis, the most common performed analysis of food commodities is due to the determination of pesticides [26], along with contaminants and compounds of interest. However, the identification and quantification of contaminants from food samples is a significant analytical challenge [77].

Contaminants and compounds of interest are usually present in low concentrations in real samples. Therefore, sample pretreatment is one of the most important steps of the whole process of analysis. ILs have attracted growing interest due to their unique physicochemical properties. There has been an increase in the application of ILs as analytical solvents, especially in many sample preparation techniques as a "green" alternative for volatile organic solvents [52]. Compared with traditional extraction solvents, ILs are safer and eco-friendly and they can be reused at least three times without a decrease in the extraction efficiency [73].

ILs can be super acidic or basic substances. They can also be water-miscible or immiscible substances, depending on its properties. Overall, the anion is applied to control the water miscibility of ILs, but the cation can also influence the hydrophobicity or hydrogen-bonding ability of the solvent [33]. Such characteristic makes possible to use ILs for different techniques and applications, such as new group of diluents in headspace gas chromatography (HS-GC) analysis [31, 77]. Therefore, this characteristic of ILs can represent an enormous advantage in food analysis, where the effective extraction of both polar and non-polar analytes is of utmost importance [26].

7.5.1   IONIC LIQUIDS (ILS) FOR DETERMINATION OF CONTAMINANTS AND COMPOUNDS OF INTEREST

An eco-friendly vortex-assisted ionic liquid-based microextraction method was reported by Bağda and Tüzen [4] for the determination of selenium in food. After microwave digestion, the method can be used to process different foods, such as pepper, wheat flour, ginger, red lentil, corn flour, cornstarch, traditional soup, and garlic, among others. Tuzen et al. [71] used 1-hexyl-3-methylimidazolium bis(trifluoromethylsulfonyl)-imide [C6MIM][NTf$_2$] as an ionic liquid for the detection of tin in foods, such as canned tuna fish, canned peas, canned olives, olive oil, black tea, green tea, cultivated mushroom, beef, chicken meat, sheep meat, herbal tea, tomato paste, cheese, butter, margarine, ice tea, orange juice, and apricot juice.

The extraction of astaxanthin from shrimp waste via aqueous biphasic systems (ABS) formed by potassium phosphate (K$_3$PO$_4$) and ILs has also been studied [22]. Also, 1-Octyl-3-methylimidazolium hexafluorophosphate ([C$_8$MIM][PF$_6$]) was used as a solution for extracting three endocrine disrupting compounds (EDCs) from the food packaging: bisphenol A (BPA), bisphenol AF and bisphenol AP. Owing to an incomplete polymerization process, residues of bisphenol-type EDCs in plastic food packages can migrate into foods, especially during storage and processing at elevated temperatures [73].

ILs were used as solvents in *in-situ* DLLME to provide rapid preconcentration of polychlorinated biphenyls (PCBs) and acrylamide from complex food samples followed by an analysis of the ionic liquid-based extraction solvent by HS-GC. The results presented a satisfactory analytical precision and linearity. The method presented detection limits lower than the ppt level for PCBs and lower than the ppb level for acrylamide in aqueous samples. The method also exhibited good matrix-compatibility with complex real-world samples. The results represent a significant advantage over the conventional solid-phase microextraction (SPME) method [77].

A $[CoCl_4^{2-}]$-based magnetic ionic liquid (MIL) was synthesized and a novel method of *in situ* DNS-Cl derivatization and DLLME method based on a magnetic ionic liquid (MIL-DLLME) combined with a chromatographic method was developed to evaluate simultaneously trace amounts of biogenic amines (Bas) in food samples. This method was combined with rapid extraction and magnetic separation techniques [10]. Following the concept of simultaneous processing, a study was developed to evaluate the online preconcentration and separation of acrylamide, asparagine, and glucose applying analyte by ionic liquid micelle collapse capillary electrophoresis (AFILMC) using 1-butyl-3-methylimidazolium bromide (BMIMBr). The process was strongly dependent on the BMIMBr concentration/conductivity in the injected sample matrix and the concentration/conductivity of the phosphate-running buffer. Short-chain (C2–C8) hydrophilic methyl imidazolium-based ILs present exciting properties applicable to the analysis of acrylamide in food samples [1].

An ultrasound-assisted ionic liquid dispersive liquid-liquid microextraction (USA-IL-DLLME) was studied to have the speciation of inorganic selenium, Se(IV) and Se(VI), in beverages and total selenium in food samples. This method presented some advantages, such as simple, rapid, low cost, high enrichment factor and green extraction [70]. Considering the physico-chemical properties and the versatility and ruggedness, polymeric ILs are being properly employed as sorbent coatings for SPME, affording advantageous analytical results in comparison to those achievable by commercially available coatings [26]. The polymeric ILs -based sorbent coatings exhibited superior selectivity and sensitivity in the extraction of PCBs and acrylamide from coffee and milk, respectively, compared to commercially available SPME coatings [77].

Electrochemical sensors based on ionic liquid or nanomaterial base have received much attention for different applications in the food industry. For example, 1-butyl-3-methyl imidazolium tetrafluoroborate ionic liquid, boron nitride and magnetite nanoparticles (Fe_3O_4NPs) based nanocomposite (IL-BN-Fe_3O_4NPs) was synthesized and applied to manufacture glassy carbon electrode (GCE) for the evaluation of ascorbic acid (AA) [7]. A high sensitive voltammetric sensor based on NiO/carbon nanotubes and 1-butyl-3-methylimidazolium hexafluorophosphate (ionic liquid) was developed as a novel sensor for the fast and selective evaluation of BPA in food samples [47].

Nowadays, various metallic nanoparticles and carbon nanotubes are applied in the combination of ILs for the fabrication of several types of electrodes [5, 7], which inevitably enhance the electrochemical sensing abilities of electrochemical sensors. Jean et al. [36] presented the synthesis and the characterization of novel fluorous ILs with fluorous anions to be used as a carrier for fluorous supported liquid membrane for future applications in the removal of metal ions. ILs with various chain lengths were synthesized: fluorous phosphates with C_8, C_{10}, and C_{12} chains and imidazolium with various chain lengths. The 1-butyl-3-decylimidazolium di(1H,1H,2H,2H-perfluorodecylphosphate) was the most appropriated ionic liquid for being applied in metal removal.

7.6   FUTURE OUTLOOK AND CHALLENGES

Recent advances in the scientific area demonstrate the potential of ILs as novel substances for extraction, fractionation, and separation, compounds synthesis, biosensors constructions, biocatalytic transformations, and as an antibacterial agent. In the future, biological activities of ILs may have exceptional attention of the scientific community to evaluate and expand their potential applications in pharmaceutics, drugs, and foods.

Applying ILs in chemical and food analysis processes has progressed within the past decade. Even though interest in ILs from natural products is still increasing, yet an enormous variety of ILs exists by combining different anions and cations. The combination represents an advantage, but sometimes it is a drawback. The main problem is related to find an adequate ionic liquid for a specific separation problem. Based on this context, the technological challenges remain to the market for less expensive ILs that can compete with traditional solvents.

7.7   SUMMARY

Research activities on the use of ILs in food-related areas for non-thermal food processing and analysis are summarized in this chapter. This chapter mainly documents the physicochemical and toxicological characteristics of ILs, especially for those produced from natural products. More detailed research is still needed to evaluate the availability of some ILs like the choline-based ones. Likewise, more studies are required on the economic evaluation of ILs in a larger scale.

ACKNOWLEDGMENTS

The authors thank the Coordination for the Improvement of Higher Education Personnel (CAPES), National Council of Technological and Scientific Development (CNPq: 409583/2018-9) and the Research Support Foundation of the State of Rio Grande do Sul (FAPERGS: 16/2551-0000522-2; 17/2551-0000893-6). M. V. Tres, R. M. Dallago, M. L. Mignoni, and G. L. Zabot (304882/2018-6) thank CNPq for the productivity grants.

KEYWORDS

- amino acids
- biodegradability
- choline-based ionic liquids
- food industry
- ionic liquid
- toxicity

REFERENCES

1. El-Hady, D. A., & Albishri, H. M., (2015). Simultaneous determination of acrylamide, asparagine and glucose in food using short-chain methyl imidazolium ionic liquid-based ultrasonic-assisted extraction coupled with analyte focusing by ionic liquid micelle collapse capillary electrophoresis. *Food Chemistry, 188*, 551–558.

2. Amoah, J., Ogura, K., Schmetz, Q., Kondo, A., & Ogino, C., (2019). Co-fermentation of xylose and glucose from ionic liquid pretreated sugar cane bagasse for bioethanol production using engineered xylose assimilating yeast. *Biomass and Bioenergy, 128*, 105283.

3. Anouti, M., Vigeant, A., Jacquemin, J., Brigouleix, C., & Lemordant, D., (2010). Volumetric properties, viscosity, and refractive index of the protic ionic liquid, pyrrolidinium octanoate, in molecular solvents. *Journal of Chemical Thermodynamics, 42*(7), 834–845.

4. Bağda, E., & Tüzen, M., (2017). A simple and sensitive vortex-assisted ionic liquid-dispersive microextraction and spectrophotometric determination of selenium in food samples. *Food Chemistry, 232,* 98–104.

5. Bavandpour, R., Karimi-Maleh, H., Asif, M., Gupta, V. K., Atar, N., & Abbasghorbani, M., (2016). Liquid phase determination of adrenaline uses a voltammetric sensor employing $CuFe_2O_4$ nanoparticles and room temperature ionic liquids. *Journal of Molecular Liquids, 213,* 369–373.

6. Berthod, A., Ruiz-Ángel, M. J., & Carda-Broch, S., (2008). Ionic liquids in separation techniques. *Journal of Chromatography A, 1184*(1/2), 6–18.

7. Bhajanthri, N. K., Arumugam, V. K., Chokkareddy, R., & Redhi, G. G., (2016). Ionic liquid-based high performance electrochemical sensor for ascorbic acid in various foods and pharmaceuticals. *Journal of Molecular Liquids, 222,* 370–376.

8. Blattner, A. R., & Hodge, C. A., (2013). Inventors; Ecolab USA Inc., Assignee. *Quaternized Alkyl Imidazoline Ionic Liquids Used for Enhanced Food Soil Removal* (p. 20). US Patent 13471532.

9. Brandt, A., Pohlmann, S., Varzi, A., Balducci, A., & Passerini, S., (2013). Ionic liquids in supercapacitors. *MRS Bull., 38*(7), 554–559.

10. Cao, D., Xu, X., Xue, S., Feng, X., & Zhang, L., (2019). An in-situ derivatization combined with magnetic ionic liquid-based fast dispersive liquid-liquid microextraction for determination of biogenic amines in food samples. *Talanta, 199,* 212–219.

11. Chen, C., Fu, Z., Zhou, W., Chen, Q., Wang, C., Xu, L., Wang, Z., & Zhang, H., (2020). Ionic liquid-immobilized NaY zeolite-based matrix solid-phase dispersion for the extraction of active constituents in *Rheum palmatum* L. *Microchemical Journal, 152,* 104245.

12. Chiappe, C., & Pieraccini, D., (2005). Ionic liquids: Solvent properties and organic reactivity. *Journal of Physical Organic Chemistry, 18*(4), 275–297.

13. Cooper, P. K., Staddon, J., Zhang, S., Aman, Z. M., Atkin, R., & Li, H., (2019). Nano- and macroscale study of the lubrication of titania using pure and diluted ionic liquids. *Frontiers in Chemistry, 7,* 1–9.

14. Dai, C., Zhang, J., Huang, C., & Lei, Z., (2017). Ionic liquids in selective oxidation: Catalysts and solvents. *Chemical Reviews, 117*(10), 6929–6983.

15. De Santis, S., Masci, G., Casciotta, F., Caminiti, R., Scarpellini, E., Campetella, M., & Gontrani, L., (2015). Cholinium-amino acid-based ionic liquids: A new method of synthesis and physicochemical characterization. *Physical Chemistry Chemical Physics, 17*(32), 20687–20698.

16. Dong, K., Zhang, S., & Wang, J., (2016). Understanding the hydrogen bonds in ionic liquids and their roles in properties and reactions. *Chemical Communications, 52*(41), 6744–6764.

17. Earle, M. J., Esperança, J. M. S. S., & Gilea, M. A., (2006). The distillation and volatility of ionic liquids. *Nature, 439*(7078), 831–834.

18. Egorova, K. S., Seitkalieva, M. M., Posvyatenko, A. V., & Ananikov, V. P., (2014). Unexpected increase of toxicity of amino acid-containing ionic liquids. *Toxicology Research, 4*(1), 152–159.

19. Filipa, A., Claudio, M., Swift, L., & Hallett, J. P., (2014). Extended scale for the hydrogen-bond basicity of ionic liquids. *Physical Chemistry Chemical Physics, 16,* 6593–6601.

20. Foulet, A., Ghanem, O. B., El-Harbawi, M., & Lévêque, J. M., (2016). Understanding the physical properties, toxicities, and anti-microbial activities of choline-amino acid-based salts: Low-toxic variants of ionic liquids. *Journal of Molecular Liquids, 221*, 133–138.
21. Galiński, M., Lewandowski, A., & Stepniak, I., (2006). Ionic liquids as electrolytes. *Electrochimica Acta, 51*(26), 5567–5580.
22. Gao, J., Fang, C., Lin, Y., Nie, F., Ji, H., & Liu, S., (2020). Enhanced extraction of astaxanthin using aqueous biphasic systems composed of ionic liquids and potassium phosphate. *Food Chemistry, 309*, 125672.
23. Garcia, M. T., Gathergood, N., & Scammells, P. J., (2005). Biodegradable ionic liquids, Part II: Effect of the anion and toxicology. *Green Chemistry, 7*(1), 9–14.
24. Gathergood, N., Scammells, P. J., & Garcia, M. T., (2006). Biodegradable ionic liquids, Part III: The first readily biodegradable ionic liquids. *Green Chemistry, 8*(2), 156–160.
25. Ghanbari-Ardestani, S., Khojasteh-Band, S., & Zaboli, M., (2019). The effect of different percentages of triethanolammonium butyrate ionic liquid on the structure and activity of urate oxidase: Molecular docking, molecular dynamics simulation, and experimental study. *Journal of Molecular Liquids, 292*, 111318.
26. Gionfriddo, E., Souza-Silva, É. A., & Ho, T. D., (2018). Exploiting the tunable selectivity features of polymeric ionic liquid-based SPME sorbents in food analysis. *Talanta, 188*, 522–530.
27. Glińska, K., Ismail, M. S. B., Goma-Camps, J., & Valencia, P., (2019). Recovery and characterization of cellulose from industrial paper mill sludge using tetrakis and imidazolium-based ionic liquids. *Industrial Crops and Products, 139*, 111556.
28. Gomes, J. M., Silva, S. S., & Reis, R. L., (2019). Biocompatible ionic liquids: Fundamental behaviors and applications. *Chemical Society Reviews, 48*(15), 4317–4335.
29. Gouveia, W., Jorge, T. F., Martins, S., & Meireles, M., (2014). Toxicity of ionic liquids prepared from biomaterials. *Chemosphere, 104*, 51–56.
30. Guangren, Y., Danchuan, Z., Lu, W., Shendu, Y., & Xiaochun, C., (2012). Viscosity of ionic liquids: Database, observation, and quantitative structure-property relationship analysis. *AIChE Journal, 58*(9), 2885–2899.
31. Ho, T. D., Yehl, P. M., Chetwyn, N. P., & Wang, J., (2014). Determination of trace level genotoxic impurities in small molecule drug substances using conventional headspace gas chromatography with contemporary ionic liquid diluents and electron capture detection. *Journal of Chromatography A, 1361*, 217–228.
32. Hou, X. D., Liu, Q. P., Smith, T. J., Li, N., & Zong, M. H., (2013). Evaluation of toxicity and biodegradability of cholinium amino acids ionic liquids. *PLoS One, 8*(3), 59145.
33. Huddleston, J. G., Visser, A. E., & Reichert, W. M., (2011). Characterization and comparison of hydrophilic and hydrophobic room temperature ionic liquids incorporating the imidazolium cation. *Green Chemistry, 3*(4), 156–164.
34. Isosaari, P., Srivastava, V., & Sillanpää, M., (2019). Ionic liquid-based water treatment technologies for organic pollutants: Current status and future prospects of ionic liquid mediated technologies. *Science of the Total Environment, 690*, 604–619.
35. Jarosik, A., Krajewski, S. R., Lewandowski, A., & Radzimski, P., (2006). Conductivity of ionic liquids in mixtures. *Journal of Molecular Liquids, 123*(1), 43–50.
36. Jean, E., Villemin, D., & Lebrun, L., (2019). Ionic liquids with fluorous anions for supported liquid membranes and characterization. *Journal of Fluorine Chemistry, 227*, 109365.

37. Jordan, A., & Gathergood, N., (2015). Biodegradation of ionic liquids: A critical review. *Chemical Society Reviews, 44*(22), 8200–8237.

38. Kapitanov, I. V., Jordan, A., & Karpichev, Y., (2019). Synthesis, self-assembly, bacterial and fungal toxicity, and preliminary biodegradation studies of a series of l-phenylalanine-derived surface-active ionic liquids. *Green Chemistry, 21*(7), 1777–1794.

39. Kuehnle, A. R., & Nolasco, N. A. B., (2011). Inventors; Kuehnle agrosystems, assignee. *Enrichment of Process Feedstock* (p. 15). USA patent WO/2011/066419.

40. Kumar, M. A., Krishna, N. V., & Selvam, P., (2019). Novel ionic liquid-templated ordered mesoporous aluminosilicates: Synthesis, characterization and catalytic properties. *Microporous and Mesoporous Materials, 275*, 172–179.

41. Lee, J. M., & Prausnitz, J. M., (2010). Polarity and hydrogen-bond-donor strength for some ionic liquids: Effect of alkyl chain length on the pyrrolidinium cation. *Chemical Physics Letters, 492*(1–3), 55–59.

42. López-Martin, I., Burello, E., Davey, P. N., Seddon, K. R., & Rothenberg, G., (2007). Anion and cation effects on imidazolium salt melting points: A descriptor modeling study. *ChemPhysChem., 8*(5), 690–695.

43. Maton, C., De Vos, N., & Stevens, C. V., (2013). Ionic liquid thermal stabilities: Decomposition mechanisms and analysis tools. *Chemical Society Reviews, 42*(13), 5963–5977.

44. Mena, I. F., Diaz, E., Palomar, J., Rodriguez, J. J., & Mohedano, A. F., (2020). Cation and anion effect on the biodegradability and toxicity of imidazolium and choline-based ionic liquids. *Chemosphere, 240*, 124947.

45. Muhammad, N., Hossain, M. I., Man, Z., & El-Harbawi, M., (2012). Synthesis and physical properties of choline carboxylate ionic liquids. *Journal of Chemical Engineering Data, 57*(8), 2191–2196.

46. Mustahil, N. A., & Baharuddin, S. H., (2019). Synthesis, characterization, ecotoxicity and biodegradability evaluations of novel biocompatible surface-active lauroyl sarcosinate ionic liquids. *Chemosphere, 229*, 349–357.

47. Nikahd, B., & Khalilzadeh, M. A., (2016). Liquid phase determination of bisphenol a in food samples using novel nanostructure ionic liquid modified sensor. *Journal of Molecular Liquids, 215*, 253–257.

48. Pacheco-Fernández, I., & Pino, V., (2020). Extraction with ionic liquids-organic compounds. Chapter 17. In: Poole, C. F., (ed.), *Liquid-Phase Extraction* (pp. 499–537). Cambridge, UK: Elsevier.

49. Passos, H., Freire, M. G., & Coutinho, J. A. P., (2014). Ionic liquid solutions as extractive solvents for value-added compounds from biomass. *Green Chemistry, 16*(12), 4786–4815.

50. Pérez, B., & Wei, W., (2016). New opportunities from ionic liquid for chemical and biochemical processes of lipids: Chapter 7. In: Xu, X., Guo, Z., & Cheong, L. Z., (eds.), *Ionic Liquids in Lipid Processing and Analysis: Opportunities and Challenges* (pp. 225–249). Cambridge, UK: Academic Press.

51. Plechkova, N. V., & Seddon, K. R., (2008). Applications of ionic liquids in the chemical industry. *Chemical Society Reviews, 37*(1), 123–150.

52. Płotka-Wasylka, J., Rutkowska, M., & Owczarek, K., (2017). Extraction with environmentally friendly solvents. *TrAC-Trends in Analytical Chemistry, 91*, 12–25.

53. Poole, C. F., & Poole, S. K., (2010). Extraction of organic compounds with room temperature ionic liquids. *Journal of Chromatography A, 1217*(16), 2268–2286.

54. Prado, R., & Weber, C. C., (2016). Applications of ionic liquids. *Application, Purification, and Recovery of Ionic Liquids, 12*, 1–58.

55. Rani, M. A. A., Brant, A., Crowhurst, L., & Dolan, A., (2011). Understanding the polarity of ionic liquids. *Physical Chemistry Chemical Physics, 13*(37), 16831–16840.

56. Reichardt, C., (2005). Polarity of ionic liquids determined empirically by means of *Solvatochromic pyridinium* N-phenolate betaine dyes. *Green Chemistry, 7*(5), 339–351.

57. Reichardt, C., & Welton, T., (2011). *Solvents and Solvent Effects in Organic Chemistry* (4th edn., p. 692). New Jersey, USA: John Wiley & Sons.

58. Restolho, J., Mata, J. L., & Saramago, B., (2012). Choline based ionic liquids: Interfacial properties of RTILs with strong hydrogen bonding. *Fluid Phase Equilibria, 322–323*, 142–147.

59. Rocha, M. A. A., Lima, C. F. R. A. C., & Gomes, L. R., (2011). High-accuracy vapor pressure data of the extended [CnC 1im][Ntf2] ionic liquid series: Trend changes and structural shifts. *The Journal of Physical Chemistry B, 115*(37), 10919–10926.

60. Romero, A., Santos, A., Tojo, J., & Rodríguez, A., (2008). Toxicity and biodegradability of imidazolium ionic liquids. *Journal of Hazardous Materials, 151*(1), 268–273.

61. Santos, A. G., Ribeiro, B. D., & Alviano, D. S., (2014). Toxicity of ionic liquids toward microorganisms interesting to the food industry. *RSC Advances, 4*(70), 37157–37163.

62. Shalabi, A., Daniels, L., Scott, M., & Mišković, Z. L., (2019). Differential capacitance of ionic liquid interface with graphene: The effects of correlation and finite size of ions. *Electrochimica Acta, 319*, 423–434.

63. Si, R., Han, Y., Wu, D., Qiao, F., Bai, L., Wang, Z., & Yan, H., (2020). Ionic liquid-organic-functionalized ordered mesoporous silica-integrated dispersive solid-phase extraction for determination of plant growth regulators in fresh *Panax ginseng. Talanta, 207*, 120247.

64. Silva, F. A., Siopa, F., & Figueiredo, B. F. H. T., (2014). Sustainable design for environment-friendly mono and dicationic cholinium-based ionic liquids. *Ecotoxicology and Environmental Safety, 108*, 302–310.

65. Sun, H., Jia, Y., Liu, B., Jin, Y., & Zhang, Q., (2019). Separation of lithium isotopes by using solvent extraction system of crown ether-ionic liquid. *Fusion Engineering and Design, 149*, 111338.

66. Tao, D. J., Cheng, Z., Chen, F. F., & Li, Z. M., (2013). Synthesis and thermophysical properties of biocompatible cholinium-based amino acid ionic liquids. *Journal of Chemical Engineering Data, 58*(6), 1542–1548.

67. Teissedre, P. L., Bornet, A., & Bruyere, J. M., (2007). *Use of a Fungal Biomass Extract as Technological Additive for Treating Food-Grade Liquids* (p. 19). French Patent WO/2007/003863.

68. Tian, J., Chen, K., Wang, H., Liu, W., & Wang, X., (2020). Determination of a thiol-based ionic liquid using ultrathin graphitic carbon nitride nanosheets as a nanofluoroprobe. *Talanta, 207*, 120291.

69. Toledo, H. A. A. C., Maximo, G. J., Costa, M. C., Batista, E. A. C., & Meirelles, A. J. A., (2016). Applications of ionic liquids in the food and bioproducts industries. *ACS Sustainable Chemistry and Engineering, 4*(10), 5347–5369.

70. Tuzen, M., & Pekiner, O. Z., (2015). Ultrasound-assisted ionic liquid dispersive liquid-liquid microextraction combined with graphite furnace atomic absorption spectrometric for selenium speciation in foods and beverages. *Food Chemistry, 188*, 619–624.

71. Tuzen, M., Uluozlu, O. D., Mendil, D., Soylak, M., Machado, L. O. R., Dos, S. W. N. L., & Ferreira, S. L. C., (2018). A simple, rapid, and green ultrasound-assisted and ionic liquid dispersive microextraction procedure for the determination of tin in foods employing ETAAS. *Food Chemistry, 245*, 380–384.

72. Verma, C., Mishra, A., Chauhan, S., & Verma, P., (2019). Dissolution of cellulose in ionic liquids and their mixed cosolvents: A review. *Sustainable Chemistry and Pharmacy, 13*, 100162.

73. Wang, L., Zhang, D., Xu, X., & Zhang, L., (2016). Application of ionic liquid-based dispersive liquid-phase microextraction for highly sensitive simultaneous determination of three endocrine disrupting compounds in food packaging. *Food Chemistry, 197*, 754–760.

74. Yim, T., Hyun, Y. L., Kim, H. J., & Mun, J., (2007). Synthesis and properties of pyrrolidinium and piperidinium bis(trifluoromethanesulfonyl)imide ionic liquids with allyl substituents. *Bulletin of the Korean Chemical Society, 28*(9), 1567–1572.

75. Yuanyuan, C., & Tiancheng, M., (2014). Comprehensive investigation on the thermal stability of 66 Ionic liquids by thermogravimetric analysis. *Industrial and Engineering Chemistry Research, 53*(20), 8651–8664.

76. Yulin, H., Yao, N., Pan, W. Y., & Bo Zheng, K., (2017). Recent advances in catalytic condensation reactions applications of supported ionic liquids. *Current Organic Chemistry, 21*, 462–484.

77. Zhang, C., & Cagliero, C., (2017). Rapid and sensitive analysis of polychlorinated biphenyls and acrylamide in food samples using ionic liquid-based *in situ* dispersive liquid-liquid microextraction coupled to headspace gas chromatography. *Journal of Chromatography A, 1481*, 1–11.

78. Zhang, J., Liu, C., Xie, Y., & Li, N., (2017). Enhancing fructooligosaccharides production by genetic improvement of the industrial fungus *Aspergillus niger* ATCC 20611. *Journal of Biotechnology, 249*, 25–33.

79. Zhao, D., Liao, Y., & Zhang, Z. D., (2007). Toxicity of ionic liquids. *Clean-Soil, Air, Water, 35*(1), 42–48.

CHAPTER 8

APPLICATION OF OZONE IN THE FOOD INDUSTRY: RECENT ADVANCES AND PROSPECTS

RAVI PANDISELVAM, NUKASANI SAGARIKA,
YARRAKULA SRINIVAS, ANJINEYULU KOTHAKOTA,
KAMALAKKANNAN GOMATHY, and KALIRAMESH SILIVERU

ABSTRACT

Ozone treatment of food products is a disinfectant and residue-free technology that has been likened as green/organic technology. Ozone has been widely adopted in water purification, wastewater treatment, sterilization of food processing equipments and food packaging materials, washing of fruits and vegetables, removal of the odors of fruits storage units, enhancement of seed germination and the preservation of marine products, the stored products, microbial destruction, enzyme inactivation, pesticide degradation and aflatoxin degradation in agro-products and also in starch modification. The main challenges associated with ozone treatment are the production of pure and high concentrated ozone, quick conversion of ozone to oxygen, necessitating the continuous supply of ozone. This chapter discusses the recent advances and challenges in ozonation in the food industry.

8.1 INTRODUCTION

The shelf life of the foods mainly depends on its moisture content and storage conditions. Some foods are perishable like meat, fish, poultry, milk, etc., while others are semi-perishable, such as fruits and vegetables. Therefore, it is necessary to develop technologies for the preservation and safe storage of food materials for a longer period and for increasing the shelf life of foods. In addition, it helps the consumers to enjoy the taste of perishable and seasonal foods throughout the year. Due to this, there will be stability in the prices of food products and the scope for supply to demand shortage will be minimized [11].

Among various thermal and non-thermal processing techniques, ozone technology has attained renewed interest in the food industry because it helps in the maintenance of safety and freshness of the produces and to improve the shelf life of the food produces [3, 49].

In the atmosphere, ozone (O_3) gas production occurs when the oxygen (O_2) gas is allowed to pass through a high electrical discharge or due to the UV irradiation [69]. The main advantage of ozone gas is its application as a gas at ambient and refrigeration temperature, and also it can be utilized in aqueous medium, including water. However, the choice of using ozone as a gaseous phase or in an aqueous phase is dependent on the type of product like solid or liquid food and also the processing method. In the USA, ozone gas is widely used in food industries, and it is generally recognized as safe (GRAS). In addition, FDA [35] approved that the ozone can be directly used as a food additive in both aqueous and gaseous phases [116].

Ozone cannot be stored, and it decomposes rapidly in the aqueous phase but in the gaseous phase, it has a relatively longer half-life. The main advantage of ozone when compared to other techniques is that it results in oxygen during decomposition and is eco-friendly. In addition, ozone is detectable at low concentrations ranging from 0.01 to 0.05 ppm [71].

As an antimicrobial agent and oxidizing agent [66, 84], ozone has several applications in the food industry because it increases the shelf life of horticultural commodities when compared to conventional antimicrobial agents including potassium sorbate and chlorine. In addition, ozone helps in problems associated with pesticide residues and mycotoxin contamination in the food industry [99, 103].

Ozone gas can be generated on-site by employing a discharge-based ozone generator enclosed with oxygen cylinder/oxygen concentrator, thus eliminating the need to store the ozone gas [56, 82]. This results in saving the cost of transportation and also energy consumption as the ozone application does not require thermal energy [57].

This chapter focuses on the application of ozone in the food processing of fruits and vegetables, fruit juice disinfection, insect control in stored products, pesticide degradation in agro-produces, degradation of aflatoxin and structural modification of native starch. The chapter also discusses the effects of ozonation on the quality of fruits, vegetables, fruit juices, and stored grains.

8.2 APPLICATIONS OF OZONATION IN FOOD INDUSTRY

Among the latest non-thermal technologies available in the food industry, ozonation is a favorable novel technology to destroy foodborne pathogens, to increase the shelf life of various food products, degradation of enzymes and pesticides, and elimination of spoilage microorganisms. It has been documented that the ozone is effective against the microorganisms for the treatment of food and water [43, 57].

In the post-harvest processing of horticultural commodities, ozone is widely used during the pre-storage period, by directly applying ozone to the harvested produces [20]. Ozone is effective in delaying the ripening of horticultural commodities by destroying the ethylene hormone, thus extending the shelf life of horticultural commodities [80, 102]. In addition, ozone acts as an antimicrobial agent and retains the quality of the agro-produces during the storage [16].

In majority of food processing operations, the most important procedure is the surface disinfection of the raw materials. It is believed that about 30% of the produce is lost due to the microbial contamination before reaching consumers [14]. Hence, ozone is useful in the food industry, because it has the ability to destroy various groups of microorganisms, thus acting as a good disinfectant [54]. The pathogenic microorganisms in foods can be easily destroyed by ozonation by affecting the cell membranes of microorganisms.

8.2.1 CONTROL OF MICROORGANISMS

Ozonation in the food industry is based on its efficacy in solid foods than the liquid foods by using either gaseous or aqueous ozone. The mechanism behind the inactivation of microorganisms is based on the fact that ozone reacts with the nucleic acid of the cell, intracellular enzymes and spore coats. During the process, oxidation of sulfhydryl groups takes place, leading to the inactivation of microorganisms. Furthermore, ozone results in the formation of aldehydes and aliphatic acids by reacting with polysaccharides. Oxidation of various cell components (such as a membrane, bound enzymes, proteins, and lipids) takes place during the ozonation process resulting in the leakage of cell contents and finally leading to the oxidative burst [28, 57, 92].

FDA has approved that ozone can be used as a food additive. Cullen et al. [28] stated that most of the fruit-juice industries in the USA started using ozone as an antimicrobial agent as it can cause 5 log reduction of pathogens in the final product.

8.2.1.1   MICROBIAL REDUCTION IN FRUITS AND VEGETABLES

The main composition of horticultural produce is water (approximately 70%), protein (around 3.5%), and fat (around 1%). These contain several vitamins, minerals, and dietary fiber, which help in preventing constipation. The good nutritional profile of horticultural produce can be an energy supplement to humans and also helps in the maintenance of body weight. Fruits and vegetables usually have living tissues, which undergo respiration throughout the post-harvest period. Respiration leads to water loss and cell softening and also the fruits and vegetables are contaminated with pathogenic microorganisms [50]. This leads to spoilage of produce and reduction in its shelf life and acceptability in the market.

The microbes mainly responsible for the spoilage of fruits and vegetables are bacteria, yeast, and molds. This deterioration leads to quality loss, nutrient loss, and off-flavor development. Hence, safety measures are needed at each stage not only in the production but also in the processing stage till the product reach the consumer in a fresh and safe form [75].

Najafi and Khodaparast [44] evaluated the efficiency of gaseous phase of ozone for reducing the microbial population in date palm fruits. Gaseous ozone (GO)-treated samples were free of microbial population compared with control samples. A minimum exposure time of one hour at 5 ppm concentration was effective in reducing the *S. aureus* and coliform of date fruits. Similarly, Brieet et al. [19] studied the effect of ozone (gaseous phase) on *Murine norovirus* (MNV-1). When compared with water treatment (washing with water), ozone treatment with a concentration of 3 ppm for 1 min inactivated the MNV-1 virus. No changes in the raspberries were observed after 3 days of post-ozone treatment. This shows that GO was highly effective against human norovirus inactivation in the raspberries.

Recently, Lone et al. [67] studied the effectiveness of ozone treatment on the sterilization of berry fruits. Three bacterial cultures isolated from kiwi fruit was subjected to ozone treatment and they found that bacterial count was reduced from 1.5×10^8 to 1.2×10^2 CFU ml^{-1} after 60 min of ozone treatment. Hence, ozone application for microbial inactivation was faster and effective when compared with other conventional techniques. Similar studies are presented in Table 8.1.

TABLE 8.1   Effects of Ozonation on Microbial Reduction in Fruits and Vegetables

Fruits and Vegetables	Treatment(s)	Results	References
Green bell pepper	Treated with constant concentration of 0.5 mg L^{-1} of ozone for 35 min exposure	After 30 min of exposure, aerobic mesophiles reduction reached to 3.27 log for continuous ozonation in contrast to 0.85 of pre-ozonation. Overall reduction of coliforms was 3.66 log. Overall reduction of yeasts/molds was in range between 2.02–2.14 log units.	[5]
Lettuce	Treated with constant concentration of 0.5 mg L^{-1} of ozone for 35 min exposure	After 30 min of exposure, 3.04 log reduction of aerobic mesophiles upon continuous ozonation in contrast to 0.67 log pre-ozonation. Overall coliforms reduction was 2.47 log.	[5]
Papaya	Application of ozone at a concentration of 0.5 mg L^{-1} for a period of 35 min	Gaseous ozone of 9.2 ± 0.2 g L^{-1} for 20 min reduced coliforms from 0.3–1.12 log10 CFU g^{-1} and mesophilic bacteria from 0.22–0.33 log 10 CFU g^{-1}.	[7]
Red bell peppers	Treated with dissolved ozone at constant concentrations of 0.3 and 2.0 ppm for 1, 2, and 3 min exposure times.	Initial loads *L. innocua* averaged 1.9×10^7 cfu/g. Changing ozone concentration from 0.3 to 2.0 ppm, the log-reductions were higher and equal to 1.9 ± 0.5, 2.4 ± 0.6 and 2.8 ± 0.5 for 1, 2, and 3 min, respectively.	[100]
Strawberries	Treated with dissolved ozone at constant concentrations of 0.3 and 2.0 ppm for 1, 2, and 3 min exposure times.	Initial load of total mesophiles was 10^7 cfu/g. Higher microbial reduction of 2.3 ± 0.4 log cycles with high concentration of ozone treatments.	[100]

TABLE 8.1   *(Continued)*

Fruits and Vegetables	Treatment(s)	Results	References
Tomatoes	Treated with 1.71, 3.43, and 6.85 mg L^{-1} ozone (gaseous form) for 2 or 4 h of exposure. In addition, the non-inoculated tomatoes were treated under the same conditions using gaseous ozone and then stored at 10°C for 21 days.	Concentration of 6.85 mg L^{-1} gaseous ozone for 2 and 4 h of exposure reduce the Salmonella by approximately 2 log CFU/fruit. Significant reductions on total plate count on fruits treated with 6.85 mg L^{-1} ozone for 2 and 4 h and 3.43 mg L^{-1} for 2 h on first and 7 days of storage.	[32]
Watercress	Treated with dissolved ozone at constant concentrations of 0.3 and 2.0 ppm for 1, 2, and 3 min of exposure time.	Initial loads of total coliforms were 10^8 cfu g^{-1}. Significant reduction of total coliforms was obtained due to the ozone treatment, particularly at the highest concentration and higher exposure time (2 and 3 min).	[100]

8.2.1.2   MICROBIAL REDUCTION IN FRUIT JUICES

Fruit juices are mostly preserved by thermal processing technologies, but ozone treatment has taken a prominent role in processing of fruit juices, such as: apple, orange, melon, etc. Dock and Floros [105] found that ozone treatment inactivated the *E. coli* O157:H7 in apple juice very effectively when compared with other technologies. In addition, Willams et al. [122] investigated the effect of ozone, hydrogen peroxide, and dimethyl dicarbonate on *E. coli* reduction in orange juice. They achieved a 5-log reduction of *E. coli* pathogen in orange juice when treated with ozone. Similarly, Patil et al. [86] also stated a 5-log reduction of *E. coli* in <7 min of ozone treatment in orange juice. Recently, Filho et al. [9] mentioned that retention of volatile compounds (p-cymene, limonene, α-terpineol, linalool, and terpinen-4-ol) was higher in ozone-treated orange juice than the plasma-treated juice. Similar studies are summarized in Table 8.2.

TABLE 8.2 Effect of Ozonation on Microbial Reduction in Fruit Juices

Fruit Juice	Treatment(s)	Results	References
Apple	Inoculated the samples with *E. coli* (pH of 3.0, 3.5, 4.0, 4.5, and 5.0) and treated with gaseous ozone of 0.048 mg min^{-1} ml^{-1} concentration for a period of 18 min.	5-log reduction was achieved at 4 min exposure of ozone with lowest pH than the highest pH with 18 min.	[18]
	Inoculated with pathogens, namely, *E. coli O157:H7, Salmonella Typhimurium, and Listeria monocytogenes* followed by treatment with ozone gas at a concentration of 2.0–3.0 g m^{-3} with 3.0 L min^{-1} flow rate and simultaneously heated at 25, 45, 50, 55°C for up to 1 min.	Heat treatment alone at 25–55°C decreased 0.20–2.54 log CFU ml^{-1} of *E. coli* O157:H7. Ozone combined with heat treatment reduced *E. coli* O157:H7 by 1.50 (at 25°C) and 1.60 log CFU ml^{-1} at 45°C) and less than one log CFU ml^{-1} at 50 and 55°C. The reduction of *L. monocytogenes* and *S. Typhimurium* followed a similar trend.	[107]
Melon	Treated with gaseous ozone at 7.0 ± 2.4 g L^{-1} of concentration for an exposure periods of 30 and 60 min	*Alicyclobacillus acidoterrestris* spores lowered 0.73±0.02 log cycles (after 30 min exposure) and 2.22±0.04 log-cycles (after 60 min exposure).	[36]

8.2.2   INSECT CONTROL IN STORED PRODUCTS

Methyl bromide is one of the toxic fumigants and pesticides. In the food industry, methyl bromide is used for space fumigation in warehouses and other storage facilities for controlling the insects and rodents, which may spoil the grains and other food commodities. However, inhalation of high concentrations of methyl bromide leads to lung infection, failure of the central nervous system and respiratory system. Not only it is dangerous to human health, but also it is harmful to the environment as it depletes the ozone layer [77]. Therefore, the scientific community and food industries are in search of new technologies that can replace this harmful fumigant. Studies have proven that ozone technology can substitute these chemicals and is safe.

Storage of food grains is essential as they are consumed throughout the year [113]. Kells et al. [56] stated that ozone has the ability to control storage insects, such as *Tribolium confusum* in wheat and *Sitophilus zeamais and Tribolium castaneum* in maize.

Tiwari et al. [113] found that ozone can effectively remove the myco-toxins in maize and wheat. Especially ozone has the ability to modify the structural, functional, and textural properties of grains. Ozone-treated wheat kernels had low germination rate and also the energy required for milling of wheat was less after the treatment [8, 30, 113]. Bonjour et al. [15] reported that ozone applied at a dose of 70 ppm over 4 days on wheat grains was effective in the control of *L. paeta* and *L. bostrychophila* at pupae stage but not on the eggs and larvae stage. Other studies on the effect of ozone on various stored food grains are summarized in Table 8.3.

TABLE 8.3   Effect of Ozonation on Insect Control in Stored Products

Product	Treatment(s)	Results	References
Popcorn kernels	Gaseous ozone at a flow rate of 0.33 m^3 min^{-1} t^{-1} and concentrations of 500, 2200, and 4500 mg L^{-1} were used for the treatment.	At 4500 mg L^{-1}, particularly, 100% mortality of the *S. zeamais* insects achieved after 2.10 h exposure. At 500 and 2200 mg L^{-1}, the mortality rates were 82% and 93% after the exposure periods of 10 and 3.12 h, respectively.	[100]
Wheat	Treated with ozone at 0.42 and 0.84 g m^{-3} concentrations.	At 0.42 and 0.84 g m^{-3}, LT_{99} (lethal time to achieve 99% mortality) values for adults on the first day were 67 and 42 h, respectively. Similarly, LD_{99} values for adults on the first day were 28 and 36 g-h.m^{-3}, respectively. The ET_{99} value for decreasing the adult progeny production was 22 h at a concentration of 0.84 g/m^3, whereas it was 39 h at 0.42 g m^{-3}. Like LD_{99} values, ED_{99} value for adult progeny reduction was lower at 0.42 g m^{-3} of ozone concentration than at 0.84 g m^{-3}.	[106]
	Treated with ozone at concentrations of 0.21 and 0.42 g m^{-3} against one phosphine-susceptible and two phosphine-resistant strains of *Rhyzopertha dominica* for up to 24 h.	At 0.42 g m^{-3}, emergence of adults from eggs, larvae, and pupae was lowered by 96.3–100%, 32.8 to 96.6%, and 19.6 to 76.5% after exposure to 72, 34, and 10 h, respectively.	[123]

8.2.3   INACTIVATION OF FOOD ENZYMES

In the horticultural produces, the inactivation of enzymes is highly essential to achieve high-quality produce. Because enzymes like polyphenol oxidase (PPO), when exposed to the atmosphere, react with oxygen and cause some undesirable changes in the food materials, e.g., changes in color, flavor, and texture of the horticultural produces. Browning of produce is a commonly observed phenomenon due to the oxidative reaction. Thermal processing of fruits, vegetables, and fruit juices is mostly preferred as the enzymes are less heat-resistant to microorganisms. However, in some cases, enzymes are more heat-resistant than microorganisms with a higher z-value [91].

Inactivation of enzymes is a more complex process because they are composed of different matrix with different paths. Some enzymes enhance the oxidation reaction by cross-linking with the cell walls. Thus, enzymes, i.e., PPO, LOX, PG, and POD are highly responsible for the quality deterioration of most of the produce [1].

Studies on enzyme inactivation by ozonation indicate good potential especially in fruits and vegetables, and fruit juices. Ozone acts as a strong oxidizing agent that suppresses the oxidation of phenols, which may cause browning of produce. Thus, ozone treatment has proved to be effective against PPO. Azevedo et al. [12] stated that the effect of ozone and temperature on the PPO is enzyme inactivation in sugarcane to produce low color sugar. The results showed that ozone at a dose rate of 150 mgL^{-1} at 105°C was effective against PPO. Some of the studies related to ozonation on enzyme inactivation are summarized in Table 8.4.

TABLE 8.4   Effect of Ozonation on Enzyme Inactivation

Product	Treatment(s)	Results	References
Mulberry	Precooled (below 5°C) samples treated with gaseous ozone at 2 ppm.	*Polyphenol oxidase* activity was lowered when ozone treatment assisted with the precooling.	[38]
Raspberries	Treated with ozone at a dose of 8–10 ppm for 30 min at every 12 h (up to 72 h of storage).	Higher activity of phenylalanine ammonia-lyase, superoxide dismutase and ascorbate peroxidase.	[87]
Wheat	Treated with gaseous ozone at an airflow rate of 5 L/min for 5–30 min exposure times.	After the exposure of 15 min *polyphenols oxidase* activity was significantly inhibited.	[64]

8.2.4   DEGRADATION OF PESTICIDES

In the food industry, sodium chloride, sodium hydroxide, sodium hypo-chlorite, sodium bicarbonate, and acid acetic solution, etc., are used for the removal of pesticide residues from horticultural produces [72, 101]. As an alternative to conventional methods (CMs), ozone was effective in the decontamination of fruits and vegetables [93]. Ozone has the ability to oxidize the organic and inorganic compounds, such as pesticides in aqueous, solid, and gaseous phases. It oxidizes pesticides either by direct attack in acidic medium or by the radical-mediated mechanism in alkaline medium. Thus, ozone can reduce surface contamination of horticultural produces and can help in removing pesticide residues by penetrating into the food [93].

Bourgin et al. [17] stated that ozone gas effectively removed 95% of pesticide imidacloprid at a dose level of 100 mg.L^{-1} for 32 min. Strawberries were fumigated with ozone gas for 1 hour at a dose of 0.8 mg.L^{-1} and it was found that the difenoconazole concentration was decreased to <0.5 mg.kg^{-1} leading to 95% reduction in the pesticide residue [46]. De Freitas et al. [29] found 91% reduction of pirimiphos-methyl residue on maize grains, when the grains were treated with ozone at a concentration of 0.86 mg L^{-1} for 1 hour.

Similarly, Souza et al. [104] stated that the removal of difenoconazole and linuron residues on carrots was increased with an increase in the concentration of ozone. More than 80% of the pesticide residues were removed when the roots were exposed to 5–10 mg.L^{-1} ozone concentration for 2 hours. After storage, around 95% reductions of both the pesticides were documented. Hence, ozone is an organic strategy for the removal of pesticide residues. Similar studies on the effect of ozone for pesticide degradation on fruits and vegetables are reported in Table 8.5.

TABLE 8.5   Effect of Ozonation on Pesticide Degradation

Product	Treatment(s)	Results	References
Grapes	Exposed to O$_3$ at a concentration of 10 m mol/mol for 60 min.	Effectively degraded the fungicide residues such as fenhexamid, cyprodinil, and pyrimethanil up to 68.5, 75.4, and 83.7% of reduction, respectively.	[53]
Lemon	Aqueous O$_3$ at a dose of 10 m mol/mol for an exposure of 5 min	Sufficiently oxidized the residues of pesticides which include chlorpyrifos ethyl, tetradifon, and chlorothalonil up to 48.5%, 59.9, and 92%, respectively.	[60]

Product	Treatment(s)	Results	References
Tomato	Immersed in water having bubbling O_3 at 1 and 3 mg L^{-1}	70 to 90% reduction of the three fungicide residues, namely, chlorothalonil, azoxystrobin, and difenoconazole was achieved at 3 mg L^{-1} ozone treatment.	[93]
Wheat	Treated with ozone concentration of 60 m mol/mol at 30–180 min exposure for the removal of pesticide residues of deltamethrin and fenitrothion	After 180 min of exposure, fenitrothion content was lowered by 66.7%. The residues of deltamethrin were decreased by 67.5, 88.1, and 89.8% after an exposure of 60, 120, and 180 min, respectively.	[98]

8.2.5 DEGRADATION OF TOXIC COMPONENTS

Mycotoxins, which belong to some species of molds, i.e., *aspergillus, penicillium,* and *fusarium,* are mostly found in food and consumption of such contaminated foods are harmful to our health. Several countries have been restricting the use of mycotoxins contaminated materials in food industries [41]. On an average, it was found that nearly 25% of the world's crop is lost due to aflatoxin contamination [96]. At present, there is a lack of technology to completely eliminate the contamination caused by aflatoxins in food products. However, ozone processing is an effective technology to manage the mycotoxins in the food [74].

Mycotoxin degradation usually follows a pseudo-first-order rate when the ozone is supplied continuously. Aflatoxins, AFB_1 and AFG_1 degrade at a higher rate compared to AFB_2 and AFG_2 because the former contains 8, 9 double bond-forming the vinyl ether at the terminal furan ring. The latter requires a higher exposure time for degradation. The efficiency of ozone on aflatoxin degradation increases with an increase in temperature and exposure time of ozone treatment [78]. Ozonation was more effective in the treatment of wheat kernels than wheat flour, because wheat kernels can be easily exposed to gaseous ozone (GO) [88].

Later, Inan et al. [47] studied the removal of aflatoxin B_1 in red pepper and found that a significant reduction of aflatoxin B_1 of 80% in red pepper following the ozone treatment. Aflatoxin B_1 removal from dried figs was studied and found that ozone (gaseous phase) was effective in the removal of aflatoxin than the aqueous ozone [127]. Similarly, ozone can effectively remove the patulin toxin without affecting the nutritional parameters of fruit juices [24]. Various studies associated with the effect of ozone on toxic component degradation are presented in Table 8.6.

TABLE 8.6   Effect of Ozonation on Degradation of Toxic Components

Product	Treatment(s)	Results	References
Alfalfa	Inoculated with Shiga toxin-producing *Escherichia coli* (STEC) and *Salmonella* were separately treated with aqueous O_3 at 5 mg L^{-1} dose for various exposure times.	After 20 min exposure, the mean log reductions for *Salmonella* and STEC were found on seeds as 2.1 ± 0.5, and 2.1 ± 0.5, respectively.	[73]
Corn	Treated with ozone at a dose of 5–40 mg L^{-1} and various reaction times (0–120 s) to detoxify ochratoxin A (OTA) and zearalenone (ZEN).	Zearalenone was not removed by ozone of 5 s treatment at 10 mg L^{-1} concentration. OTA was degraded to 65.4% after 120 s of treatment. Higher ozone dose and fumigation time increased the degradation rates of ZEN and OTA. When corn (19.6% MC) treated with ozone at a dose of 100 mg.L^{-1} for 180 min, ZEN, and OTA were reduced by 90.7% and 70.7%, respectively.	[90]
Wheat	Treated with ozone at a concentration of 60 µmol/mol for 40–120 min exposure	Inhibited *F. graminearum* and *P. citrinum* colonies.	[98]
	Treated with ozone concentrations of 0–80 mg L^{-1}, reaction times of 0, 2, 4, 8, and 12 h at moisture contents of 9.52% and 17.04%	When 10 mg.L^{-1} gaseous ozone exposed to 1 µg ml^{-1} of deoxynivalenol solution in wheat, the rate of degradation was found as 93.6%.	[63]

8.2.6   STARCH MODIFICATION

Starch is mostly available in roots, tubers, pulses, and cereals [95]. In the food industry, starch is used in the preparation of various food products, such as pasta, bread, dairy products, breakfast cereals, confectionery, etc., [117]. The main purpose of utilization of starch in food industries is to control the viscosity of foods and maintain their stability during processing as well as storage [55].

Sometimes the properties of starch can be affected during its application in the food industry. Thus, modification of starch is another alternative to improve its functional properties. Several methods are used to modify starch, such as: physical, chemical, genetic, and enzymatic methods but chemical method have been very effective in the food industry because it enhances the textural properties of gel, improves film adhesion, etc., [23]. However, the CMs are time-consuming, leaving toxic wastes and deposition of chemical residues on treated foods. To avoid these problems, ozone is an emerging technology employed in the food industry for the modification of starch [83].

The main advantage of modification of starch using ozone is that ozone reacts with starch even at room temperature. Controlled environment and catalysts are not necessary for an effective modification of starch. In addition, ozonation has a positive impact on the modified starch properties. Studies have stated that absorption capacity, swelling power, and solubility of wheat flour were increased after ozonation. Rise in the properties like color, viscosities, paste clarity, greater cooking stability, low retrogradation potential, and higher strength of dough, etc., are various positive impacts of ozonation on the starch properties [85].

Ozone application leads to high recovery rates compared with CMs [25]. A comparative study on wheat starch indicated that application of ozone for a period of 30 and 45 min did not show any variations in granular morphology between ozonated and control samples [97]. In addition, oxidation of cassava starch by ozone in aqueous solutions at different pH levels of 3.5, 6.5, and 9.5 indicated that ozone treatment was more effective at a pH of 3.5 to reduce the final viscosity and breakdown of cassava starch [59]. Similar studies are reported in Table 8.7.

TABLE 8.7   Effect of Ozonation on Starch Modification of Food Products

Product	Treatment(s)	Results	References
Cassava	Treated at a concentration of 41 mg of ozone L^{-1} and a gas flow of 1 L min^{-1} for an exposure of 15 and 30 min	Increased young's modulus, tensile strength, hydrophilic surface and lowered solubility, and elongation were found as in the case of biodegradable films that were produced with ozonated cassava starch. In addition, increased exposure time enhanced water vapor transmission and oxygen permeation.	[61]

TABLE 8.7   *(Continued)*

Product	Treatment(s)	Results	References
Potato	Processed with ozone concentration of 47 mg L^{-1} (at the flow rate of 5 L min^{-1}) for 15, 30, 45 and 60 min exposure periods.	Carbonyl, carboxyl, and reducing sugar contents were increased, but pH, molecular size, and apparent amylose content was decreased.	[21]
Wheat	Treated with gaseous O_3 at a dose of 0.00042 g dissolved ozone/100 g for an exposure of 15, 30 and 60 min at 5°C.	Cooking stability and the shear-thinning behavior were increased but retrogradation tendency and pasting temperature were decreased.	[22]

8.3   EFFECT OF OZONIZATION ON FOOD QUALITY

Ozone is very effective in providing safe food to the consumer in terms of the overall quality of the final product. Several studies were conducted to know the ozonation effect on the quality attributes of horticultural produces. A limited dosage of ozone was effective for processing of foods in food industry. Ozone application beyond certain concentration leads to oxidation of foods. Hence, the concentration and time must be optimized to have an effective utilization of ozone.

8.3.1   FRUITS AND VEGETABLES

Among the sensory attributes of fresh produce (form. Color, taste, texture, appearance, and aroma), color is an important parameter of agro-products as it is a sign of senescence. During the storage of agricultural products, such as, apples, papaya, tomato, etc., ozone treatment did not show any significant effect on the color of the produce. Ozone treatment at a dose of 38 to 95 µg L^{-1} for exposure time of 10 min during the storage period delayed the formation of red color in tomatoes [124]. In case of papaya, there was a change in the color of papaya peel when exposed to GO at a dose of 4.5 µg L^{-1} for 96 h [7]. These results reveal that the possible cause for discoloration might be due to the reaction of ozone with carotenoids in the food products.

Other important quality parameters of food products are antioxidant compounds, such as flavonoids, polyphenols, fatty acids, and vitamins. Ozone with its oxidizing activity can cause loss of antioxidant compounds.

However, there are no significant changes in the phenolic content profile in lettuce when treated with ozone [13]. Zhang et al. [125] stated that there was not significant variation in the ascorbic acid (AA) (vitamin C) content in the celery sand lettuce samples, which were treated with ozonated water. In addition, slight increment in the ascorbic acid content was observed in spinach, pumpkin leaves, and strawberries under ozone treatment. In addition, a slight decrease in ascorbic acid level was observed in broccoli florets. The increase in vitamin C level of the product was due to the presence of an antioxidative property, which can synthesize vitamin C [52].

Kiris et al. [58] reported that olives ozonated for 5 min has no effect on the fatty acid profile in olives. Minor changes in the anthocyanin content of strawberries and blackberries were observed when ozonated at various concentrations. Blackberries were ozonated at 0.1 and 0.3 ppm of ozone and stored in ambient conditions, and it was observed that anthocyanin content was almost stable in blackberries at 0.1-ppm concentration, whereas it fluctuated at 0.3 ppm concentration [114].

Ozone at low concentrations of 0.04 ml L^{-1} was more effective in the shelf-life extension of broccoli and seedless cucumbers under refrigerated conditions. In addition, ozone at a concentration of 0.4 ml. L^{-1} was effective in removing ethylene gas during the storage of apples and pears, leading to no quality loss during the whole storage period [102].

Wang et al. [120] stated that coriander leaves retained their fresh appearance and color during the storage period when treated with ozone. The overall acceptability of ozone-washed coriander leaves was highest compared with non-ozonated coriander leaves. The results on the impact of ozonation on apple bioactive compounds during short-term cold storage [68] showed that the total phenols content in apple was decreased slightly after one month of storage. In addition, no significant changes in TSS and acidity were observed in the ozone-treated apples during the storage period. Similar studies are reported in Table 8.8.

8.3.2 FRUIT JUICES

Fruit juices contain bioactive compounds with minimal loss in taste, flavor, vitamins, etc., during the processing and their storage. Ozone treatment of fruit juices causes a non-significant color loss due to oxidation of chemicals. For example, strawberry juice when subjected to ozone treatment of concentration 7.8% w/w for 10 min resulted in 98.2% reduction of pelargonidin-3-glucoside content [114]. Also, under the same treatment conditions of blackberry juice, >90% reduction in the cyanidin-3-glucoside content was observed. The reduction in anthocyanin content is mainly due to the direct reaction with ozone or indirect reaction with secondary oxidants (such as HO^2, OH, O_2^-, and O_3^- thus leading to nucleophilic and electrophilic reactions with aromatic ring.

TABLE 8.8   Effect of Ozonation on Quality of Horticultural Produces

Produce	Treatment(s)	Results	References
Carrot	Treated with gaseous ozone at a concentration of 0–5 mg L^{-1} and aqueous ozone of 0–10 mg.L^{-1} for 60 min at 14°C and also the quality evaluated during 5 days storage.	The average weight loss of carrots at the end of 5 days was 20% and 14% with ozone gas and dissolved gas in water was 20% and 14%, respectively. ΔE and chroma, showed a degree of saturation, purity or intensity of color, were also not changed. No change in pH values (remained 6) in ozonated water treated samples but a slight drop as in case of ozone gas treated samples. From day 4 of storage, solids content increased.	[104]
Mulberry	Pre-cooled (below 5°C) and then treated with ozone (gaseous phase) at a dose of 2 ppm.	Increase in the level of total soluble solids (TSS), titratable acidity, better retention in color and firmness, and lowered decay rate and respiratory intensity.	[45]
Papaya	Treated with ozone of 9.2 ± 0.2 l/L at 10 to 30 min	The total phenolic content of the treated sample with gaseous ozone of 9.2 ± 0.2 l/L for 20 min increased by 10.3% while the ascorbic acid level decreased by 2.3%.	[7]

Torres et al. [112] studied the impact of ozone treatment on the phenolic content, rheological properties and color of apple juice, in which apple juice samples were treated at ozone concentration of 1–4.8% w/w and processing time of 0–10 min. Significant quality changes in the juice were observed during the ozonation process. Gabriela et al. [37] conducted ozone processing of peach juice at a dose rate of 0.06–2.48 g.L^{-1} and they indicated no significant changes in the pH, °Brix, and titratable acidity (TA) during the ozonation process. Similar studies are also presented in Table 8.9.

TABLE 8.9   Effect of Ozone Treatment on Quality of Fruit Juices

Fruit Juice	Treatment(s)	Results	References
Apple	Treated with ozone at various concentrations up to 4.8% w/w for different processing time up to 10 min.	Significant decrease in quality parameters includes rheological properties, color, and phenolic content.	[112]
Apple	Ozone with different concentrations of 2.8 and 5.3 mg L^{-1} at 4 and 22°C was exposed up to 40 min.	At a dose of 5.3 mg L^{-1}, total phenolic content was reduced but at 2.8 mg L^{-1}, there was no significant reduction.	[110]
Melon	Exposed to ozone gas at a dose of 7.0 ± 2.4 g L^{-1} for 30 and 60 min.	There was a significant increase in total phenolics content, but color, total carotenoids, antioxidant activity, and vitamin C were altered after the exposures of 30 and 60 min.	[36]

8.3.3   EFFECTS ON STORED GRAINS

Fumigation of storage rooms with ozone was highly efficient to control the storage pests. The efficiency of ozone depends on the concentration of ozone, moisture content of grains, exposure time of grains to ozone, velocity of ozone, and insect species. The mechanism for the control of insects in grain storage is based on the attack by ozone on the respiratory system of insects. Due to this, insects respire discontinuously, leading to the alteration in pulmonary function, oxidation of cell membranes and DNA strand breaks. The high concentration of ozone is unfavorable for the reproduction of insects under low humidity conditions. This leads to the opening of spiracles of insects, resulting in rapid loss of water from their body. Thus, ozone is highly efficient in controlling insects/pests in storage structures [85].

Ozonation of stored grains results in changes in quality attributes of grains (Table 8.10). Mendez et al. [70] stated that the ozone treatment (@ of 50 ppm) of grains for a period of about 30 days caused significant changes in the popping volume of popcorn. Similarly, fatty acid profile, amino acid composition, and milling characteristics of wheat, maize, and soybean were not affected during the ozone treatment. In addition, no detrimental changes were found in the stickiness of rice and baking characteristics of wheat. Mostly in cereals, ozone reacts with sulfhydryl-containing proteins leading to changes in product quality. However, Prudente and King [89] reported that ozone-treated corn indicated no significant changes in the fatty acid composition. Similarly, the impact of ozone treatment (50 ppm at a flow rate of 8 L min^{-1} for 24 to 48 h) on the quality of corn grains revealed that lipid content of corn grain and corn oil showed little changes [33].

The ozonation at 50 ppm had no effect on the germination of corn. Similarly, long exposure of grains to ozone (at a concentration of 0.98 mg.g^{-1} for 45 min) did not affect the germination. However, the rate of germination was affected when a higher dose of ozone was used for a shorter time. Finally, the efficiency of ozonation on stored grains not only depends on the nature of grains but also on the conditions of ozonation, such as time, temperature, moisture content, etc. [118].

TABLE 8.10   Effect of Ozone Treatment on Quality of Stored Grains

Product	Treatment(s)	Results	References
Corn	Ozonated at 100 mg L^{-1} for an exposure of 180 min at different moisture level of 14.1% and 19.6%	After 180 min, the moisture content of initial corn and humidified corn lowered from 14.1% to 13.2%, and 19.6% to 15.3%, respectively. There was a significant change in color. The whiteness and yellowness increased and decreased with increasing time, respectively. There was a significant increase in fatty acid value.	[90]
Peanuts	Ozonated at 6.0 mg L^{-1} for an exposure period of 30 min	No significant differences in peroxide value, acid value, resveratrol, and polyphenols	[26]
Popcorn kernels	Gaseous ozone or atmospheric air (control) at a flow rate of 0.33 m^3 min^{-1} and concentrations of 500, 2200, and 4500 mg L^{-1} were used.	An increase in ozone concentration did not alter the water content of grains. No significant difference between control and the ozone-treated kernels at 500 mg L^{-1}. The expansion volume of grains treated at 2200 and 4500 mg L^{-1} decreased by 4.3%	[34]

8.4 CURRENT CHALLENGES

Ozone processing of foods is one of the novel techniques in which product quality can be improved compared with other processing techniques with a significant saving in cost and energy consumption. Some challenging issues are:

- At the outset, an effective ozone generation system must be designed since it is a highly complex process and the half-life of ozone is too short. The equipment for ozone generation should be corrosion-free.
- The concentration of ozone and time of exposure must be optimized for various food products, because exposure to the high concentration of ozone leads to negative effects not only on the final quality of products but also on the human health. It is the maturity index and variety of fruit or vegetable, which determines the optimum dosage of ozone in processing. Hence, those factors are more important in optimizing the ozone treatment of fruits and vegetables.
- Consumers are still not aware of the practical utilization of this technique due to some misconceptions on the toxic properties of ozone. Hence, to achieve consumer acceptance, qualitative research on ozone has to be conducted, and information on ozone-based technologies as well as benefits of ozone should be provided to the consumers.

8.5 FUTURE RESEARCH WORK

GO is highly effective compared to aqueous ozone for microbial inactivation on fruits and vegetables, but new operating procedures are to be identified for inactivation of harmful viruses like HAV on fruits. For complex food systems, there should be detailed information about the effect of various food ingredients on the final quality and also inactivation of harmful pathogens.

Mostly in fruit juice processing, care should be taken on the by-products of organic compounds obtained during the processing. These should be identified and analyzed for their health benefits. Suitable models should be developed by employing various software to describe complex reactions and inactivation of microorganisms, which helps in controlling the quality aspects of ozonated foods. Research on evaluating the molecular weight of different starch sources for finding out the alterations in starch properties is also a priority. Moreover, further studies should be carried out by utilizing various starch modification methods, and also the combination of technologies should be studied for the effective processing of foods.

8.6 SUMMARY

This chapter includes discussions on: effects of ozonation on various fruits and vegetables, fruit juices, and stored grains; enzyme inactivation of foods using ozone and toxic compounds like aflatoxin degradation; role of ozone in the modification of starch and its use in food industry; and effects of ozone on the quality of horticultural produce. Ozone generation is relatively easy, eco-friendly, and requires no storage facility. Ozone concentration, exposure time and other treatment parameters should be optimized for effective use of ozone in the food industry. Preliminary trials should be conducted before commercializing the ozonation process for any product. Research is needed on ozone processing of foods in combination with ionization and pulsed light (PL) to provide safe foods to the consumers.

KEYWORDS

- aflatoxin
- enzyme inactivation
- food quality
- food safety
- ozone
- pesticide degradation

REFERENCES

1. Agcam, E., Akyildiz, A., & Dundar, B., (2018). Thermal pasteurization and microbial inactivation of fruit juices: Chapter 17. In: Rajauria, G., & Kumar, T. B., (eds.), *Fruit Juices-Extraction, Composition, Quality and Analysis* (pp. 309–339). Apple Academic Press Inc.
2. Ahmad, S., Raghunathan, S., Davoodbasha, M., Srinivasan, H., & Lee, S., (2019). An investigation on the sterilization of berry fruit using ozone: Option to preservation and long-term storage. *Biocatalysis and Agricultural Biotechnology*.
3. Ahvenainen, R., (1996). New approaches in improving the shelf-life of minimally processed fruits and vegetables. *Trends in Food Science and Technology, 7,* 179–187.

4. Alexandre, E. M. C., & Santos-Pedro, D. M., (2011). Influence of aqueous ozone, blanching and combined treatments on microbial load of red bell peppers, strawberries, and watercress. *Journal of Food Engineering, 105*(2), 277–282.

5. Alexopoulos, A., Plessas, S., Ceciu, S., & Lazar, V., (2013). Evaluation of ozone efficacy on the reduction of microbial population of fresh-cut Lettuce (*Lactuca Sativa*) and green bell pepper (*Capsicum Annuum*). *Food Control, 30*(2), 491–496.

6. Alexopoulos, A., Plessas, S., Kourkoutas, Y., & Stefanis, C., (2017). Effect of ozone upon the microbial flora of commercially produced dairy fermented products. *International Journal of Food Microbiology, 246*, 5–11.

7. Ali, A., Kying, M., & Forney, C. F., (2014). Effect of ozone pre-conditioning on quality and antioxidant capacity of papaya fruit during ambient storage. *Food Chemistry, 142*, 19–26.

8. Allen, B., Wu, J. N., & Doan, H., (2003). Inactivation of fungi associated with barley grain by gaseous ozone. *Journal of Environmental Science and Health Part B-Pesticides, Food Contaminants and Agricultural Wastes, 38*, 617–630.

9. Alves, F. E. G., Rodrigues, T. H. S., & Fernandes, F. A. N., (2019). Untargeted chemometric evaluation of plasma and ozone processing effect on volatile compounds in orange juice. *Innovative Food Science and Emerging Technologies, 53*, 63–69.

10. Aparecida, A. R. Z., & De Queiroza, M. E. L. R., (2019). Use of ozone and detergent for removal of pesticides and improving storage quality of tomato. *Food Research International, 125*, 108626.

11. Ashiya, K., (2019). *What is the Importance of Food Preservation?* www.preservearticles.com/articles/what-is-the-importance-of-food-preservation/5187 (accessed on 16 January 2021).

12. Azevedo, Borba, D. A. C., & Honorato, D. S. F. L., (2019). Enzymatic *Polyphenoloxidase* inactivation with temperature and ozone in sugarcane variety RB 92579 to produce lower color Sugar. *Brazilian Journal of Food Technology, 22*, 2018043.

13. Beltran, D., Selma, M. V., Marin, A., & Gil, M. I., (2005). Ozonated water extends the shelf life of fresh-cut lettuce. *Journal of Agricultural and Food Chemistry, 53*(14), 5654–5663.

14. Beuchat, L. R., (1992). Surface disinfection of raw produce, *Dairy, Food, and Environmental Sanitation, 12*(1), 6–9.

15. Bonjour, E. L., Opit, G. P., Hardin, J., Jones, C. L., Payton, M. E., & Beeby, R. L., (2011). Efficacy of ozone fumigation against the major grain pests in stored wheat. *Journal of Economic Entomology, 104*(1), 308–316.

16. Botondi, R., Bartoloni, S., Baccelloni, S., & Mencarelli, F., (2015). Biodegradable polylactic acid hinged trays keep quality of fresh-cut and cooked spinach. *Journal of Food Science and Technology, 52*, 5938–5945.

17. Bourgin, M., Albet, J., & Violleau, F., (2013). Study of the degradation of pesticides on loaded seeds by ozonation. *Journal of Environmental Chemical Engineering, 1*, 1004–1012.

18. Bourke, P., Patil, S., Valdramidis, V. P., Cullen, P. J., & Frias, J., (2010). Inactivation of *Escherichia coli* by ozone treatment of apple juice at different pH levels. *Food Microbiology, 27*, 835–840.

19. Brie, A., Boudaud, N., Mssihid, A., Loutreul, J., Bertrand, I., & Gantzer, C., (2018). Inactivation of murine norovirus and hepatitis a virus on fresh raspberries by gaseous ozone treatment. *Food Microbiology, 70*, 1–6.

20. Carletti, L., Botondi, R., Moscetti, R., Stella, E., Monarca, D., & Cecchini, M., (2013). Use of ozone in sanitation and storage of fruits and vegetables. *Journal of Food Agriculture Environment, 11,* 585–589.
21. Castanha, N., Matta, J. M. D., & Augusto, P. E. D., (2017). Potato starch modification using the ozone technology. *Food Hydrocolloids, 66,* 343–356.
22. Çatal, H., & Ibanoglu, S., (2014). Effect of aqueous ozonation on the pasting, flow and gelatinization properties of wheat starch. *LWT-Food Science and Technology, 59*(1), 577–582.
23. Catal, H., & Ibanoglu, S., (2012). Ozonation of corn and potato starch in aqueous solution: Effects on the thermal, pasting, and structural properties. *International Journal of Food Science and Technology, 47*(9), 1958–1963.
24. Cataldo, F., (2008). Ozone decomposition of patulin: Mycotoxin and food contaminant. *Ozone Science Engineering, 30,* 197–201.
25. Chan, H. T., Fazilah, A., Bhat, R., Leh, C. P., & Karim, A. A., (2012). Effect of deproteinization on degree of oxidation of ozonated starch. *Food Hydrocolloids, 26*(2), 339–343.
26. Chen, R., Ma, F., Li, P. W., Zhang, W., & Ding, X. X., (2014). Effect of ozone on aflatoxins detoxification and nutritional quality of peanuts. *Food Chemistry, 146,* 284–288.
27. Christ, D., Renata, S., Coelho, M., & Aparecida, C., (2019). Evaluation of grain and oil quality of packaged and ozonized flaxseed. *Journal of Stored Products Research, 83,* 311–316.
28. Cullen, P. J., Tiwari, B. K., & O'Donnell, C. P., (2009). Modeling approaches to ozone processing of liquid foods. *Trends in Food Science and Technology, 20*(3), 125–136.
29. De Freitas, R. S., & Faroni, L. R. D. A., (2017). Degradation kinetics of pirimiphos-methyl residues in maize grains exposed to ozone gas. *Journal of Stored Products Research, 74,* 1–5.
30. Desvignes, C., Chaurand, M., Dubois, M., & Sadoudi, A., (2008). Changes in common wheat grain milling behavior and tissue mechanical properties following ozone treatment. *Journal of Cereal Science, 47,* 245–51.
31. Diva, F., Almeida, L., & Cavalcante, R. S., (2015). Effects of atmospheric cold plasma and ozone on prebiotic orange juice. *Innovative Food Science and Emerging Technologies, 32,* 127–135.
32. Fan, X., Wang, L., Sokorai, K., & Sites, J., (2019). Quality deterioration of grape tomato fruit during storage after treatments with gaseous ozone at conditions that significantly reduced populations of salmonella on stem scar and smooth surface. *Food Control, 103,* 9–20.
33. Faroni, L. R. D., Pereira, A. M., & Sousa, A. D., (2007). Influence of corn grain mass temperature on ozone toxicity to *Sitophilus zeamais* (*Coleoptera: Curculionidae*) and quality of oil extracted from ozonized grains. In: *IOA Conference and Exhibition* (pp. 1–6). Valencia, Spain.
34. Faroni, L. R. A., Silva, M. V. A., Sousa, A. H., Prates, L. H. F., & Abreu, A. O., (2019). Kinetics of the ozone gas reaction in popcorn kernels. *Journal of Stored Products Research, 83,* 168–175.
35. FDA (Food and Drug Administration), (2001). Secondary direct food additives permitted in food for human consumption. *Rules Regulations, 64,* 44122, 44123.

36. Fundo, J. F., Miller, F. A., Tremarin, A., & Garcia, E., (2018). Quality assessment of cantaloupe melon juice under ozone processing. *Innovative Food Science and Emerging Technologies, 47*, 461–466.

37. Gabriela, M. J., Analia, B. G. L., Paula, L. G., & Stella, M. A., (2017). Ozone processing of peach juice: Impact on physicochemical parameters, color, and viscosity. *Ozone: Science and Engineering, 40*(4), 305–312.

38. Gao, H., Han, Q., Chen, H., Fang, X., & Wu, W., (2017). Precooling and ozone treatments affects postharvest quality of black mulberry (*Morus nigra*) fruits. *Food Chemistry, 221*, 1947–1953.

39. García-Martín, J. F., Olmo, M., & María, (2018). Effect of ozone treatment on postharvest disease and quality of different citrus varieties at laboratory and at industrial facility. *Journal of Postharvest Biology and Technology, 137*, 77–85.

40. Gibson, K. E., Almeida, G., Jones, S. L., Wright, K., & Lee, J. A., (2019). Inactivation of bacteria on fresh produce by batch wash ozone sanitation. *Food Control, 106*, 106747.

41. Gilbert, J., & Anklam, E., (2002). Validation of analytical methods for determining mycotoxins in foodstuffs. *Trends in Analytical Chemistry, 21*, 468–486.

42. Granella, S. J., Christ, D., Werncke, I., Bechlin, T. R., & Coelho, R. M., (2018). Effect of drying and ozonation process on naturally contaminated wheat seeds. *Journal of Cereal Science,* 205–211.

43. Guzel-Seydim, Z. B., Greene, A. K., & Seydim, A. C., (2004). Use of ozone in the food industry. *LWT-Food Science and Technology, 37*(4), 453–460.

44. Habibi-Najafi, M. B., & Haddad-Khodaparast, M. H., (2009). Efficacy of ozone to reduce microbial population in date fruits. *Food Control, 20*, 27–30.

45. Han, Q., Gao, H., Chen, H., Fang, X., & Wu, W., (2017). Precooling and ozone treatments affects postharvest quality of black mulberry (*Morus Nigra*) fruits. *Food Chemistry, 221*, 1947–1953.

46. Heleno, F. F., Queiroz, M. E. L., & Neves, A. A., (2014). Effects of ozone fumigation treatment on the removal of residual difenoconazole from strawberries and on their quality. *Journal of Environmental Science and Health, 49*, 94–101.

47. Inan, F., Pala, M., & Doymaz, I., (2007). Use of ozone in detoxification of aflatoxin B1 in red pepper. *Journal of Stored Products Research, 43*, 425–429.

48. Isikber, A. A., & Athanassiou, C. G., (2014). The use of ozone gas for the control of insects and micro-organisms in stored products. *Journal of Stored Products Research,* 1–7.

49. Jegadeeshwar, H., & Vijay, S. S., (2017). *A Review on Utilization of Ozone in Food Preservation.* Garuda Graphics Publisher; Guideline 21.

50. Kader, A. A., (1997). Quality in relation to marketability of fruits and vegetables. In: Studman, C. J., (ed.), *Proceeding of the 5th Fruit Nut and Vegetable Engineering Symposium* (Vol. 4, No. 3, pp. 243–272). University of California, Davis.

51. Keat, W., Ali, A., & Forney, C. F., (2014). Postharvest biology and technology effects of ozone on major antioxidants and microbial populations of fresh-cut papaya. *Postharvest Biology and Technology, 89*, 56–58.

52. Karaca, H., & Velioglu, Y. S., (2007). Ozone applications in fruit and vegetable processing. *Food Reviews International, 23*, 91–106.

53. Karaca, H., Walse, S. S., & Smilanick, J. L., (2012). Effect of continuous gaseous ozone exposure on fungicide residues on table grape berries. *Postharvest Biology and Technology, 64*(1), 154–159.

54. Katz, D., (1986). *Ozone Technology* (pp. 28–36). Golf Course Management.
55. Kaur, B., Ariffin, F., Bhat, R., & Karim, A., (2012). A. progress in starch modification in the last decade. *Food Hydrocolloids, 26,* 398–404.
56. Kells, S. A., Mason, L. J., & Maier, D. E., (2001). Efficacy and fumigation characteristics of ozone in stored maize. *Journal of Stored Products Research, 37,* 371–382.
57. Khadre, M. A., Yousef, A. E., & Kim, J. G., (2001). Microbiological aspects of ozone applications in food: Review. *Journal of Food Science, 66,* 1242–1252.
58. Kiris, S., Velioglu, Y. S., & Tekin, A., (2017). Effect of ozonated water treatment on fatty acid composition and some quality parameters of olive oil. *Ozone: Science and Engineering, 31*(2), 91–96.
59. Klein, B., Vanier, N. L., Moomand, K., & Pinto, V. Z., (2014). Ozone oxidation of cassava starch in aqueous solution at different pH. *Food Chemistry, 155,* 167–173.
60. Kusvuran, E., Yildirim, D., Mavruk, F., & Ceyhan, M., (2012). Removal of chlorpyrifos ethyl, tetradifon and chlorothalonil pesticide residues from citrus by using ozone. *Journal of Hazardous Materials, 241–242,* 287–300.
61. La, C. I. A., Tamyris, A., Souza, D., Tadini, C. C., Esteves, P., & Augusto, D., (2019). Ozonation of cassava starch to produce biodegradable films. *International Journal of Biological Macromolecules, 141,* 713–720.
62. Li, Q., Bao, X., Lu, C., Zhang, X., Zhu, J., Jiang, Y., & Liang, W., (2012). Soil microbial food web responses to free-air ozone enrichment can depend on the ozone-tolerance of wheat cultivars. *Soil Biology and Biochemistry, 47,* 27–35.
63. Li, M. M., Guan, E. Q., & Bian, K., (2015). Effect of ozone treatment on deoxynivalenol and quality evaluation of ozonized wheat. *Food Additives and Contaminants: Part A., 32,* 544–553.
64. Li, M., Peng, J., Zhu, K., Guo, X., Zhang, M., Peng, W., & Zhou, H., (2013). Delineating the microbial and physical-chemical changes during storage of ozone-treated wheat flour. *Innovative Food Science and Emerging Technologies, 20,* 223–229.
65. Lin, S., Chen, C., Luo, H., Xu, W., Zhang, H., Tian, J., Ju, R., & Wang, L., (2019). The combined effect of ozone treatment and polyethylene packaging on postharvest quality and biodiversity of *Toona Sinensis. Postharvest Biology and Technology, 154*(5), 1–10.
66. Loeb, B. L., (2011). Ozone: Science and engineering: Thirty-three years and growing. *Ozone: Science and Engineering, 33,* 329–342.
67. Lone, S. A., Raghunathan, S., & Davoodbasha, M. A., (2019). Investigation on the sterilization of berry fruit using ozone: Option to preservation and long-term storage. *Biocatalysis and Agricultural Biotechnology, 20,* 101212.
68. Lv, Y., & Tahir, I. I., (2019). Effect of ozone application on bioactive compounds of apple fruit during short-term cold storage. *Scientia Horticulturae, 253,* 49–60.
69. Mahapatra, A. K., Muthukumarappan, K., & Julson, J. L., (2005). Applications of ozone, bacteriocins and irradiation in food processing: Review. *Critical Reviews in Food Science and Nutrition, 45,* 447–461.
70. Mendez, F., Maier, D., Mason, L., & Woloshuk, C. P., (2003). Penetration of ozone into columns of stored grains and effects on chemical composition and processing performance. *Journal of Stored Products Research, 39,* 33–44.
71. Miller, G. W., Rice, R. G., Robson, C. M., Scullin, R. L., Kuhn, W., & Wolf, H., (1978). *An Assessment of Ozone and Chlorine Dioxide Technologies for Treatment of Municipal Water Supplies* (pp. 13–43). US Environmental Protection Agency Report No. EPA-600/2-78-147; Washington, DC: US Government Printing Office.

72. Mirani, B. N., Sheikh, S. A., Nizamani, S. M., & Mahmood, N., (2013). Effect of household processing in removal of lufenuron in tomato. *International Journal of Agricultural Science Research, 3,* 235–244.

73. Mohammad, Z., Kalbasi-ashtari, A., Riskowski, G., & Castillo, A., (2019). Reduction of salmonella and Shiga toxin-producing *Escherichia coli* on alfalfa seeds and sprouts using an ozone generating system. *International Journal of Food Microbiology, 289,* 57–63.

74. Naitou, S., & Takahara, H., (2006). Ozone contribution in food industry in Japan. *Ozone Science Engineering, 28,* 425–429.

75. Obetta, S. E., Nwakonobi, T. U., & Adikwu, O., (2011). A. microbial effects on selected stored fruits and vegetables under ambient conditions in Makurdi-Nigeria. *Research Journal of Applied Sciences, Engineering and Technology, 3*(5), 393–398.

76. Odilichukwu, C., Okpala, R., Bono, G., & Falsone, F., (2016). Aerobic microbial inactivation kinetics of shrimp using a fixed minimal ozone discharge: Facts during iced storage ? *Italian Oral Surgery, 7,* 47–52.

77. Olmez, H., & Dogan, H., (2002). Applications of ozone in food industry: An alternative to methyl bromide and chlorine. In: *Priority Thematic Area of Research FP 6* (p. 23). Marmara Research Center: Turkey.

78. Otniel, F. S., & Armando, V., (2010). Ozone applications to prevent and degrade mycotoxins: A review. *Drug Metabolism Reviews, 42*(4), 612–620.

79. Öztekin, S., Evliya, I. B., Zorlugenç, B., & Kırog, F., (2008). The influence of gaseous ozone and ozonated water on microbial flora and degradation of aflatoxin b1 in dried figs. *Food and Chemical Toxicology, 46,* 3593–3597.

80. Palou, L., Smilanick, J. L., Crisosto, C. H., & Mansour, M., (2001). Effect of gaseous ozone exposure on the development of green and blue molds on cold-stored citrus fruit. *Plant Disease, 85,* 632–638.

81. Patil, S., Valdramidis, V. P., Cullen, P. J., Frias, J., & Bourke, P., (2010). Inactivation of *Escherichia coli* by ozone treatment of apple juice at different pH levels. *Food Microbiology, 27*(6), 835–840.

82. Pandiselvam, R., Kothakota, A., & Thirupathi, V., (2016). Numerical simulation and validation of ozone concentration profile in green gram Vigna radiate bulks. *Ozone: Science and Engineering, 39*(1), 54–60.

83. Pandiselvam, R., Sunoj, S., & Manikantan, M. R., (2017). Application and kinetics of ozone in food preservation. *Ozone: Science and Engineering, 39*(2), 115–126.

84. Pandiselvam, R., Thirupathi, V., & Anandakumar, S., (2015). Reaction kinetics of ozone gas in paddy grains. *Journal of Food Process Engineering, 38,* 594–600.

85. Pandiselvam, R., Thirupathi, V., & Mohan, S., (2019). Gaseous ozone: A potent pest management strategy to control *Callosobruchus maculatus* (Coleoptera: Bruchidae) infesting green gram. *Journal of Applied Entomology,* 1–9.

86. Patil, S., Bourke, P., & Frias, J. M., (2009). Inactivation of *Escherichia coli* in orange juice using ozone. *Innovative Food Science and Emerging Technologies, 10*(4), 551–557.

87. Piechowiak, T., & Balawejder, M., (2019). Impact of ozonation process on the level of selected oxidative stress markers in raspberries stored at room temperature. *Food Chemistry, 298,* 125093.

88. Proctor, A. D., Ahmedna, M., Kumar, J. V., & Goktepe, I., (2004). Degradation of aflatoxins in peanut kernels/flour by gaseous ozonation and mild heat treatment, *Food Additives and Contaminant Part A, 21,* 786–793.

89. Prudente, A. D., & King, J. M., (2002). Efficacy and safety evaluation of ozonation to degrade aflatoxin in corn. *Journal of Food Science, 67*(8), 2866–2872.

90. Qi, L., Li, Y., Luo, X., Wang, R., & Zheng, R., (2016). Detoxification of zearalenone and ochratoxin *A* by ozone and quality evaluation of ozonized corn. *Food Additives and Contaminant Part A, 33*, 1700−1710.

91. Ramaswamy, H. S., & Marcotte, M., (2006). *Food Processing: Principles and Applications* (p. 592). Boca Raton, CRC Press.

92. Restaino, L., Frampton, E. W., Hemphill, J. B., & Palnikar, P., (1995). Efficacy of ozonated water against various food-related microorganisms. *Applied and Environmental Microbiology, 61*(9), 3471–3475.

93. Rodrigues, A. A. Z., & De Queiroz, M. E. L. R., (2019). Use of ozone and detergent for removal of pesticides and improving storage quality of tomato. *Food Research International, 125,* 108626.

94. Rodrigues, A. A. Z., & De Queiroz, M. E. L. R., (2017). Pesticide residue removal in classic domestic processing of tomato and its effects on product quality. *J. Environ Sci. Health B, 52*(12), 850–857.

95. Rohit, T., Deepak, K., & Annapure, U. S., (2017). Cold plasma: An alternative technology for the starch modification. *Food Biophysics, 12*(1), 129–139.

96. Saad, N., (2001). *Aflatoxins: Occurrence and Health Risks.* http://poisonousplants. ansci.cornell.edu/toxicagents/aflatoxin/aflatoxin.html (accessed on 08 February, 2021).

97. Sandhu, H. P., Manthey, F. A., & Simsek, S. K., (2011). Ozone gas affects the physical and chemical properties of wheat (*Triticum aestivum* L.) starch. *Carbohydrate Polymers, 87*(2), 1261–68.

98. Savi, G. D., & Scussel, V. M., (2014). Reduction in residues of deltamethrin and fenitrothion on stored wheat grains by ozone gas. *Journal of Stored Product Research,* 1–5.

99. Schomer, H. A., & McColloch, L. P., (1948). *Ozone in Relation to Storage of Apples* (Vol. 765, pp. 1–23). USDA Circular.

100. Silva, M. V. A., Faroni, L. R. A., & Sousa, A. H., (2019). Kinetics of the ozone gas reaction in popcorn kernels. *Journal of Stored Products Research, 83*, 168–175.

101. Sheikh, S. A., Panhwar, A. A., Mirani, B. N., & Nizamani, S. M., (2014). Effectiveness of traditional processing techniques on residual removal in chilies sprayed with various pesticides. *Journal of Biodiversity and Environmental Sciences, 5*, 365–370.

102. Skog, L. J., & Chu, C. L., (2001). Effect of ozone on qualities of fruits and vegetables in cold storage. *Canadian Journal of Plant Science, 81*, 773–778.

103. Smilanick, J. L., Margosan, D. M., & Gabler, F. M. 2002). "Impact of Ozonated Water on the Quality and Shelf-Life of Fresh Citrus Fruit, Stone Fruit and Table Grapes," *Ozone Sci. Eng., 24*(5), 343–356.

104. Souza, L. P. D., Faroni, L. R. D., & Heleno, F. F., (2018). Ozone treatment for pesticide removal from carrots: Optimization by response surface methodology. *Food Chemistry, 243*, 435–441.

105. Steenstrup, D. L., & Floros, J. D., (2004). Inactivation of *E. coli* O157:H7 in apple cider by ozone at various temperatures and concentrations. *Journal of Food Processing Preservation, 28*, 103–116.

106. Subramanyam, B., & Xinyi, E., (2017). Efficacy of ozone against *Rhyzopertha dominica* adults in wheat. *Journal of Stored Products Research, 70*, 53–59.

107. Sung, H., Song, W., Kim, K., Ryu, S., & Kang, D., (2014). Combination effect of ozone and heat treatments for the inactivation of *Escherichia Coli* O157: H7, *Salmonella Typhimurium*, and *Listeria Monocytogenes* in apple juice. *International Journal of Food Microbiology, 171*, 147–153.

108. Tabakoglu, N., & Karaca, H., (2018). Effects of ozone-enriched storage atmosphere on postharvest quality of black mulberry fruits (*Morus Nigra* L.). *LWT-Food Science and Technology, 92*, 276–281.

109. Taylor, P., Li, M. M., Guan, E. Q., & Bian, K., (2015). Effect of ozone treatment on deoxynivalenol and quality evaluation of ozonized wheat. *Food Additives and Contaminants: Part A, 32*(4), 544–53.

110. Torlak, E., (2014). Efficacy of ozone against *Alicyclobacillus Acidoterrestris* spores in apple juice. *International Journal of Food Microbiology, 172*, 1–4.

111. Torlak, E., & Ulca, P., (2013). Efficacy of gaseous ozone against *Salmonella* and microbial population on dried oregano. *International Journal of Food Microbiology, 165*, 276–280.

112. Torres, B., Tiwari, B. K., & Patras, A., (2011). Effect of ozone processing on the color, rheological properties, and phenolic content of apple juice. *Food Chemistry, 124*(3), 721–726.

113. Tiwari, B. K., Brennan, C. S., Curran, T. P., & Gallagher, E., (2010). Application of ozone in grain processing. *Journal of Cereal Science, 51*(3), 248–255.

114. Tiwari, B. K., O'Donnell, C. P., Patras, A., Brunton, N., & Cullen, P. J., (2009). Effect of ozone processing on anthocyanins and ascorbic acid degradation of strawberry juice. *Food Chemistry, 113*(4), 1119–1126.

115. Torres, B., Tiwari, B. K., Patras, A., & Wijngaard, H. H., (2011). Effect of ozone processing on the color, rheological properties, and phenolic content of apple juice. *Food Chemistry, 124*, 721–726.

116. Toti, M., & Carboni, C., (2017). Postharvest gaseous ozone treatment enhances quality parameters and delays softening in cantaloupe melon during storage at 6°C. *Journal of the Science of Food and Agriculture, 98*(2), 487–494.

117. Vanier, N. L., & El Halal, S. L. M., (2017). Molecular structure, functionality, and applications of oxidized starches: A review. *Food Chemistry, 221*, 1546–59.

118. Vikash, C. H., (2018). Applications and investigations of ozone in cereal grain storage and processing: Benefits and potential drawbacks. *International Journal of Current Microbiology and Applies Sciences, 7*, 5034–5041.

119. Wang, L., Fan, X., Sokorai, K., & Sites, J., (2019). Quality deterioration of grape tomato fruit during storage after treatments with gaseous ozone at conditions that significantly reduced populations of *Salmonella* on stem scar and smooth surface. *Food Control, 103*, 9–20.

120. Wang, H., Feng, H., & Luo, Y. G., (2004). Microbial reduction and storage quality of fresh-cut cilantro washed with acidic electrolyzed water and aqueous ozone. *Food Research International, 37*(10), 949–956.

121. Watson, I., Kamble, P., & Shanks, C., (2019). Decontamination of chili flakes in a fluidized bed using combined technologies: Infrared, UV and ozone. *Innovative Food Science and Emerging Technologies,* 102248.

122. Willams, R. C., Sumner, S. S., & Golden, D. A., (2005). Inactivation of *Escherichia Coli* O157:H7 and *Salmonella* in apple cider and orange juice treated with combinations

of ozone dimethyl dicarbonate and hydrogen peroxide, *Journal of Food Science, 70,* 197–201.

123. Xinyi, E., Li, B., & Subramanyam, B., (2019). Efficacy of ozone against adults and immature stages of phosphine susceptible and resistant strains of *Rhyzopertha Dominica. Journal of Stored Products Research, 83,* 110–116.

124. Zambre, S. S., Venkatesh, K. V., & Shah, N. G., (2010). Tomato redness for assessing ozone treatment to extend the shelf life. *Journal of Food Engineering, 96*(3), 463–68.

125. Zhang, L., Lu, Z., Yu, Z., & Gao, X., (2005). Preservation fresh-cut celery by treatment of ozonated water. *Food Control, 16,* 279–283.

126. Zhu, F., (2018). Effect of ozone treatment on the quality of grain products. *Food Chemistry, 264,* 358–366.

127. Zorlugenç, B., KiroğluZorlugenç, F., Oztekin, S., & Evliya, I. B., (2008). The influence of gaseous ozone and ozonated water on microbial flora and degradation of aflatoxin B1 in dried figs. *Food and Chemical Toxicology, 46,* 3593–3597.

CHAPTER 9

ADVANCES IN OSMOTIC DEHYDRATION

BARINDERJEET SINGH TOOR, HARINDERJEET KAUR BHULLAR, and AMARJEET KAUR

ABSTRACT

Osmotic dehydration is used for the reduction of the moisture content in food products by immersing in hypertonic solution. Mass transfer during osmotic dehydration takes place at ambient conditions avoiding phase change. The rate of mass transfer across the food structure is affected by temperature and time, osmotic agent, solution to sample ratio, solute concentration, structure of the food matrix, and agitation. In recent times, it has been coupled with ohmic heating, irradiation, high hydrostatic pressure (HHP), pulsed electric field (PEF), ultrasound, vacuum, centrifugal force, freezing, and micro-wave to enhance the cell permeability, thus making the process faster and cost-efficient.

9.1 INTRODUCTION

The osmotic dehydration includes the immersion of the food products in the hypertonic solution. The food cellular matrix behaves as a semi-permeable membrane, and the potential difference across the membrane results in the movement of the water from the food matrix into the surrounding solution. Concurrently, there is diffusion of the solute particles from solution into the cellular structure of the food [22, 41, 77].

The permeability of the membrane is not very specific, which leads to the leaching out of various food components, including organic acids, soluble sugars, salt, pigments, and flavor compounds. It results in the modification of the food composition and the composition of osmotic solution. Various osmotic agents have been tested in the dehydration process, including glucose, sucrose, fructose, maltodextrin, sodium chloride, etc. However, sucrose, and sodium chloride are prime choices in the food industry. Osmotic agents can be used individually or in combination depending on the requirement [62, 86].

The major application of osmotic dehydration is in fruits and vegetables, where it helps in the reduction of water activity of the products. Lower water activities are generally related with high microbial stability and better shelf life [15]. The process is generally carried out at ambient or mild conditions, which ensures the maximum retention of the nutrients as well as desired organoleptic properties. In addition, the mild processing conditions do not cause any phase conversion, which helps in the reduction of energy consumption thus makes, the process economically feasible. The low moisture content of the end products, result in lower weight and volume, which reduces the cost of packaging, storage, as well as transportation [2, 92].

The process of osmotic dehydration is influenced by type and concentration of osmotic agent, time, and temperature combination, solution to food matrix ratio, agitation speed, as well as type of agitation and food geometry. The variations of these factors can be beneficial in terms of better mass transfer rates and superior quality, however, after a certain extent the further variation can lead to low efficiency and inferior quality of products. Therefore, the parameters must be adjusted carefully to achieve the desired results [22, 62].

An intensive research has been going on to enhance the rate of osmosis and processing time. The vacuum, pulse electric field, high-pressure processing, ultrasonic waves, centrifugation, ohmic heating, microwave, and edible coating have been coupled with osmotic dehydration to serve the purpose. Pulsed electric field (PEF) include application of high voltage pulses (10–80 KV/cm) for few seconds resulting in electroporation, which enhances the rate of water loss and solid gain [90]. Similarly, ohmic heating relies on the phenomenon of electroporation to provide the better mass transfer rate [50].

Power ultrasound through its cavitation phenomenon accelerates the diffusion of solute and water particles across the semi-permeable membrane [24]. High-pressure processing (100–700 MPa) modifies the permeability of cell membrane and enhance water loss rates [59]. Similarly, gamma radiation brings sin changes in the cellular structure to improve the efficiency [74]. Application of vacuum during osmosis accelerates the degassing of tissues and ultimately enhances the overall rates [89]. These novel technologies also help in retaining the nutritional value of the food and ensure the better organoleptic properties.

This chapter focuses on various aspects of the osmotic dehydration processing with major emphasis on the recent advancements to enhance the processing rates. The chapter also covers the effects of various factors on the overall process and final product quality.

9.2 MECHANISM OF OSMOTIC DEHYDRATION

Basically, osmotic dehydration includes the movement of water and solute particles across the semi-permeable membrane, when any food matrix is dipped into the concentrated solution. Since the food matrix and solution exhibit concentration gradient, it acts as a driver for mass transfer. After dipping, the first stage includes the diffusion of water from the outermost layer of the food to the surrounding medium resulting in the shrinkage of the food structure. A potential gradient between the outermost layer and second layer leads to the diffusion of the water, resulting in the shrinkage of the second layer.

The process follows the same pattern till the equilibrium moisture content has reached. Concurrently, there is diffusion of solute particles into food particles, which is generally expressed in terms of solid gain [62, 77]. In addition, some soluble components (for instance, color pigments, organic acids, soluble sugars, and salts) can also leach out into solution due to partial selectivity of the cell membrane [20]. Figure 9.1 depicts the movement of different components during osmosis.

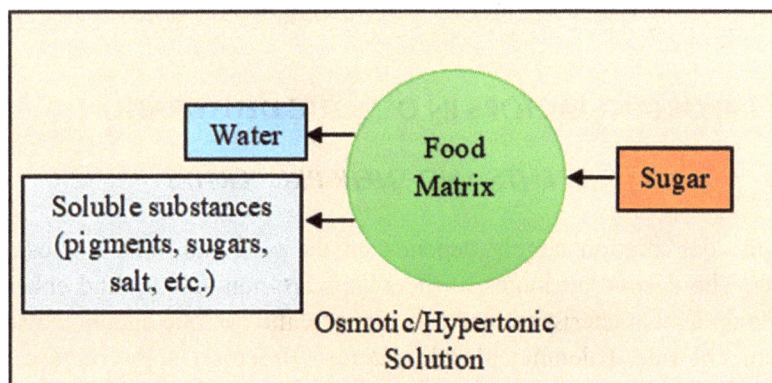

FIGURE 9.1   Mass flow during osmotic dehydration.

The rate of water movement into the surrounding medium (osmotic solution) is quick in the initial few hours (1 to 2 hours), which decreases after the defined time and then ultimately become constant. However, the rate of solute diffusion follows a different trend, where the rate is low in the initial hours and then increase with the loss of water. The movement of water follows the different mechanism for the different type of structures. In non-porous structure, the movement takes place through diffusion, while capillary action is responsible in porous structures [68, 69].

Different modeling techniques have been used to comprehend the osmotic dehydration kinetics and the changes induced during processing. According to simplified semi-empirical models in any osmotic dehydration process, the cells involved are parenchymatous cells. These cells mainly consist of intercellular and extracellular volume, separated by the cell membrane. The extracellular part consists of cell wall and voids among the cells, whereas cytoplasm and vacuole are present in the intercellular volume. A potential gradient between the hypertonic solution and cellular material induces the movement of mass through a semi-permeable membrane.

The process of mass transfer continues until the moisture content of the solution and food matrix reach an equilibrium point. When the surrounding of the cells has lower osmotic pressure than the cells, then there is migration of the water into cells. It results in the swelling of the cells up to a certain extent, which induces the movement of solute particles into the extracellular volume of the cell. Further movement includes the diffusion of the solute particles into the intracellular region with or without penetrating the cell membrane. The penetration of the solute particles into tissue produce cell membrane potential, which leads to the movement of water into the extracellular region and ultimately into the surrounding medium [1, 93].

9.3   IMPORTANT FACTORS IN OSMOTIC DEHYDRATION

9.3.1   OSMOTIC AGENTS AND THEIR PROPERTIES

Osmotic dehydration largely depends on the type and nature of osmotic agents. The desired attributes of the agents are non-toxicity and enhanced organoleptic characteristics. For this purpose, the osmotic agents in use are sodium chloride, calcium chloride, sucrose, fructose, high fructose corn syrup, honey, invert syrup, maltodextrin, and sorbitol [86]. Sucrose and sodium chloride are primary choices in the processing of vegetables and fruits [85]. Extensive use of sucrose is attributed to its ability to produce high efficiency and desired organoleptic properties. However, drying can result in crystallization of sucrose rather than forming amorphous solids.

Sometimes during processing, the natural sugars present in the food materials combines with the sucrose resulting in the formation of an amorphous layer on the food tissues. Sucrose can be applied in the dry or powder form or can be used in the syrup form of specific concentration depending upon the type of the food. Even though the dried forms are very effective and possess excellent efficiencies, yet the difficulties related with the disposal of residual sugar at the end of processing make the syrup form to be prime choice. Sometimes sucrose is also used in combination with other ingredients (e.g., sodium chloride, glycerol, maltodextrin, or corn syrup) to modify the properties of the solution according to the requirement of the process [22].

Sodium chloride has been successfully utilized in the osmotic processing of various types of fruits and vegetables, such as apple, mango, carrot, onion, potato, etc. It is a low molecular weight substance, which produces a greater impact on the chemical potential compared to sugars and this is basically due to its ability to get ionized in the solution. The resulting ions can easily penetrate or diffuse into the food matrix to yield high solid gains. The major drawback in the utilization of sodium chloride is the development of salty taste in the final product. Therefore, it is generally combined with sugars to achieve the desired results with better efficiencies [22, 46, 48].

Recently, honey has been utilized in the osmotic dehydration of fruits. It shows better osmotic behavior compared to sucrose due to the presence of a variety of saccharides including glucose, fructose, sucrose, and maltose. The combination of such type of saccharides leads to a large pressure gradient and enhanced water removal rates. Honey contributes to better texture properties as it does not crystallize on drying and it also imparts enhanced organoleptic properties to the end product [41].

Efficiency and effectiveness of any osmotic dehydration process largely depends on the rate of diffusion/penetration of osmotic solution, removal of water from matrix and solid gain. All these attributes are correlated to each other, and they further depend upon the certain characteristics of the osmotic agent. Technically, the penetration rate of osmotic agent is inversely proportional to its molecular weight. Therefore, sucrose has lower penetration rates than the simple sugars (glucose/fructose), invert sugar, and high fructose corn syrup; and better penetration rates than corn syrup and maltodextrin [37, 42]. In some instances, the leaching of acids into osmotic solution leads to hydrolysis of sugars into simpler forms, which eventually enhances the rate of the process [13].

Higher diffusion capacity generally leads to better water removal and higher solid gains. Examples are the application of glucose, sucrose, and corn syrup for the dehydration of the fruits like apple, kiwi, papaya, and banana, where the glucose provides the highest values for water loss and solid gain, whereas corn syrup shows the least water loss and solid gain values. Sometimes, the conventional osmotic agents are coupled with other ingredients to bring out specific changes. The selection of these additional ingredients is based upon the certain factors like cost, compatibility, and contribution to organoleptic properties, overall effectiveness, and efficiency.

Addition of glycerol during sucrose-based osmotic dehydration of peeled mandarins significantly enhanced the overall efficiency due to the low molecular weight of the glycerol. This is because of the rise in the osmatic hydration by the concentration of low molecular weight compounds. Thus the osmotic pressure gradient gets enhanced, which results in more water loss and solid gain [22, 60].

The osmotic solution concentration significantly influences the overall process. Generally, the increase in concentration enhances the water loss and solid gain and contributes in developing products with low water activity values [2, 62]. In this context, the processing of watermelon slabs in sucrose solution shows enhanced water loss and solid gains at 60°Brix compared to 40°Brix and 50°Brix solutions [23].

Similar trends can also be observed during the osmotic processing of pomegranate arils and apricot in sucrose solution, where the 60°Brix solution gives the higher processing efficiencies than solution of lower concentration [32, 52]. However, the direct relationship between concentration and overall efficiency is valid to a certain level only. As the solution concentrates, the viscosity increases, which reduce its mobility, thus, lower the rate of diffusion and water removal. This is the reason, why concentrated solutions take more time to achieve the equilibrium stage. For instance, the water loss rate of mango increases significantly with the increase in concentration of sucrose solution from 35 to 55°Brix and reduces on achieving the concentration of 65°Brix [30, 62]. In addition, more power is required to pump concentrated solution, which again affects the overall efficiencies.

9.3.2 PROCESSING TEMPERATURE

The change in the processing temperature significantly modifies the flow properties of the osmotic solution. With the increase in the temperature, the solution viscosity gets reduced, which enhances the rate of diffusion through food matrix, thus the water loss activity is improved. At high temperature values, the enhanced rate of solute diffusion can also be attributed to the elimination of the entrapped air from the porous structure of the food matrix. Generally, the water loss is rapid in the early stages of the processing. However, the increase in the temperature does not cause any major change in the solid gain. High temperature can also be combined with higher solution concentration to maximize the processing efficiencies as they have synergetic effects [53, 67].

The application of the osmosis at high temperature shows an improved rate of water removal from the fruits and vegetables like papaya, peach, tomato, cranberry ad tomato [69]. During osmosis processing of watermelon slabs in 50°Brix solution, the 40°C showed the better water removal activity compared to processing at 20–30°C [23]. Processing of apples slices in the sucrose solution (60%) showed higher efficiencies, when the temperature was raised from 20 to 50°C [21, 35]. Similarly, the osmotic dehydration of apricots and pomegranates in sucrose solution was significantly enhanced with increase in the temperature from 25 to 45°C and 35 to 55°C, respectively, and thereby improving the overall efficiency [32, 52].

The high temperature values during osmosis are beneficial to a certain level (60°C) only, because beyond this level, the processing rate reduces due to changes in cell wall permeability. It is also very important to mention that the temperature above 50°C can significantly reduce the vitamin C and chlorophyll content and deterioration of organoleptic properties can also takes place [15, 22, 44].

9.3.3 PROCESSING TIME

The rate of mass transfer across the semi-permeable membrane is greatly influenced by the duration of contact between osmotic solution and food matrix. Generally, with the passage of processing time, the rate of solid gain and water loss reduces. Major water loss occurs during the first hour of the processing, beyond which the rate of water loss reduces by many folds. Sometimes, it falls down to 10% of the original rate after 2–3 hours of processing. After 1–2 hours of processing, the cell geometry tends to change; thereby, the permeability of the cell membrane also gets affected, which is suggested to be the potential reason for the reduction of osmosis rate. On the other hand, the moisture content of the product reduces with the passage of time [41, 66, 92].

9.3.4   AGITATION

Agitation can successfully enhance the rate of mass transfer during osmosis due to the elimination of the saturated areas around the food product, which helps in maintaining the effective mass transfer forces during processing. The requirement of agitation generally depends upon the properties of the osmotic agent and food sample.

Agitation is more effective in the case of concentrated solutions as they have limited mobility. Also, in concentration solutions, the food pieces tend to float, resulting in smaller contact areas with osmotic solution, which can be countered by agitating the solution. However, the agitation of osmotic agents can cause damage to the food product resulting in reduction of overall quality.

Centrifugal pumps can be used to circulate the solution around the sample so that the damage to the food can be minimized. Examples include osmotic dehydration of apple pieces combined with the circulation of syrup resulting in enhanced water loss and solid gain. Agitation speed should also be considered as gentle agitation does not produce any major changes. In contrast, few studies show that the overall change induced by agitation or circulation is negligible or very small, which puts an extra burden on the overall cost of production. Therefore, it is better to avoid agitation in such cases [49, 67, 92].

9.3.5   RATIO OF OSMOTIC SOLUTION TO FOOD MATRIX

The ratio of solution to food matrix has considerable effect on the parameters, such as osmotic pressure, rate of dehydration, solid gain, and final product quality. Generally, it is advisable to keep the ratio high (e.g., 30:1 or 60:1) to prevent the dilution of working solution with continuous removal of water from the matrix into the surrounding solution [28, 81]. In addition, a higher volume of solution ensures maximum dehydration within 1–2 hours of the operation and also prevents the surface oxidation of the food sample [15, 41]. However, it is very challenging to maintain such high ratios at the industrial level, since it is difficult to handle high volumes of the osmotic solution. At laboratory scale, the ratios of 3:1 or 4:1 are being used to analyze the effect on the mass transfer with the change in the concentration of the osmotic solution [3, 63].

9.3.6   CHARACTERISTICS OF FOOD MATRIX

Food characteristics like porosity, surface area, shape, and cell arrangement and chemical composition have pronounced effects on the dehydration process [15]. It is always recommended to keep the surface area of food as large as possible to ensure maximum removal of the water within 1 to 2 hours of the process [55]. Technically, the relation of surface area and rate to mass transfer is expressed as A/L (surface area/ half thickness) ratio. With the increase in A/L ratio, the rate of water loss and solid gain enhances [2]. Foods with small surface area take a longer time to achieve the equilibrium stage attributed to long distance travel by water from inside of the food to the surrounding medium.

9.4   BENEFITS OF OSMOTIC DEHYDRATION IN FRUITS AND VEGETABLES [1, 15, 86, 94]

- Osmotic dehydration significantly reduces water activity of the food matrix, thus provides the better shelf life and microbiological stability.
- The process is carried out at mild temperature conditions of 30 to 50°C, which ensures elimination of any major heat damage.
- The hypertonic solution used in processing preserves the color of the product by avoiding the reactions like enzymatic browning.
- The hypertonic solution inhibits the activity of polyphenol oxidase (PPO), thus can avoid the utilization of chemical preservatives.
- The effect of processing on nutritional quality is negligible.
- Osmotic product tends to have better organoleptic and aromatic properties.
- The process lowers the moisture content of the product leading to reduction in weight and volume, thus cutting down the packaging and transportation costs.
- The solid gain during processing protects the texture of the end product, which is attributed to its cryo-protectant effects.
- The process is carried out at mild conditions; therefore, no phase change is involved resulting in lower consumption of energy as compared to the other dehydration techniques.
- The used solution can be utilized in the other industries activities; especially beverage industry, which contribute to the economics of the process.

9.5   LIMITATIONS OF OSMOTIC DEHYDRATION

The foremost limiting component is the osmotic solution, which has a significant effect on the rate of osmosis and properties of the end product. However, as the process proceeds, the concentration of the solution starts changing due to the exchange of water and solute particles between solution and food matrix. The used or resulted solution tends to have a lower concentration of solute, undesirable sensory attributes and lower acidity values [17].

The lower concentration of the solution makes the process slow and longer processing times increase the chances of microbial contamination and nutrition loss. Therefore, the used solution has to be processed before reuse. The recycling becomes more difficult if the used solution contains more than one type of osmotic agent.

The lower acidity levels negatively affect the organoleptic properties of the food matrix [86]. Usually, the end product of osmosis has some residue of syrup left on its surface, which is not desirable. However, it can be removed by rinsing the product with water, but the additional step increases the overall cost [1].

After processing, the food particles left in the solution makes the conditions favorable for the growth of the harmful microorganisms. Thus, the solution must be processed immediately after the completion of the process to enable its reuse. Otherwise, the solution can deteriorate the product quality resulting in the undesirable sensory attributes [17, 68]. Sometimes lack of processing control and inadequate design can affect the overall quality. The continuous flow of solution around the food matrix and agitation during batch processing can cause breakage of the food product. Highly viscous solution is difficult to pump and agitate and food products tend to float in such solution, which reduces the efficiency of the process and also it increases the chances of enzymatic browning. Osmotic dehydration followed by freezing results in large shrinkage of the product, which depicts inferior quality of the product [1, 41].

9.6   RECENT ADVANCEMENTS IN OSMOTIC DEHYDRATION

Osmotic drying is a time taking process compared to other drying methods. Thus, in order to enhance the drying time, mass transfer rate should be increased by using different concentrations of different osmotic agents (e.g., sodium chloride, maltodextrin, glucose, glycerol, fructose, and corn syrup).

Mass transfer rate for water loss and solid gain depends on the concentration of solute in the solution as a combination of salt and sucrose solution is best for the adequate osmotic drying of tomatoes and can also enhance the mass transfer rate with the increase in salt concentration [84]. For instance, osmotic solution of 50% sucrose and 15% salt concentration agitated at 250 rpm led to 54.5% water loss and 12.30% solid gain, which was 12.37% and 23%, respectively higher than osmotic solution of 65% sucrose concentration [6].

Agitation reduces external resistance against water removal during osmotic drying that elevates water loss and solid gain. Agitation of osmotic solution resulted in 52% of the water loss in carrot, which was higher than water loss in apple and banana treated under similar conditions due to higher water content in carrots than apple and banana [9]. Solute gain during the osmotic drying helps in the production of confectionery products, whereas high water loss and less solid gain are useful for development of dietetic products [6].

Mass transfer can also be increased by pre-treatment of food matrix, thereby modifying the cellular composition by certain methods and increasing the cell-wall permeability. Pre-treatment methods employed for osmotic dehydrated products are pulsed electric fields, high-pressure processing, ultrasonication, gamma irradiation, freezing, and microwave.

9.6.1 PULSE ELECTRIC FIELD

Pulsed electric field (PEF) increases cell permeability by cell-wall electroporation due to application of electric field for short duration. Application of external electric field induces plasmolysis due to the compositional changes of cell membrane (removal of native electrolytes as Ca^{2+} and Na^{2+}) and symplastic transport occurs during osmotic dehydration of pulse electric field pretreated product [87]. Pre-treatment of osmotic dehydrated products with pulse electric field increases the product solids and decreases the sugar concentration of osmotic solution. Treatment of apples with pulse electric field at 0.90 kV/cm and 750 pulses prior to immersion in 44.5% of sucrose solution reduced the sugar concentration of osmotic solution from 44.5% to 39.0% after 4 hours [8].

Pulse electric field in combination with other treatment methods gives better results than individual application. As pre-treatment of agro-food tissues with pulse electric field followed by agitation of solution during osmotic dehydration led to 21.11%, 26.61% and 15.72% more water loss in apple, carrot, and banana, respectively; and 84.4% and 17.23% more solid gain in apple and carrot, respectively than the pulse electric field individually treated samples [9]. Amami et al. [6] found the remarkable effect of agitation in combination with pulsed electric field on solute uptake and moisture loss. The rate of mass transfer is affected largely by pulse number and electric field strength and slightly by duration of pulses. For example, water loss and uptake of sucrose molecules was increased with a substantial increase in the strength of electric field from 1 to 2 kV/cm and pulse number from 8 to 16 in osmo-dehydrated apples [54].

Preliminary exposure of blueberries to pulse electric field of 3 kV/cm voltage for 5 minutes increased the water and solute diffusivity and reduced the dehydration time from 130 to 48 hours [95]. Osmotic dehydration in combination with pulsed electric field pre-treatment improved the nutritional quality of the blueberries by deactivation of PPO activity. Degradation loss of phenols and antioxidants was 27.9% and 58.5%, respectively, lower than non-treated osmotically dried blueberries [95].

9.6.2   HIGH HYDROSTATIC PRESSURE (HHP)

High hydrostatic pressure (HHP) promotes mass transfer by increasing cell permeability and by causing irreversible damage to the cell membrane due to the compression of the cellular structure by high pressure [70]. Disruption of cell wall under pressure results in water flow from vacuoles to the cytoplasm and extracellular spaces resulting in higher water loss during dehydration of agro-food tissues [39]. During osmosis, the water loss depends on the type of pre-treatment, pre-treatment conditions and conditions of the osmotic solution. Therefore, an increase in pressure, temperature, and time of immersion in the osmotic solution remarkably affects the solute concentration and water loss and solid gain.

Diffusion coefficients of osmotically dried strawberries were enhanced with the increase in the pressure from 200 to 400 MPa for 10 minutes [58]. High pressure (50, 100, 200, 400 MPa for 10 minutes) and osmotic dehydration synergistically affect the initial rate of solute gain and water loss thus resulting in higher effective diffusion coefficient of pressure-treated dried wumei (3.76×10^{-10}) than the control (1.53×10^{-10}) [39]. Osmotic dehydration (50°Brix at 40°C) of pineapple under elevated pressure (700 MPa for 5 minutes) showed significant increase in mass transfer rate [97]. Taiwo et al. [83] observed that osmotic drying of pre-frozen, high-pressure, and pulse electric field pretreated strawberries caused 96–270%, 40–160%, and 50–62%, respectively, more solid gain than untreated osmotically dried strawberries. In contradiction to this, several researchers observed no significant effect of pressure on the mass transfer during osmotic drying [18, 57].

Mobility of water bound to cell wall and membrane reduces beyond 400 MPa pressure due to the starch gelation or release of hydrophilic groups of protein and phenols during cell-wall disruption that hinders the mass transfer rate [79]. In addition, the high-pressure of the system and the concentration of hypertonic solution cause remarkable variation in the browning index (BI) of the fruits. BI is directly proportional to pressure and inversely proportional to concentration and immersion time. It increases by high-pressure due to acceleration of non-enzymatic browning and interaction of PPO with phenols during compression of cellular structure. On the contrary, a decrease in BI shows that sugar has a protective effect on the ascorbic acid (AA) of citrus fruit that reduces the enzymatic browning [88].

Generally, pre-treatment of osmotically dehydrated products by pressure resulted as a promising technique to obtain good quality product. However, the rehydration capacity of the pressure-treated product is poor due to the interaction of divalent ions to de-esterified pectin resulting in the development of gel network thus reducing the solute diffusion out of the tissue. Moreover, the irreversible damage to the cell wall causes reduction in water uptake [71].

9.6.3   OHMIC HEATING

Blanching is usually done by steam that damages the cell structure due to expulsion of air during expansion of tissue by heat thereby increases intercellular spaces and accelerates mass transfer. Generation of heat in conventional blanching depends upon the thermal conductivity of aqueous medium; thereof liquid overheating may take place that might affect the quality of dehydrated products [62]. Therefore, blanching by ohmic heating can be a potential alternate to enhance the rate of osmotic dehydration process.

In ohmic heating, the food material placed between two electrodes acts as resistance, and the passage of alternative electric current through food leads to the generation of heat. During ohmic heating, uniform heating takes place, which has less impact on the nutritive as well as organoleptic properties of food thus provide product with superior quality than conventional blanching.

Ohmic heating solubilizes the pectin, thus affects the biological integrity of tissue and provide electroporation of cell membrane, which can be confirmed by scanning electron microscopic image of ohmic heating assisted osmotically dried pears [50, 64].

The fruit structure, duration of ohmic heating and temperature affects the electrical conductivity of the sample and eventually the kinetics of solid gain and water loss. Longer the duration of ohmic heating higher would be the electrical conductivity as ohmic heating damages the tissue and releases intercellular fluid that increases electrical conductivity thereby increases water loss [5]. Similar results were reported by Allali et al. [4] for ohmic heating assisted osmo-dehydration of strawberries.

Additionally, combination of ohmic heating with osmotic dehydration reduces the drying time, thus increases the process efficiency. For example, ohmic heating (electric field intensity of 100 V/cm) of osmotic solution up to 50°C reduced the drying time by 50% in osmo-dehydrated raspberries. In addition, coupling of ohmic heating with osmotic drying significantly improves the firmness and color of fruits and vegetables [62].

9.6.4   CENTRIFUGATION

Pre-treatment by centrifugation results in better dehydration by limiting the solute uptake thereby the osmotic dried products are best suited for dietetic people. Amami et al. [6] reported that centrifugation in combination with pulsed electric field led to 21.5% less solid gain in osmotic dehydrated product and more water loss. Decrease in solute uptake is because of the centrifugal pressure exerted by the centrifugal force that inhibits the flow of solutes to and from the solution whereas, water loss increases as centrifugal force increases the movement of hypotonic intracellular water from sample to make equilibrium with hypertonic solution during osmosis.

Various parameters as centrifugation force and centrifugation time have significant effect on mass kinetics and product quality. In reference to this, upon centrifugation prior to osmotic dehydration of bamboo shoot, there was rise in water loss from 12.56 to 39.18% and decline in solid gain, hardness, and L value by 1.13 to 25.68%, 9.74 to 44.61%, and 1.22 to 10.82%, respectively [10]. Similarly, Barman and Badwaik [11] reported that with an increase in centrifugation time, water loss and rehydration capacity were increased from 4.28 to 11.02% and 1.42 to 8.05%, respectively, but uptake of sucrose molecules was reduced from 5.05 to 48.19%, without any remarkable effect on the texture. Therefore, centrifugation before drying produces the product with better quality.

9.6.5 EDIBLE COATING

Various research studies have been conducted to control the solute uptake by modifying the size of the product, using the ternary solution or osmotic agents of high molecular weight, or applying an edible coating before osmotic dehydration [62]. Among these, coating treatments showed good barrier effects on solid gain. Coating prior to osmotic dehydration decreases the excessive solute uptake without major impact on the water loss. High water loss in coated samples is due to the hydrophilic nature of coating agents, whereas a decrease in solid gain is caused by the barrier created by the coating to the solute movement into the product. Coating also helps in the retention of nutrients and flavors and maintains the product integrity as well as physical strength during packaging and transportation of the product [36, 45].

Polysaccharides, lipids, protein, and resins can be used individually or in combination as a coating material. Coating material must be edible in nature and should possess good mechanical strength, film-forming capability, and high water diffusivity and should act as a barrier to mass transfer and flavor loss, and must remain intact in osmotic solution. Most common polysaccharides as chitosan, cornstarch, potato starch, maltodextrin, sodium alginate, carrageenan guar gum, low methoxy pectin, high methoxy pectin and carboxymethylcellulose can be used as coating agents [62].

During osmosis of fruits, the concentration of coating solution, dipping time and drying time of the coating significantly affect the solute uptake, water loss and performance ratio (water loss/solid gain). In certain cases, coating agent acts as a barrier to the water removal as observed during the osmotic dehydration of pineapples pre-coated with sodium alginate of concentration greater than 3% [78]. Furthermore, double coating of strawberries by 0.5% sodium alginate reduced solute uptake by 44%, increased performance ratio by 123% and resulted in minimal drip loss after freeze/thawing of osmo-dehydrated strawberries [45].

During osmotic dehydration, the ripening stage of fruits has a remarkable effect on the rate of mass transfer. Garcia et al. [26] reported higher solid gain and water loss in green papaya than ripened papaya due to the high porosity of green papaya. Furthermore, an increase in temperature and concentration of salt in hypertonic solution revealed a considerable increase in the mass transfer rate in coated product. This behavior can be due to high-pressure gradients with increase in salt concentration and high molecular energy with the rise in temperature [96].

Coating of fruits and vegetable before drying improves organoleptic attributes, increases the retention of color, flavor, and nutrients, and also extends the shelf life of the product by acting as a barrier to the entry of microorganisms. In reference to this, edible coating of potato by calcium alginate extended the shelf life up to 4 days by delaying microbial growth at refrigerated conditions than non-coated potatoes [47]. Therefore, coating prior to osmotic dehydration can be used to develop value-added product for dietetic people.

9.6.6   ULTRASOUND

Power ultrasound (20–100 kHz) induces cavitation of cell membrane that forms microscopic channels through which osmotic solution flows into the product. Alternative compression and expansion of cell wall provide sponge effect that increases mass diffusion rate and reduces the dehydration time [34]. This fact is justified by scanning electron microscopic image of ultra-sonic pre-treated kiwi fruit slices that shows the alteration of cell structure and formation of micro-channels through the cell membrane upon increase in the time of ultrasonic pre-treatment [66]. Preliminary exposure to ultrasonic waves increases cell permeability to water loss and solute uptake, thereby increases the sucrose concentration, drying rate, and rehydration capacity of the product and also improve the color as well as the ascorbic acid content [56, 91] but reduces the phenolic content [14].

Additionally, an increase in the ultrasonic treatment time and immersion time in osmotic solution enhanced the solute uptake and moisture loss and reduced the dehydration time by several hours [7, 14]. Moreover, ultrasound-assisted osmotic dehydration increases mass transfer rate without much effect on the solution temperature. Therefore, the application of ultrasound can help in the preservation of heat-sensitive compounds, flavors, and color during osmotic dehydration.

9.6.7   IRRADIATION

Gamma irradiation alters and damage the internal tissue structure thereby increases cell permeability and enhances the mass transfer rate. Gamma irradiation prior to osmotic dehydration significantly influences the quality characteristics and increases the kinetics of water loss and solid gain [72]. Scanning electron microscope image of gamma-irradiated osmo-dehydrated potatoes revealed the swelling and aggregation of starch granules and disruption of cell wall, resulting in the formation of micro-channels that decreased the hardness (48.7–84.1%) and increased the solid gain and water diffusivity.

Fruits and vegetables are packed in polyethylene pouches for exposure to irradiation and are then osmotically dried. Increase in the dose of irradiation from 3 to 9 kGy significantly increased moisture diffusion coefficient of potato from 2.38×10^{-9} to 3.51×10^{-9} m²/s and of carrot from 1.63×10^{-9} to 2.64×10^{-9} m²/s [74]. Furthermore, coupling of gamma irradiation with osmotic dehydration extended the shelf life, increases the microbial resistance, and improves the quality characteristics of guava, therefore decreases the need of thermal treatments [80].

9.6.8 VACUUM

Application of vacuum as pre-treatment has significant effect on the dehydration rates and product-related properties. Mostly it is applied as short pulses, which results in the compression and expansion of the gases present inside the porous structure of the food matrix. This results in the exchange of the gas molecules with the surrounding liquid medium. In other words, the application of the vacuum causes the degassing of the pores, and when the vacuum is removed, the pores tend to get filled with osmotic solution.

The advantages of the vacuum treatment include the enhanced water loss rate at ambient temperature conditions and the higher solid gain. The higher rates of the osmosis reduce the time requirement, which helps in the production of quality food products. Certain studies show that the impact of vacuum treatment amplifies in the presence of a high concentration of syrup solution, which can be attributed to a large potential gradient across cell membrane [22, 25, 73].

In sucrose solution of 40 to 60°Brix, the application of the vacuum pulses (100 mbar) for 0 to 15 minutes significantly enhanced the rate of water loss of guavas slices, and the similar results were also obtained for solid gain [16]. In the osmotic dehydration of melon, the application of vacuum (0 to 600 mbar) for 10 to 360 minutes showed better mass transfer rates and the results were intensified, for instance, the water loss and solid gain values of 191.6 mg/g and 36.4 mg/g after 10 minutes of combined treatment of vacuum (300 mbar) and osmosis (60°C, 40°Brix) increased to 683 mg/g and 79.4 mg/g after 360 minutes [43]. In figs, application of the vacuum (130 mbar) in the first 15 minutes of the process improved the drying rates and reduced the overall processing time, thus help in making the process economically feasible [76]. Table 9.1 indicates the effects of different treatments on osmotic dehydration.

9.6.9 FREEZING

Osmotic dehydration can be used as pre-treatment for the freezing of fruits and vegetables, and the combined process can be called osmotic dehydro-freezing. This combined technique is found beneficial in the production of quality food products with better textural properties as compared to conventional freezing.

In addition, the technique prevents the collapsing of cellular structure as well as drip loss during thawing of frozen products. Certain studies demonstrate that the application of osmotic dehydration before freezing can help in the retention of ascorbic acid content [1, 73].

The application of osmotic dehydration prior to freezing can be applied to a variety of food products. In peas, the combined application minimized the changes in textural properties, color, and ascorbic acid content [29]. Similar results were also found in the processing of strawberries and muskmelon [40]. In tomatoes, the application of osmotic dehydrofreezing provides better quality attributes compared to traditional freezing. The beneficial effects also include the better retention of ascorbic acid content during storage period [19]. In addition, this technique has the ability to increase the solid gain and water loss values by a certain extent; for instance, the water loss and solid gain values increased by 1.4 and 3.5 times, respectively, during the osmotic dehydration of pre-frozen pomegranate seeds. Thus, the overall rate of processing can be improved by pre-treatment of freezing [12].

9.6.10 MICROWAVE

Microwaves have been utilized for cooking, drying, baking, etc. Its combination with osmotic dehydration has a significant impact on the color, texture, taste, and nutrients of food matrix and solid gain and water loss values. In addition, the described combination ensures uniform heating, lesser

TABLE 9.1 Effect of Different Treatments on Osmotic Dehydration

Product	Treatment(s)	Conditions	Osmotic Solution	Findings	References
Apple	Agitation	1000 rpm	65% Sucrose solution	36.7% Water loss	[9]
Apple	Pulse electric field	0.90 kV/cm + 750 pulses	44.5% Sucrose solution at 25°C for 4 hour	40% Water loss; 6% Solid gain	[8]
Apple	Pulse electric field + Agitation	0.90 kV/cm + 750 pulses + 1500 rpm	65% Sucrose solution for 4 hour	7.34% Water loss; 53.07% Solid gain	[9]
Apple	Coating	1% Low Methoxy Pectin	60% sucrose solution at 30°C for 3 hour	Reduces 7.64% Water loss; Reduces 31.06% Solid gain	[33]
Apple	Coating	2% Low Methoxy Pectin	60% sucrose solution at 30°C for 3 hour	5.30% Water loss; Reduces 36.12% Solid gain	[33]
Apple	Coating	3% Low Methoxy Pectin	60% sucrose solution at 30°C for 3 hour	11.56% Water loss; Reduces 44.52% Solid gain	[33]
Apple	Coating	0.5% Carboxymethylcellulose	60% sucrose solution at 30°C for 3 hour	Reduces 14.15% Water loss; Reduces 36.64% Solid gain	[33]
Apple	Coating	1% Carboxymethylcellulose	60% sucrose solution at 30°C for 3 hour	Reduces 8.76% Water loss; Reduces 42.70% Solid gain	[33]
Apple	Coating	2% Carboxymethylcellulose	60% sucrose solution at 30°C for 3 hour	17.32% Water loss; Reduces 54.91% Solid gain	[33]
Apple	Coating	1% Corn Starch	60% sucrose solution at 30°C for 3 hour	Reduces 2.41% Water loss; Reduces 6.95% Solid gain	[33]

TABLE 9.1   *(Continued)*

Product	Treatment(s)	Conditions	Osmotic Solution	Findings	References
Apple	Coating	2% Corn Starch	60% sucrose solution at 30°C for 3 hour	0.09% Water loss; Reduces 16.71% Solid gain	[33]
Apple	Coating	3% Corn Starch	60% sucrose solution at 30°C for 3 hour	4.88% Water loss; Reduces 18.54% Solid gain	[33]
Bamboo Shoots	Centrifugation	1600 rpm	12% sucrose solution at 37°C for 2 h	12.56% Water loss; Reduces 1.14% Solid gain; Reduces 9.74% hardness; Reduces 1.22% L value	[10]
Bamboo Shoots	Centrifugation	2800 rpm	12% sucrose solution at 37°C for 2 h	39.69% Water loss; Reduces 20.91% Solid gain; Reduces 39.63% hardness; Reduces 9.92% L value	[10]
Bamboo Shoots	Centrifugation	3200 rpm	12% sucrose solution at 37°C for 2 h	39.17% Water loss; Reduces 25.68% Solid gain; Reduces 44.61% hardness; Reduces 10.82% L value	[10]
Banana	Agitation	1000 rpm	65% Sucrose solution	43.5% Water loss	[9]
Banana	Pulse electric field + Agitation	0.30 kV/cm + 500 pulses + 1500 rpm	65% Sucrose solution for 4 hour	45.76% Water loss	[9]
Banana	High hydrostatic pressure	200 MPa	40% Sucrose solution at 40°C for 6 hour	18.36% Water loss; 152% Solid gain	[88]
Banana	High hydrostatic pressure	400 MPa	40% Sucrose solution at 40°C for 6 hour	36.92% Water loss; 174% Solid gain	[88]

TABLE 9.1 *(Continued)*

Product	Treatment(s)	Conditions	Osmotic Solution	Findings	References
Banana	High hydrostatic pressure	200 MPa	60% Sucrose solution at 40°C for 6 hour	6.62% Water loss; 33.89% Solid gain	[88]
Banana	High hydrostatic pressure	400 MPa	60% Sucrose solution at 40°C for 6 hour	11.22% Water loss; 92.35% Solid gain	[88]
Banana	High hydrostatic pressure	200 MPa	40% Sucrose solution at 40°C for 8 hour	1.49% Water loss; 64.89% Solid gain	[88]
Banana	High hydrostatic pressure	400 MPa	40% Sucrose solution at 40°C for 8 hour	18.70% Water loss; 83.74% Solid gain	[88]
Banana	High hydrostatic pressure	100 MPa	50% Sucrose solution at 50°C for 7 hour	3.22% Water loss; 4.96% Solid gain	[88]
Banana	High hydrostatic pressure	500 MPa	50% Sucrose solution at 50°C for 7 hour	9.77% Water loss; 35.03% Solid gain	[88]
Blueberries	Pulse electric field	3 kV/cm + 200 pulses	70% cane sugar solution at 40°C	58.82% increase in phenolics; 100% increase in antioxidants	[95]
Carambola Slices	Centrifugation	2800 rpm for 30 min	70% sucrose solution at 50°C for 3 hours	8.24% Water loss; 1.42% Rehydration capacity; Reduces 4.75% Solid gain	[11]
Carambola Slices	Centrifugation	2800 rpm for 60 min	70% sucrose solution at 50°C for 3 hours	11.02% Water loss; 8.05% Rehydration capacity; Reduces 48.19% Solid gain	[11]
Carrot	Agitation	1000 rpm	65% Sucrose solution	52% Water loss	[9]
Carrot	Agitation	250 rpm	65% Sucrose solution	48.5% Water loss; 10% Solid gain	[6]

TABLE 9.1   *(Continued)*

Product	Treatment(s)	Conditions	Osmotic Solution	Findings	References
Carrot	Agitation	250 rpm	50% Sucrose solution + 15% Salt solution for 4 hour	54.5% Water loss; 12.30% Solid gain	[6]
Carrot	Pulse electric field + Agitation	0.60 kV/cm + 500 pulses + 1500 rpm	65% Sucrose solution for 4 hour	58.50% Water loss; 13.40% Solid gain	[9]
Carrot	Pulse electric field + Agitation	0.60 kV/cm + 500 pulses + 250 rpm	65% Sucrose solution for 4 hour	50% Water loss; 10.8% Solid gain	[6]
Carrot	Pulse electric field + Agitation	0.60 kV/cm + 500 pulses + 250 rpm	50% Sucrose solution + 15% Salt solution for 4 hour	58.8% Water loss; 13.9% Solid gain	[6]
Carrot	Pulse electric field + Centrifugation	0.60 kV/cm + 500 pulses + 2400 g	50% Sucrose solution + 15% Salt solution for 4 hour	61.9% Water loss; 10.9% Solid gain	[6]
Green Papaya	Coating	1% Chitosan-Tween 80	40% Sucrose solution at 25°C for 24 hour	22% Water loss; Reduces 32.63% Solid gain; 78% weight reduction	[27]
Green Papaya	Coating	1% Chitosan-Tween 80 and oleic acid	40% Sucrose solution at 25°C for 24 hour	14.17% Water loss; Reduces 43.13% Solid gain; 65.89% weight reduction	[27]
Guava	Irradiation	0.25 kGy; 2.75 kGy per hour	40% sucrose solution at 30°C for 9 hours	33.62% Water loss, 88% Weight Reduction; Reduces 29.9% Solid gain	[80]
Melon	Vacuum	600 mbar	50% sucrose solution at 60°C for 10 min	0.67% Water loss; Reduces 10.26% Solid gain	[43]

TABLE 9.1 *(Continued)*

Product	Treatment(s)	Conditions	Osmotic Solution	Findings	References
Melon	Vacuum	600 mbar	50% sucrose solution at 60°C for 360 min	13.27% Water loss; Reduces 41.52% Solid gain	[43]
Papaya	Ultrasound	25 kHz for 30 min	25% sucrose solution at 30°C	16.37% Water loss; 10.26% Solid gain	[75]
Pears	Ohmic Heating	50/60 Hz; 100 V	65% Sucrose solution at 40°C for 5 hour	13.33% Solid gain	[58]
Persimmon Fruit	Ultrasound	35 kHz for 30 min	45% sucrose solution at 30°C for 4 hours	10.45% Water loss; 5.83% Solid gain	[14]
Pineapple	Ultrasound	35 kHz for 30 min	35% sucrose solution	8.3% Water loss; 18.2% Solid gain	[14]
Potato	Irradiation	6 kGy; 8.80 kGy per hour	50% sucrose solution at 25°C for 5 hours	11.57% Solid gain	[72]
Potato	Irradiation	12 kGy; 8.80 kGy per hour	50% sucrose solution at 25°C for 5 hours	22.62% Solid gain	[72]
Ripped Papaya	Coating	1% Chitosan-Tween 80	40% Sucrose solution at 25°C for 24 hour	14.13% Water loss; Reduces 13.16% Solid gain; 72.98% weight reduction	[27]
Ripped Papaya	Coating	1% Chitosan-Tween 80 and oleic acid	40% Sucrose solution at 25°C for 24 hour	14.53% Water loss; Reduces 19.56% Solid gain; 73.11% weight reduction	[27]
Scallop Adductors	Coating	1% Chitosan-Tween 80	20% salt solution at 35°C for 5 hour	Reduces 2.87% Water loss; Reduces 15.52% Solid gain	[96]
Scallop Adductors	Coating	2% Low Methoxy Pectin	20% salt solution at 35°C for 5 hour	18.97% Water loss; Reduces 11.93% Solid gain	[96]

TABLE 9.1   *(Continued)*

Product	Treatment(s)	Conditions	Osmotic Solution	Findings	References
Scallop Adductors	Coating	2% Sodium Alginate	20% salt solution at 35°C for 5 hour	11.18% Water loss; Reduces 14.12% Solid gain	[96]
Scallop Adductors	Coating	1% Chitosan-Tween 80	30% salt solution at 35°C for 5 hour	24.92% Water loss; Reduces 13.25% Solid gain	[96]
Scallop Adductors	Coating	2% Low Methoxy Pectin	30% salt solution at 35°C for 5 hour	17.58% Water loss; Reduces 7.39% Solid gain	[96]
Scallop Adductors	Coating	2% Sodium Alginate	30% salt solution at 35°C for 5 hour	12.51% Water loss; Reduces 15.58% Solid gain	[96]
Strawberries	High hydrostatic pressure	400 MPa	50% Sucrose solution	61.99% Water loss; 26.51% Solid gain	[58]
Strawberries	Ultrasound	40 kHz for 30 min	50% sucrose solution	31% Solid gain	[27]

shrinkage, high processing rates and enhanced rehydration capacity and porosity. Uniform heating produces products with better sensorial attributes.

During combined processing, microwaves raise the temperature of system, which enhances the evaporation rates of moisture from cellular structure resulting into greater potential gradient between food matrix and surrounding medium. This phenomenon leads to better mass flow rates and shorter processing time [82]. Microwave-assisted osmotic dehydration of fruits accelerates the water removal rate and shortens the overall processing time, resulting in the product with better quality and sensory attributes [61, 65]. In tomato processing, the combination enhanced the shelf life of the product compared to the conventional method [31]. The application of osmotic dehydration prior to microwave drying minimized the shrinkage of cellular structure and preserved the sensorial attributes of the pineapples [73].

9.6.11 FRYING

The application of osmotic dehydration as pre-treatment to frying signifi-cantly affect the various attributes of the final product, for instance, color, texture, and oil absorption. Studies show that osmotic dehydrated products absorb less oil as compared to non-treated products which is attributed to increase in solid matrix of food product during dehydration. The combined treatment of osmotic dehydration and frying shows a great potential in the production of low fat and high-quality food products [38, 51].

9.7 SUMMARY

Osmotic dehydration (OH) plays an important role in shelf-life stability. The technique includes the immersion of fruit and vegetable products in the hypertonic solution resulting in the mass transfer across the semi-permeable membrane. The force behind the water loss and solid gain is the potential gradient across the membrane. Treatments like ultrasound, centrifugation, and HHP, ohmic heating, PEF, microwaves, irradiation, etc., in combination with OH have shown great potential in the design of novel food products with better quality attributes.

KEYWORDS

- dehydration
- hypertonic
- membrane
- osmosis
- semi-permeable
- solid gain
- water loss

REFERENCES

1. Ahmed, I., Qazi, I. M., & Jamal, S., (2016). Developments in osmotic dehydration technique for the preservation of fruits and vegetables. *Innovative Food Science and Emerging Technologies, 34,* 29–43.
2. Akbarian, M., Nila, G., & Fatemeh, M., (2014). Osmotic dehydration of fruits in food industrial: A review. *International Journal of Biosciences, 4,* 42–57.
3. Alam, M. M., & Islam, M. N., (2013). Effect of process parameters on the effectiveness of osmotic dehydration of summer onion. *International Food Research Journal, 20,* 391–396.
4. Allali, H., Marchal, L., & Vorobiev, E., (2010). Blanching of strawberries by ohmic heating: Effects on the kinetics of mass transfer during osmotic dehydration. *Food and Bioprocess Technology, 3,* 406–414.
5. Allali, H., Marchal, L., & Vorobiev, E., (2009). Effect of blanching by ohmic heating on the osmotic dehydration behavior of apple cubes. *Drying Technology, 27,* 739–746.
6. Amami, E., Fersi, A., Vorobiev, E., & Kechaou, N., (2007). Osmotic dehydration of carrot tissue enhanced by pulsed electric field, salt, and centrifugal force. *Journal of Food Engineering, 83,* 605–613.
7. Amami, E., Khezami, W., Mezrigui, S., & Badwaik, L. S., (2017). Effect of ultrasound-assisted osmotic dehydration pre-treatment on the convective drying of strawberry. *Ultrasonics Sonochemistry, 36,* 286–300.
8. Amami, E., & Vorobiev, E., (2005). Effect of pulsed electric field on the osmotic dehydration and mass transfer kinetics of apple tissue. *Drying Technology, 23*(3), 581–595.
9. Amami, E., Khezami, L., Jemai, A. B., & Vorobiev, E., (2014). Osmotic dehydration of some agro-food tissue pre-treated by pulsed electric field: Impact of impeller's Reynolds number on mass transfer and color. *Journal of King Saud University-Engineering Sciences, 26,* 93–102.
10. Badwaik, L. S., Choudhury, M., Dash, K. K., Borah, P. K., & Deka, S. C., (2014). Osmotic dehydration of bamboo shoots enhanced by centrifugal force and pulsed vacuum using salt as osmotic agent. *Journal of Food Processing and Preservation, 38,* 2069–2077.

11. Barman, N., & Badwaik, L. S., (2017). Effect of ultrasound and centrifugal force on carambola (*Averrhoa carambola L.*) slices during osmotic dehydration. *Ultrasonics Sonochemistry, 34,* 37–44.

12. Bchir, B., Besbes, S., Attia, H., & Blecker, C., (2011). Osmotic dehydration of pomegranate seeds (*Punica Granatum* L.): Effect of freezing pre-treatment. *Journal of Food Process Engineering, 35*(3), 335–354.

13. Bolin, H. R., Huxsoll, C. C., Jackson, R., & Ng, K. C., (1983). Effect of osmotic agents and concentration of fruit quality. *Journal of Science, 48,* 202–205.

14. Bozkir, H., Ergün, A. R., Serdar, E., Metin, G., & Baysal, T., (2019). Influence of ultrasound and osmotic dehydration pre-treatments on drying and quality properties of persimmon fruit. *Ultrasonics Sonochemistry, 54,* 135–141.

15. Chandra, S., & Kumari, D., (2013). Recent development in osmotic dehydration of fruit and vegetables: A review. *Critical Reviews in Food Science and Nutrition, 55,* 552–561.

16. Corrêa, J., Rodrigues, L., Vieira, G., & Hubinger, M., (2010). Mass transfer kinetics of pulsed vacuum osmotic dehydration of guavas. *Journal of Food Engineering, 96,* 498–504.

17. Dalla, R. M., & Giroux, F., (2001). Osmotic treatments and problems related to the solution management. *Journal of Food Engineering, 49,* 223–236.

18. Dermesonlouoglou, E., Boulekou, S., & Taoukis, P., (2008). Mass transfer kinetics during osmotic dehydration of cherry tomatoes pretreated by high hydrostatic pressure. In: *IV International Symposium on Applications of Modeling as an Innovative Technology in the Agri-Food-Chain* (pp. 127–133).

19. Dermesonlouoglou, E., Giannakourou, M., & Taoukis, P., (2007). Stability of dehydro-frozen tomatoes pretreated with alternative osmotic solutes. *Journal of Food Engineering, 78,* 272–280.

20. Derossi, A., De Pilli, T., Severini, C., & McCarthy, M. J., (2008). Mass transfer during osmotic dehydration of apples. *Journal of Food Engineering, 86,* 519–528.

21. Devic, E., Guyoi, S., Daudin, J., & Bonazzi, C., (2010). Effect of temperature and cultivar on polyphenol retention and mass transfer during osmotic dehydration of apples. *Journal of Agricultural and Food Chemistry, 58,* 606–614.

22. Falade, K. O., & Igbeka, J. C., (2007). Osmotic dehydration of tropical fruits and vegetables. *Food Reviews International, 23,* 373–405.

23. Falade, K. O., Igbeka, J. C., & Ayanwuyi, F. A., (2007). Kinetics of mass transfer and color changes during osmotic dehydration of watermelon. *Journal of Food Engineering, 80,* 979–985.

24. Fernandes, F. A. N., Gallão, M. I., & Rodrigues, S., (2009). Effect of osmosis and ultrasound on pineapple cell tissue structure during dehydration. *Journal of Food Engineering, 90,* 186–190.

25. Fito, P., & Chiralt, A., (2000). Vacuum impregnation of plant tissues. In: Alzamora, S. M., Tapia, M. S., & Lopez-Malo, A., (eds.), *Minimally Processed Fruits and Vegetables* (pp. 189–205). Gaithersburg, MD: Aspen Publishers.

26. Garcia, M., Diaz, R., Martinez, Y., & Casariego, A., (2010). Effects of chitosan coating on mass transfer during osmotic dehydration of papaya. *Food Research International, 43,* 1656–1660.

27. Garcia-Noguera, J., Oliveira, F. I. P., & Gallão, M. I., (2010). Ultrasound-assisted osmotic dehydration of strawberries: Effect of pre-treatment time and ultrasonic frequency. *Drying Technology, 28*(2), 294–303.

28. Gheybi, F., Rahman, R. A., Bakar, J. B., & Aziz, S. H. A., (2013). Optimization of osmotic dehydration of honeydew using response surface methodology. *International Journal of Agriculture and Crop Sciences, 5*, 2308–2317.

29. Giannakourou, M. C., & Taoukis, P. S., (2003). Stability of dehydro frozen green peas pretreated with non-conventional osmotic agents. *Journal of Food Science, 68*, 2002–2012.

30. Giraldo, G., Talens, P., Fito, P., & Chiralt, A., (2003). Influence of sucrose solution concentration on kinetics and yield during osmotic dehydration of mango. *Journal of Food Engineering, 58*, 33–43.

31. Heredia, A., Barrera, C., & Andrés, A., (2007). Drying of cherry tomato by a combination of different dehydration techniques. Comparison of kinetics and other related properties. *Journal of Food Engineering, 80*, 111–118.

32. Ispir, A., & Togrul, T. I., (2009). Osmotic dehydration of apricot: Kinetics and the effect of process parameters. *Chemical Engineering Research and Design, 87*, 166–180.

33. Jalaee, F., Fazeil, A., Fatemain, H., & Tavakolipour, H., (2011). Mass transfer coefficient and the characteristics of coated apples in osmotic dehydrating. *Food and Bioproducts Processing, 89*, 367–374.

34. Kapturowska, A., Stolarzewicz, I., & Chmielewska, I., (2011). Ultrasound a tool to inactivate yeast and to extract intracellular protein. *Żywnosc Nauka Technology Jakość* [*Food Science Technology Quality*], *4*(77), 160–171.

35. Kaymak-Ertekin, F., & Sultanoglu, M., (2000). Modeling of mass transfer during osmotic dehydration of apples. *Journal of Food Engineering, 46*, 243–250.

36. Khin, M. M., Zhou, W., & Perera, C., (2007). Impact of process conditions and coatings on the dehydration efficiency and cellular structure of apple tissue during osmotic dehydration. *Journal of Food Engineering, 79*, 817–827.

37. Kowalska, H., Lenart, A., & Leszczyk, D., (2008). The effect of blanching and freezing on osmotic dehydration of pumpkin. *Journal of Food Engineering, 86*, 30–38.

38. Krokida, M. K., & Oreopoulou, V., (2001). Effect of osmotic dehydration pre-treatment on quality of French fries. *Journal of Food Engineering, 49*, 339–345.

39. Luo, W., Tappi, S., Wang, C., Yu, Y., Zhu, S., & Rocculi, P., (2018). Study of the effect of high hydrostatic pressure (HHP) on the osmotic dehydration mechanism and kinetics of Wumei fruit (*Prunus mume*). *Food and Bioprocess Technology, 11*, 2044–2054.

40. Maestrelli, A., Lo Scalzo, R., Lupi, D., Bertolo, G., & Torreggiani, D., (2001). Partial removal of water before freezing: Cultivars and pre-treatments as quality factors of frozen muskmelon (*Cucumis Melo*). *Journal of Food Engineering, 49*, 255–260.

41. Maftoonazad, N., (2010). Use of osmotic dehydration to improve fruits and vegetables quality during processing. *Recent Patents on Food, Nutrition and Agriculture, 2*, 233–242.

42. Marani, C. M., Agnelli, M. E., & Mascheroni, R. H., (2007). Osmo-frozen fruits: Mass transfer and quality evaluation. *Journal of Food Engineering, 79*, 1122–1130.

43. Martinez, B., Abud-Archila, M., & Ruiz-Cabrera, M., (2011). Pulsed vacuum osmotic dehydration kinetics of melon (*Cucumis melo*) var. cantaloupe. *African Journal of Agricultural Research, 6*, 3588–3596.

44. Matusek, A., & Merez, P., (2002). Modeling of Sugar transfer during osmotic dehydration of carrots. *Periodica Polytechnica Series of Chemical Engineering, 46*, 83–93.

45. Matuska, M., Lenart, A., & Lazarides, N. H., (2006). On the use of edible coatings to monitor osmotic dehydration kinetics for minimal solids uptake. *Journal of Food Engineering, 72*, 85–91.

46. Mayor, L., Moreira, R., Chenlo, F., & Sereno, A. M., (2006). Kinetics of osmotic dehydration of pumpkin with sodium chloride solutions. *Journal of Food Engineering, 74*, 253–262.

47. Mitrakas, G. E., Koutsoumanis, K. P., & Lazarides, H. N., (2008). Impact of edible coating with or without antimicrobial agent on microbial growth during osmotic dehydration and refrigerated storage of a model plant material. *Innovative Food Science and Emerging Technologies, 9,* 550–555.

48. Moreira, R., Chenlo, F., Chaguri, L., & Vazquez, G., (2011). Air drying and color characteristics of chestnuts to osmotic dehydration with sodium chloride. *Food and Bio Products Processing, 89*, 109–115.

49. Moreira, R., Chenlo, F., Torres, M. D., & Vazquez, G., (2007). Effect of stirring in the osmotic dehydration of chestnut using glycerol solutions. *LWT-Food Science and Technology, 40*, 1507–1514.

50. Moreno, J., Simpson, R., Sayas, M., Segura, I., Aldana, O., & Almonacid, S., (2011). Influence of ohmic heating and vacuum impregnation on the osmotic dehydration kinetics and microstructure of pears (cv Packham's Triumph). *Journal of Food Engineering, 104*, 621–627.

51. Moreno, M. C., & Bouchon, P., (2008). A different perspective to study the effect of freeze, air, and osmotic drying on oil absorption during potato frying. *Journal of Food Science, 73*, 122–128.

52. Mundada, M., Hathan, S. B., & Maske, S., (2011). Mass transfer kinetics during osmotic dehydration of pomegranate arils. *Journal of Food Science, 76*, 31–39.

53. Narang, G., & Pandey, J. P., (2013). Optimization of osmotic dehydration process of grapes using response surface methodology. *Focusing Modern Food Industry, 2*, 78–85.

54. Nazari, A., Salehi, M. A., & Souraki, B. A., (2019). Experimental investigation of effective factors of pulsed electric field in osmotic dehydration of apple. *Heat and Mass Transfer, 55*, 2049–2059.

55. Nieuwenhuijzen, N. H., Zareifard, M. R., & Ramaswamy, H. S., (2001). Osmotic drying kinetics of cylindrical apple slices of different sizes. *Drying Technology, 19*, 525–45.

56. Nowacka, M., Tylewicz, U., Laghi, L., Dalla, R. M., & Witrowa-Rajchert, D., (2014). Effect of ultrasound treatment on the water state in kiwifruit during osmotic dehydration. *Food Chemistry, 144,* 18–25.

57. Nuñez-Mancilla, Y., & Pérez-Won, M., (2013). Osmotic dehydration under high hydrostatic pressure: Effects on antioxidant activity, total phenolics compounds, vitamin C and color of strawberry (*Fragaria vesca*). *LWT- Food Science and Technology, 52*, 151–156.

58. Nuñez-Mancilla, Y., & Perez-Won, M., (2011). Modeling mass transfer during osmotic dehydration of strawberries under high hydrostatic pressure conditions. *Innovative Food Science and Emerging Technologies, 12*, 338–343.

59. Núñez-Mancilla, Y., & Vega-Gálvez, A., (2014). Effect of osmotic dehydration under high hydrostatic pressure on microstructure, functional properties, and bioactive compounds of strawberry (*Fragaria Vesca*). *Food and Bioprocess Technology, 7*(2), 516–524.

60. Pattanapa, K., Therdthai, N., Chantrapornchai, W., & Zhou, W., (2010). Effect of sucrose and glycerol mixtures in the osmotic solution on characteristics of osmotically dehydrated mandarin (cv. *Sai-Namphaung*). *International Journal of Food Science and Technology, 45*, 1918–1924.

61. Pereira, N. R., Marsaioli, J. A., & Ahrné, L. M., (2007). Effect of microwave power, air velocity, and temperature on the final drying of osmotically dehydrated bananas. *Journal of Food Engineering, 81*(1), 79–87.

62. Phisut, N., (2012). Factors affecting mass transfer during osmotic dehydration of fruits. *International Food Research Journal, 19*, 7–18.

63. Pisalkar, P. S., Jain, N. K., & Jain, S. K., (2011). Osmo-air drying of aloe vera gel cubes. *Journal of Food Science and Technology, 48*, 183–189.

64. Praporsoic, I., Lebovka, N. I., Ghnimi, S., & Vorobiev, E., (2006). Ohmic heated, enhanced expression of juice from soft engineering. *Biosystems Engineering, 93*, 199–204.

65. Prothon, F., Ahrne, L. M., & Funebo, T., (2001). Effects of combined osmotic and microwave dehydration of apple on texture, microstructure and rehydration characteristics. *LWT-Food Science and Technology, 34*, 95–101.

66. Ramaswamy, H. S., (2007). Osmotic drying: Principles, techniques and modeling. In: *Proc 5ᵗʰ Asia-Pacific Drying Conference* (pp. 49–59). Hong Kong.

67. Ramya, V., & Jain, N. K., (2017). A review on osmotic dehydration of fruits and vegetables: An integrated approach. *Journal of Food Process Engineering, 40*, 1–22.

68. Raoult-Wack, A. L., (1994). Recent advances in the osmotic dehydration of foods. *Trends in Food Science and Technology, 5*, 255–260.

69. Rastogi, N. K., Angersbach, A., & Knorr, D., (2000). Evaluation of mass transfer mechanisms during osmotic treatment of plant materials. *Journal of Food Science, 65*, 1016–1021.

70. Rastogi, N. K., Angersbach, A., & Knorr, D., (2000). Synergistic effect of high hydrostatic pressure pre-treatment and osmotic stress on mass transfer during osmotic dehydration. *Journal of Food Engineering, 45*, 25–31.

71. Rastogi, N. K., Angersbach, A., Niranjan, K., & Knorr, D., (2000). Rehydration kinetics of high-pressure treated and osmotically dehydrated pineapple. *Journal of Food Science, 65*, 838–841.

72. Rastogi, N. K., & Raghavarao, K. S. M. S., (2004). Increased mass transfer during osmotic dehydration of γ-irradiated potatoes. *Journal of Food Science, 69*, 259–263.

73. Rastogi, N. K., Raghavarao, K. S. M. S., & Niranjan, K., (2014). Recent developments in osmotic dehydration. In: Sun, D. W., (ed.), *Emerging Technologies for Food Processing* (1ˢᵗ edn., pp. 181–212). London, UK: Academic Press.

74. Rastogi, N. K., Suguna, K., & Nayak, C. A., (2006). Combined effect of gamma-irradiation and osmotic pre-treatment on mass transfer during dehydration. *Journal of Food Engineering, 77*, 1059–1063.

75. Rodrigues, S., & Oliveira, F. I. P., (2009). Effect of immersion time in osmosis and ultrasound on papaya cell structure during dehydration. *Drying Technology, 27*, 220–225.

76. Şahin, U., & Ozturk, H., (2016). Effects of pulsed vacuum osmotic dehydration (PVOD) on Drying Kinetics of Figs (*Ficus carica* L). *Innovative Food Science and Emerging Technologies, 36*, 104–111.

77. Shi, J., & Xue, J. S., (2008). Application and development of osmotic dehydration technology in food processing. In: Ratti, C., (ed.), *Advances in Food Dehydration* (pp. 205–226). Boca Raton, FL-USA: CRC Press.

78. Singh, C., Sharma, K. H., & Sarkar, C. B., (2010). Influence of process conditions on the mass transfer during osmotic dehydration of coated pineapple samples. *Journal of Food Processing and Preservation, 34,* 700–714.

79. Sopanangkul, A., Ledward, D. A., & Niranjan, K., (2002). Mass transfer during sucrose infusion into potatoes under high-pressure. *Journal of Food Science, 67,* 2217–2220.

80. Srijaya, M., & Priya, S. B., (2017). Impact of Gamma irradiation and osmotic dehydration on quality characteristics of guava (*Psidium guajava*) slices. *Asian Journal of Dairy and Food Research, 36,* 197–205.

81. Sutar, N., & Sutar, P. P., (2013). Developments in osmotic dehydration of fruits and vegetables: A review. *Trends in Postharvest Technology, 1,* 20–36.

82. Sutar, P. P., & Prasad, S., (2011). Modeling mass transfer kinetics and mass diffusivity during osmotic dehydration of blanched carrots. *International Journal of Food Engineering, 7*(4), 1–5.

83. Taiwo, K. A., & Eshtiaghi, M. N., (2003). Osmotic dehydration of strawberry halves: Influence of osmotic agents and pre-treatment methods on mass transfer and product characteristics. *International Journal of Food Science and Technology, 38,* 693–707.

84. Telis, V. R. N., Murari, R. C. B. D. L., & Yamashita, F., (2004). Diffusion coefficients during osmotic dehydration of tomatoes in ternary solutions. *Journal of Food Engineering, 61,* 253–259.

85. Telis, V. R. N., Romanelli, P. F., Gabas, A. L., & Telis-Romero, J., (2003). Salting kinetics and salt diffusivities in farmed Pantanal caiman muscle. *Pesquisa Agropecuária Brasileira* [*Brazilian Agricultural Research*], *38*(4), 529–535.

86. Tortoe, C., (2010). A review of osmo-dehydration for food industry. *African Journal of Food Science, 4,* 303–324.

87. Traffano-Schiffo, M. V., Laghi, L., & Castro, G. M., (2017). Osmotic dehydration of organic kiwifruit pretreated by pulsed electric fields: Internal transport and transformations analyzed by NMR. *Innovative Food Science and Emerging Technologies, 41,* 259–266.

88. Verma, D., Kaushik, N., & Rao, P. S., (2014). Application of high hydrostatic pressure as a pre-treatment for osmotic dehydration of banana slices (*Musa cavendishii*) finish-dried by dehumidified air-drying. *Food and Bioprocess Technology, 7,* 1281–1297.

89. Viana, A. D., Corrêa, J. L., & Justus, A., (2013). Optimization of the pulsed vacuum osmotic dehydration of cladodes of fodder palm. *International Journal of Food Science and Technology, 49*(3), 726–732.

90. Wiktor, A., & Witrowa-Rajchert, D., (2012). Applying pulsed electric field to enhance plant tissue dehydration process. *Żywnosc Nauka Technology Jakość* [*Food Science Technology Quality*], *19*(2), 22–32.

91. Xin, Y., Zhang, M., & Adhikari, B., (2014). Freezing characteristics and storage stability of broccoli (*Brassica oleracea* L. var. *botrytis* L.) under osmodehydro freezing and ultrasound-assisted osmodehydro freezing treatments. *Food and Bioprocess Technology, 7*(6), 1736–1744.

92. Yadav, A. K., & Singh, S. V., (2014). Osmotic dehydration of fruits and vegetables: A review. *Journal of Food Science and Technology, 51,* 1654–1673.

93. Yao, Z., & Le Maguer, M., (1996). Mathematical modeling and simulation of mass transfer in osmotic dehydration process, Part I: Conceptual and mathematical models. *Journal of Food Engineering, 29,* 349–360.

94. Yetenayet, B., & Hosahalli, R., (2010). Going beyond conventional osmotic dehydration for quality advantage and energy savings. *Ethiopian Journal of Applied Sciences and Technology, 1*, 1–15.

95. Yu, Y., Jin, T. Z., Fan, X., & Wu, J., (2018). Biochemical degradation and physical migration of polyphenolic compounds in osmotic dehydrated blueberries with pulsed electric field and thermal pre-treatments. *Food Chemistry, 239*, 1219–1225.

96. Yuan, T., Ya, Z., & Qilong, S., (2017). Appropriate coating pre-treatment enhancing osmotic dehydration efficiency of scallop adductors. *Transactions of the Chinese Society of Agricultural Engineering, 32*(17), 266–273.

97. Yucel, U., Alpas, H., & Bayindirli, A., (2010). Evaluation of high-pressure pre-treatment for enhancing the drying rates of carrot, apple, and green bean. *Journal of Food Engineering, 98*, 266–272.

INDEX

For Product Safety Concerns and Information please contact our EU
representative GPSR@taylorandfrancis.com
Taylor & Francis Verlag GmbH, Kaufingerstraße 24, 80331 München, Germany

www.ingramcontent.com/pod-product-compliance
Lightning Source LLC
Chambersburg PA
CBHW060330220326
41598CB00023B/2661

9 781774 638514